IRB, Human Research Protections, and Data Ethics for Researchers

Robin Throne
University of the Cumberlands, USA

IGI Global
Publishing Tomorrow's Research Today

Vice President of Editorial	Melissa Wagner
Managing Editor of Acquisitions	Mikaela Felty
Managing Editor of Book Development	Jocelynn Hessler
Production Manager	Mike Brehm
Cover Design	Phillip Shickler

Published in the United States of America by
IGI Global Scientific Publishing
701 East Chocolate Avenue
Hershey, PA, 17033, USA
Tel: 717-533-8845
Fax: 717-533-8661
E-mail: cust@igi-global.com
Website: https://www.igi-global.com

Library of Congress Cataloging-in-Publication Data

Names: Throne, Robin, 1960- editor.
Title: IRB, human research protections, and data ethics for researchers / edited by Robin Throne.
Description: Hershey, PA : IGI Global Scientific Publishing, 2025. | Includes bibliographical references and index. | Summary: "This book offers guidance and the current scholarship surrounding institutional review boards policies and procedures, human research protections for researchers and principal investigators, data privacy, data ethics, and other areas of interest related to human subjects research"-- Provided by publisher.
Identifiers: LCCN 2024052910 (print) | LCCN 2024052911 (ebook) | ISBN 9798369338483 (hardcover) | ISBN 9798369351611 (paperback) | ISBN 9798369338490 (ebook)
Subjects: LCSH: Institutional review boards (Medicine)--Case studies. | Human experimentation in medicine--Evaluation--Case studies. | Medicine--Research--Evaluation--Case studies. | Medical ethics--Case studies.
Classification: LCC R852.5 .I73 2025 (print) | LCC R852.5 (ebook) | DDC 174.2/8--dc23/eng/20250105
LC record available at https://lccn.loc.gov/2024052910
LC ebook record available at https://lccn.loc.gov/2024052911

British Cataloguing in Publication Data
A Cataloguing in Publication record for this book is available from the British Library.

Table of Contents

Detailed Table of Contents

This chapter introduces human subjects research protections basics from a United States regulatory perspective. An introduction to the institutional review board (IRB) office provides descriptions of the types of research activities considered human subjects research (HSR) and not human subjects research (NHSR). Guidance is offered to new investigators for best practices in preparing for the ethical and responsible protection of human subjects in research and their data. Considerations for IRB policy and procedures, including suggestions for improving research protocol submissions for IRB review, are also offered.

This chapter presents how a comprehensive education program for Principal Investigators (PI) leads to an efficient IRB process providing in-depth guidance on data ethics, procedures, and curriculum, reduction in PI time investment, and elevating the quality of research protocols and applications through IRB education and training. The authors emphasize inclusivity and adaptability, catering to the needs of diverse stakeholders ranging from novice doctoral students to seasoned researchers and IRB professionals. The education program offers a range of services including 1:1 office hours, tailored learning sessions, individualized feedback, and live webinars to offer real-time engagement and interactive learning opportunities. Resources include a dynamic IRB website offering guidance, templates, FAQs, and other PI support materials. Regular newsletters provide updates, best practices, and upcoming events. By providing comprehensive training on human research ethics, regulatory requirements, and best practices, the program equips PIs to conduct studies responsibly and ethically.

This chapter provides key takeaways for investigators and clinical researchers in healthcare settings doing minimal-risk research and quality improvement. Following an in-depth overview of the current regulations for informed consent (IC) with a focus on the 2018 regulatory revisions, it then describes the current state of IC in quality improvement activities commonly employed in learning healthcare systems and the ethical challenges they present. This chapter does not address the complexities of IC in clinical trials or in research with greater than minimal risk because the IC processes in such activities are well established. Instead, the focus is on what the regulations state regarding the requirements of informed consent; when and how consent can be waived; how research and quality improvement activities differ from the viewpoint of IC; and what information potential research subjects need to make decisions about participating in minimal-risk research and quality improvement in healthcare settings.

This chapter discusses ethical challenges and barriers unique to graduate-level qualitative information technology (IT) research faced by computer science, systems, and management researchers. IT research presents several challenges stemming from a discrepancy between the researcher's approach to ethical research design and the characteristics of qualitative methodology, bias, and discrimination. The review of the ethical considerations considers current implementations and recommendations in safeguards and risk mitigation strategies to protect research participants' rights, privacy, and well-being. The chapter explores ethical reflections in research design, the regulatory landscape governing IT research, and current studies addressing the general approach to applied ethics, focusing on education and knowledge of compliance requirements and the consequences of non-compliance in human subject research (HSR). Ultimately, this chapter underscores the need for researchers to prioritize ethical considerations throughout the IT research lifecycle, fostering ethical research practices.

In today's digital age, ethical issues are a principal concern during responsible data research which requires thoughtful consideration and approaches for managing them. As technological advances materialize to increase the standard of living for people across the world, ethics must be prioritized by cybersecurity professionals to ensure these technologies are not being misused or inflicting harm to the general population, especially to underrepresented and underprivileged communities. Cybersecurity subject matter experts must develop awareness regarding ethical ramifications of their research endeavors to ensure security is balanced with moral standards. Observance of and adherence to ethical based policies, principles, and security best practices will delineate cybersecurity professionals from threat actors.

Past scholarship has noted the growing concerns and complex challenges surrounding the emergent use of ubiquitous digital technologies in medicine and healthcare, including artificial intelligence (AI) and machine learning (ML). The scholarship has also noted the challenges that human research protection programs (HRPP) and institutional review boards (IRB) have faced in an evolving regulatory climate to review human subjects research involving AI and ML for medical devices. This chapter presents the results of a trends analysis of the current scholarship surrounding compliance trends for AI ethics specific to the use of protected health information (PHI) and compliance trends specifically surrounding AI ethics in software as a medical device (SaMD) for new investigators seeking IRB review for AI/ML research activities. The trends offered in the current scholarship also present many opportunities for policy implications needed to assure ethical oversight for the protection of human subjects in research involving AI/ML for IRBs and HRPPs to review this research.

Recently, there has been increased awareness of the health benefits of animal-assisted interventions. When it comes to using animals in research, this field is unique since

Policy in the context of human subjects' research and PII to address the mounting data ethical issues surrounding AI, ML, and DM, and the implications for IRBs and HRPPs for the protection of human subjects' data.

This chapter explores the Belmont Report, a cornerstone in the ethics of human subject research. It details the report's historical development, emphasizing the three pivotal principles: Respect for People, Beneficence, and Justice. These principles advocate for autonomy through informed consent, the minimization of harm coupled with the maximization of benefits, and fairness in research burden distribution. The chapter discusses the operational challenges in applying these principles across modern research settings and examines the evolving role of Institutional Review Boards (IRBs) in enforcing these standards. Additionally, it addresses the limitations of the Belmont Report in contemporary issues like digital data privacy and the globalization of research practices. The chapter concludes by emphasizing the necessity for dynamic ethical frameworks that adapt to ongoing technological and societal shifts, urging continued ethical vigilance in human subject research.

With the growth of ubiquitous digital technologies, such as artificial intelligence and machine learning, specialized training and preparation are needed to best guide social science researchers for human research protections that involve protections for data with personally identifiable information. Some human research protection (HRP) officers have called for a Belmont 2.0 that offers more aligned guidance for HRP programs and institutional review boards (IRB) to address data ethics in this new era. This chapter presents an analysis of the shifting climate of HRP data ethics, Belmont Principles, and IRB and HRP implications for artificial intelligence, machine learning, data mining/scraping, and other ubiquitous technologies.

This chapter presents the best practices used by institutional review boards (IRBs) and human research protections programs (HRPPs) to prepare online graduate student investigators for human research protections specific to research within online graduate degree programs or where research supervisors are not proximal to graduate student investigators and their research protocols. In recent years, advances in artificial intelligence (AI), machine learning (ML), and other data mining/scraping forms have adversely impacted individual privacy and the unintended sharing of personally identifiable information (PII). With this growth of ubiquitous digital technologies, such as AI, ML, and data mining/scraping, used across online graduate degree programs, specialized training and preparation are needed to best prepare graduate student researchers for human research protections involving data with PII.

It is imperative for companies to adopt ethical policies for sustainable marketing. The objective of the study is to analyze the marketing ethics policies adopted by companies and their roles in creating a sustainable organization. The methodology adopted is a conceptual analysis of the corporate marketing ethics policies. Various controversies and rules and regulations in ethical marketing practices are discussed. In the digital age, consumer privacy is under threat. The policies may help in developing a sustainable company. Initiatives adopted by companies towards remaining ethical and adopting sustainable marketing are highlighted. Academicians may analyze the existing policies and suggest relevant and effective marketing ethics policies. Practicing managers may evaluate the existing marketing ethics policies in their organizations and adopt policies that are effective. These strategies allow companies to adopt ethical policies and sustainable marketing research objectives and for the ethical and responsible protection of human subjects and their data.

Preface

"It is important to adhere to ethical principles in order to protect the dignity, rights and welfare of research participants. As such, all research involving human beings should be reviewed by an ethics committee to ensure that the appropriate ethical standards are being upheld."

~World Health Organization

The evolving global landscape of research in the social, behavioral, educational, and biomedical sciences necessitates that researchers remain informed about the ethical standards and protocols guiding human subjects research. The integrity of research hinges on adherence to these principles, ensuring the protection of human participants, respect for their privacy, and compliance with the ethical obligations outlined by institutional review boards (IRBs) and Human Research Protections Programs (HRPPs) in the United States and various forms of ethics review committees globally.

This edited reference book, *IRB, Human Research Protections, and Data Ethics for Researchers*, is designed to serve as a comprehensive guide for researchers and new investigators navigating the complexities of IRB policies and human research protections. Each chapter has been carefully curated to address the key areas essential to conducting ethical research with human subjects. These topics include IRB guidelines, data ethics, data privacy, and the responsible conduct of research. Our contributors bring both depth and diversity of perspective, offering guidance rooted in current scholarship, practical experience, and evolving best practices.

The collection is particularly useful not only for established researchers and principal investigators but also for new investigators, such as doctoral students, dissertation mentors, and research supervisors. As they embark on their scholarly journeys, these emerging researchers will find invaluable insights into the ethical dimensions of research involving human subjects.

The aim of this book is not only to inform but also to inspire researchers to engage with ethical standards as a critical element of their work. As research methodologies and technologies evolve, so too must the ethical frameworks that guide them.

We hope this resource supports you in conducting ethical, impactful research that respects human participants' dignity, rights, and privacy.

ORGANIZATION OF THE BOOK

Chapter 1

This chapter introduces the fundamentals of human subjects research (HSR) protections from a United States regulatory standpoint. It begins with an overview of the institutional review board (IRB) office, delineating the types of research activities categorized as human subjects research (HSR) versus those classified as non-human subjects research (NHSR). New investigators are provided with practical guidance on ethical research practices and the responsible protection of human subjects and their data. The chapter also explores considerations for IRB policies and procedures, offering suggestions for improving protocol submissions and ensuring compliance with ethical standards.

Chapter 2

The second chapter highlights how a comprehensive education program for Principal Investigators (PIs) can streamline the IRB process and improve the quality of research protocols. Focusing on inclusivity and adaptability, the authors outline services ranging from personalized learning sessions to real-time engagement through webinars. These educational initiatives help reduce the time investment required by PIs while enhancing the quality of their research applications. The chapter emphasizes the importance of continuous education on human research ethics, regulatory requirements, and best practices, equipping investigators at all stages of their careers with the knowledge necessary to conduct ethical research.

Chapter 3

This chapter provides valuable insights for investigators and clinical researchers conducting minimal-risk research and quality improvement (QI) initiatives in healthcare settings. A particular focus is given to the 2018 revisions to informed consent (IC) regulations, with an emphasis on IC processes in quality improvement activities. The chapter contrasts the ethical challenges of informed consent in minimal-risk research with those in more established clinical trials. It also explores regulatory requirements for consent waivers and offers a detailed examination of the information participants need to make informed decisions about involvement in healthcare research.

Chapter 4

The fourth chapter addresses the ethical challenges faced by graduate-level researchers in the fields of computer science and information technology (IT), particularly in qualitative research. The authors explore the intersection between qualitative methodologies and ethical research design, discussing common issues such as bias, discrimination, and privacy risks. A review of current ethical frameworks and regulatory considerations offers recommendations for safeguarding the rights and well-being of research participants. The chapter underscores the importance of ethical reflection throughout the research lifecycle, advocating for compliance with ethical standards in IT research.

Chapter 5

As the digital age accelerates, ethical concerns around responsible data use become increasingly prominent. This chapter focuses on the ethical responsibilities of cybersecurity professionals, particularly in ensuring that technological advancements do not harm vulnerable populations. The authors discuss the need for heightened awareness of ethical ramifications in research and emphasize the importance of balancing security measures with moral standards. By following ethics-based policies and best practices, cybersecurity researchers can differentiate themselves from malicious actors and promote responsible innovation.

Chapter 6

The use of artificial intelligence (AI) and machine learning (ML) in healthcare and medical devices presents new ethical and regulatory challenges. This chapter offers a trends analysis of compliance issues related to AI ethics, particularly regarding the use of protected health information (PHI) in medical research. It examines the regulatory landscape for AI and software as a medical device (SaMD), focusing on the implications for IRB reviews of AI/ML research. The chapter highlights opportunities for policy development to ensure the protection of human subjects in this rapidly evolving field.

Chapter 7

Animal-assisted interventions are a growing field of research that introduces unique ethical considerations. This chapter explores the responsibilities of the Institutional Animal Care and Use Committee (IACUC), which oversees ethical research involving animals, similar to how IRBs manage human subjects research. Researchers working with animals as therapeutic agents are advised on the ethical guidelines and protocols required to protect animal welfare while conducting studies. The chapter offers essential guidance for researchers and practitioners on maintaining ethical standards in human-animal interactions for therapeutic purposes.

Chapter 8

Tracing the evolution of research administration, this chapter examines the shift from its modest origins to its current strategic role in advancing scientific research. Key policy developments, including the Bayh-Dole Act and the Belmont Report, are analyzed for their impact on the professionalization of research administration. The chapter also discusses contemporary challenges such as technological integration, the rise of specialized roles, and the growing emphasis on compliance. Emerging technologies like AI and automation are evaluated for their potential to further transform research administration.

Chapter 9

This chapter delves into the significant challenges posed by the widespread use of AI, machine learning (ML), and data mining (DM) in human subjects research. With a focus on the protection of personally identifiable information (PII), the authors present an analysis of the White House's "Blueprint for the AI Bill of Rights" and its implications for IRBs and human research protection programs (HRPPs). The chapter provides recommendations for addressing the ethical issues surrounding the use of AI, ML, and DM in federally regulated research and stresses the importance of protecting human subjects' data in this context.

Chapter 10

The Belmont Report remains a cornerstone in the ethics of human subject research, and this chapter revisits its key principles: Respect for Persons, Beneficence, and Justice. The authors discuss the challenges of applying these principles in modern research, particularly in areas such as digital data privacy and globalized research practices. The chapter calls for more dynamic ethical frameworks that can adapt to technological advancements and the shifting landscape of human subjects research, emphasizing the continued relevance of the Belmont principles.

Chapter 11

With the growth of AI, machine learning, and data mining in research, human research protections need to evolve to meet new ethical challenges. This chapter explores calls for a "Belmont 2.0" to provide updated guidance on data ethics for IRBs and HRPPs. The authors analyze current issues surrounding AI and data privacy and propose new ethical frameworks to guide research involving pervasive digital technologies. This chapter is particularly relevant for researchers and IRB professionals navigating the complexities of data ethics in the digital age.

Chapter 12

Online graduate programs present unique challenges for human research protections, especially when research supervisors are geographically distant from student investigators. This chapter offers best practices for preparing graduate students in these programs for ethical research involving personally identifiable information (PII). With a focus on digital technologies such as AI, ML, and data mining, the authors provide specialized training recommendations to ensure that student researchers are equipped to protect human subjects and their data in an increasingly digital research environment.

Chapter 13

In the context of sustainable marketing, this chapter explores the ethical policies adopted by companies and their role in promoting ethical corporate sustainability. Through a conceptual analysis, the authors examine marketing practices that respect consumer privacy and adhere to ethical standards. The chapter highlights the importance of ethical marketing in building sustainable organizations and offers practical strategies for academics and practitioners to evaluate and improve marketing ethics policies.

Chapter 14

This chapter presents a compelling discussion of healthcare data ethics and the need for data safety and privacy. Data ownership debates public or private ownership and details an ethical society's global divergent perspectives and philosophical foundations. The building of a consistent data governance model may be beneficial to all and may address the issues of data bias, achieve data justice, and articulate individual autonomy and individual data rights.

CONCLUSION

As we bring this edited reference book, *IRB, Human Research Protections, and Data Ethics for Researchers*, to readers, we reflect on the crucial themes and insights explored throughout its chapters. This collection was designed to be more than just a guide; it is a call to action for researchers at every stage of their careers to engage deeply with the ethical dimensions of their work. The evolving landscape of research, particularly with the rise of digital technologies and novel methodologies, underscores the need for researchers to remain vigilant and adaptive to the ethical challenges they face.

The chapters in this volume have collectively addressed the multifaceted responsibilities that investigators, IRBs, HRPPs, and animal protection professionals must uphold in safeguarding human participants in research and their data. From understanding the regulatory frameworks guiding IRB policies to grappling with the ethical complexities presented by artificial intelligence, machine learning, and cybersecurity, each chapter has contributed to building a robust foundation of knowledge. Equally, the book emphasizes that ethical research is not static—it must evolve in response to the growing intricacies of research environments and the technologies that shape them.

For seasoned research professionals and emerging researchers alike, this reference aims to provide actionable insights, practical recommendations, and a thorough understanding of the ethical considerations that must govern the research process. Whether one is navigating minimal-risk healthcare studies, exploring human-animal interactions, or contributing to the rapidly advancing field of information technology, the importance of protecting human subjects and adhering to ethical research practices remains constant.

In closing, the authors of this collection offer this book as both a reference and an inspiration for human research protections. It encourages researchers to maintain a strong ethical compass, fostering a culture of integrity that respects the dignity, rights, and privacy of those who participate in research. As technology and research methodologies continue to progress, we must ensure that ethical standards evolve alongside them, ensuring that the protection of human subjects remains at the forefront of scientific and scholarly inquiry.

Robin Throne
University of the Cumberlands, USA

Acknowledgment

In honor and appreciation of the 50th anniversary of the enactment of the U.S. National Research Act of 1974, which included the creation of federal rules to protect human participants in research and led to the establishment of the National Commission for the Protection of Human Subjects of Biomedical and Behavioral Research that established ethical standards to guide human research.

Robin Throne
University of the Cumberlands, USA

Chapter 1
Human Research Protections Basics for New Investigators

Rochelle Siapno
Mayo Clinic, USA

Laura Smith
https://orcid.org/0009-0002-8367-3608
Stanley Manne Children's Research Institute, USA

Robin Throne
https://orcid.org/0000-0002-3015-9587
University of the Cumberlands, USA

ABSTRACT

This chapter introduces human subjects research protections basics from a United States regulatory perspective. An introduction to the institutional review board (IRB) office provides descriptions of the types of research activities considered human subjects research (HSR) and not human subjects research (NHSR). Guidance is offered to new investigators for best practices in preparing for the ethical and responsible protection of human subjects in research and their data. Considerations for IRB policy and procedures, including suggestions for improving research protocol submissions for IRB review, are also offered.

DOI: 10.4018/979-8-3693-3848-3.ch001

INTRODUCTION

"The IRB is the researcher's friend. It helps each of us to ensure that the research will protect human subjects as well as possible."
~Nina Kammerer, Brandeis University[1]

In past work, the authors, among others, have provided best practices for the preparation of new investigators for the protection of human subjects involved in research for privacy, confidentiality, fairness, and equity in research recruitment and to ensure risks are minimized for all research involving human subjects and their data, and to make certain no physical or psychological harm comes to the research participants[2] (Lewis & Throne, 2021; Throne et al., 2023). For many new investigators, submitting the application to the institutional review board (IRB) for their research is a rite of passage, and like research, it can be a procedure improved upon with experience. The IRB provides oversight to ensure the rights and welfare of the human subjects involved in the research are protected, as required under federal requirements. Despite these assurances, some research supervisors continue to describe the IRB as a bureaucratic nuisance and one to overcome rather than understand (Reisig et al., 2022). Still others may blame their research supervisors or organizations for inadequate preparation or resources to be well informed on the knowledge foundation needed for human subjects protections and the IRB submission process, or even a revamp of IRB policies and procedures (Ciampa & Wolfe, 2019; Slovin & Semenec, 2019).

Nevertheless, new investigators must understand the purpose, function, and procedures for IRB submission and review of research that involves human subjects and their data to ensure they are engaged in these protections as members of the research community. This chapter presents the essential elements of the IRB office, its policies, and typical expectations new investigators may experience as they submit their human subjects research for IRB review.

BACKGROUND

Institutional review boards (IRB), often placed within human research protection programs (HRPP), are one of the many entities charged with ensuring the protection of human research subjects that include governmental agencies, research organizations and institutions, academic institutions, research sponsors, and research investigators (U.S. Government Accountability Office [GAO], 2023). While IRBs continue to oversee research compliance in an evolving regulatory climate, many HRP professionals have called for improved practices and procedures to oversee human subjects' protection in research (Rosenfeld et al., 2023). The ongoing emergence of ubiquitous digital technologies, such as artificial intelligence in research (Throne et al., 2023), and the calls for a revisit of ethical principles, such as in *The Belmont Report* (Throne & Hendon, 2024), continue to be debated among federal agencies and IRB professionals. Despite the need for ongoing improvement, a soundly structured and proactive HRPP and IRB staff within a research organization provides new investigators with the training, education, and resources needed to balance the risks and benefits of research activities for their participants and to ensure ethical and responsible research is conducted (Rosenfeld et al., 2023; Throne et al., 2023).

THE INSTITUTIONAL REVIEW BOARD OFFICE

The Institutional Review Board office is critical in research institutions, particularly those involved in human subjects research. The Institutional Review Board office is an administrative board established within an institution or organization to ensure the protection of the rights and welfare of human research subjects recruited to participate in research activities conducted under the organization or institutional affiliation. The functions and responsibilities of the institutional review board office consist of reviewing research proposals, risk assessments, and monitoring compliance. A biomedical IRB typically operates under various regulatory frameworks and guidelines, including the *Belmont Report*, FDA Regulations (21 CFR parts 50 and 56), and The Common Rule. A study investigated whether students in two experimental and one nonexperimental research methods class would gain greater knowledge of research ethics and its practical application in daily life by engaging in various aspects of IRB protocol procedures. Participation in mentored IRB-like activities was ultimately associated with an increased understanding of ethical research issues, consequently enhancing awareness of ethical principles. (Ritchie, 2021) Overall, investigators must familiarize themselves with the IRB's policies and procedures before submitting research for review.

HUMAN SUBJECTS RESEARCH SUBJECT TO IRB OVERSIGHT

The Institutional Review Board is a vital entity of any research organization where research involving human subjects is conducted. The IRB must review and approve all research involving human subjects before any study can begin or determine that the research is exempt from IRB oversight to ensure ethical standards are met for responsible research (Office of Human Subjects Research Protections, n.d.). Research involving human subject vulnerable populations allows for broader research to be conducted, explicitly allowing for studies on healthcare disparities among marginalized communities. Researchers must consider these populations respectfully, culturally sensitive, and inclusively while focusing on social justice. Vulnerable populations include groups who are at increased risk of harm or exploitation, such as age, pregnant women, prisoners, and tribal members. IRBs can also negotiate with immigrants, refugees, and undocumented students in social and behavioral research. In studies involving these participants, informed consent is crucial and needs to be ongoing. This ensures that participants, considering their vulnerabilities, still want to be part of the study. In today's political landscape, social and behavioral researchers need to examine issues related to vulnerable populations, including those who might be undocumented (Jach et al., 2020). Research teams and IRBs need to be aware of their institution's guidelines regarding the inclusion of these groups.

WHAT IS HUMAN SUBJECTS RESEARCH?

First, a research study must meet the federal definition of *Research*[3]. Secondly, the research must involve human subjects research (HSR), which involves a living individual about whom data or biospecimens are obtained/used/studied/analyzed through interaction/intervention, or identifiable, private information is used (see Figure 1). Typically, there are three types of IRB review: Exempt, expedited, and full board review. It is essential to follow the IRB's policies and procedures for each type of submission and to understand the IRB application to be used for the submission. Many IRB offices offer sample forms and templates to assist researchers in preparing the IRB application, so it is essential to consult the IRB staff to ensure the study team understands the process prior to submission of the IRB application for exempt, expedited, or research protocols that may require review by the convened IRB.

Figure 1. What is human subjects research?

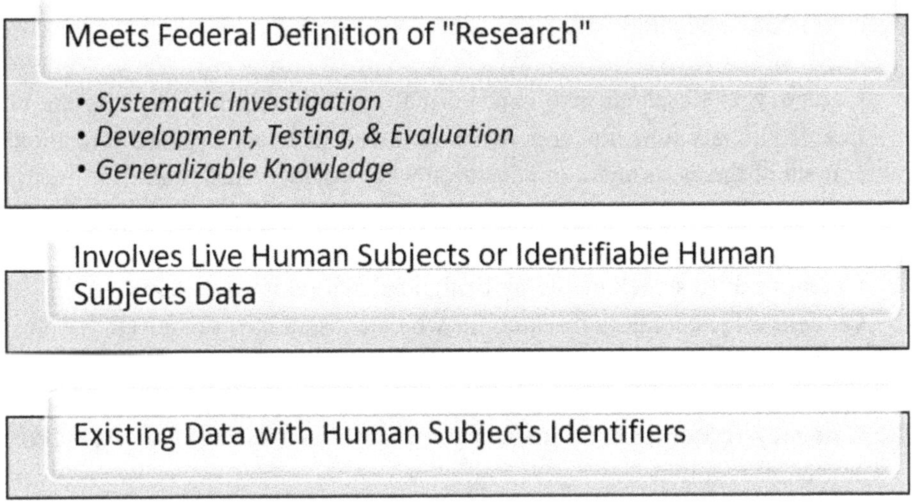

Three types of review are typically conducted by IRBs:

Exempt Review

Exempt review involves human subjects research across eight defined categories that may be considered exempt from IRB oversight (U.S. Department of Health and Human Services [HHS], 2021). The IRB may review these applications to determine whether the protocol meets the exempt criteria and ensure the research meets ethical standards. These categories include

i. The research involves no more than minimal risk to participants
ii. The selection of participants is equitable
iii. If there is a recording of identifiable information, adequate safeguards are in place to maintain the confidentiality of the information.
iv. If there are participant interactions, someone with appropriate qualifications and expertise should determine whether there should be a consent process that will disclose the following:
   a. The activity involves research
   b. A description of the procedures
   c. The participation is voluntary
   d. Name and contact of the researcher

Research activities in which the only involvement of human subjects pertains to one or more of the eight categories below are considered exempt research and not subject to IRB oversight:

Category 1: Research involving normal educational practices that are not likely to impact students' opportunity to learn adversely required educational content or the assessment of education who provide instruction
Category 2: Research involving tests, surveys, interviews, or observations of public behavior
Category 3: Research involving benign behavioral interventions
Category 4: Secondary research for which consent is not required
Category 5: Federal research and demonstration projects
Category 6: Taste and food quality evaluation and consumer acceptance studies
Category 7: Storage or maintenance for secondary research for which broad consent is required
Category 8: Secondary research for which broad consent is required: research involving the use of identifiable private information or identifiable biospecimens

To start exempt research, written documentation of the determination of exempt research should be obtained before the initiation of any research activity. Please note that if it is determined that the project does not need the above categories, the researcher should submit the protocol for IRB review and approval.

A regulatory file must be maintained with current and accurate records of all study documentation, including the finalized IRB protocol, the IRB exempt determination letter, and approved study documents.

Expedited Research

Expedited review is a procedure through which certain kinds of research may be reviewed and approved without convening a meeting of the IRB. The research must involve no more than minimal risk to human subjects and involve one or more of the following categories:

Category 1: Clinical studies of drugs and medical devices
Category 2: Collection of blood samples by finger stick, heel stick, ear stick, or venipuncture
Category 3: Prospective collection of biological specimens for research purposes by noninvasive means

Category 4: Collection of data through noninvasive procedures (not involving general anesthesia or sedation) routinely employed in clinical practice, excluding x-rays or microwaves
Category 5: Research involving materials (data, documents, records, or specimens) that have been collected or will be collected for non-research purposes
Category 6: Collection of data from voice, video, digital, or image recording made for research purposes
Category 7: Research on individual or group characteristics or behavior (including but not limited to research on perception, cognition, motivation, identity, language, communication, cultural beliefs or practices, and social behavior)
Category 8: Continuing research review of previously approved by the convened IRB:

a. Research is permanently closed to the enrollment of new subjects; all subjects have completed all research-related interventions, or research remains active only for long-term follow-up
b. Where no subjects have been enrolled, and no additional risk has been identified
c. Where the remaining activities are limited to data analysts

Category 9: Continuing research reviews not conducted under an investigational new drug application or investigational device where categories two through either do not apply. However, the IRB has determined and documented in the convened meeting that the research involves no more than minimal risk, and no additional risks have been identified.

Once all issues noted by the IRB staff during the review have been addressed, the submission is forwarded to an IRB reviewer for expedited review. The goal of the expedited review is to streamline IRB procedures without reducing the protection for human participants.

Full Board Review

A convened IRB board review is required for research proposals that involve greater than minimal risk to participants, a minor increase over minimal risk, or the study includes an investigational device that is non-exempt and requires a risk determination and is otherwise not found to be exempt from regulatory requirements.

The IRB review is typically conducted at a convened meeting of IRB members. For a convened board review to proceed, there are quorum requirements, which refer to a majority of the voting members present, including one nonscientist and layperson representative of the larger community.

The IRB cannot determine the necessary risk without a detailed research protocol description. Therefore, for the initial applications, all documents submitted to the IRB (protocol, consent documents, recruitment materials, Investigators Brochure) will need to be available for review by all members of the IRB via the electronic IRB system. For Renewals and Modifications that require convened IRB review, the protocol, current consents, proposed consents, and status report that includes the research study elements will need to be available for review by all members of the IRB via the electronic IRB system. The initial application is assigned to reviewers based on each reviewer's expertise.

If the IRB is satisfied with the application in the IRB system, the IRB can approve the study outright. Investigators cannot begin any study activities until they have received the formal letter from the IRB and stamped consent documents, as applicable. In instances where approval is given with minor directed contingencies, the IRB reviewer can include questions or requests to make amendments within the application or the consent forms for clarity if the changes are not substantive. Suppose responses to the contingencies on the Initial Application are not returned for IRB review after 45 days. In that case, the submission may be administratively withdrawn, and the PI will be required to re-submit the study as an Initial application.

The IRB approval study period is no more than one year. Typically, the approval period expires on the last day of the month prior to the anniversary of the approval. Depending on the degree of the study, the approval period may be shortened for more frequent review.

- Any clinical research testing new or non-FDA-approved drugs or biologics must be reviewed and approved by the FDA, which is done by filing an Investigational New Drug (IND) application. There are three types of IND: Investigator IND, Emergency Use IND, and Treatment IND. (Clark et al., 2019).
    - o Investigator Use IND is the use of an investigational drug submitted by a physician

    - Emergency Use IND is the use of an investigational drug on an emergency basis without enough time to file for a full IND
    - Treatment IND is the use of a clinically experimental drug
- A medical device being tested for research. There are three types of medical devices: (a) non-significant risk devices, (b) significant risk devices, and (c) humanitarian use devices. (Clark et al., 2019).
    - For non-significant risk devices, an investigational device exemption is not required. Although IRB approval is required.
    - Significant Risk Devices typically present a potential risk to participants and require an Investigational Device Exemption approval by the FDA or the IRB.
    - Humanitarian Use Device is a medical device that benefits patients with rare conditions. The IRB is required to review DSMB/C reports yearly.

When submitting a research protocol involving investigating a medical device or drug, it is important to include several supporting documents. The inclusion of supporting documents helps the IRB evaluate the submission. These may include brochures, schematics, or other documentation from the manufacturer or developer.

Once the study has been approved, the IRB may perform regular reviews to ensure adherence to ethical standards and monitor the safety of participants (Carlson & Hecht, 2024). The IRB will be issued an approval period depending on the risk associated with the study. During this period, the PI and study team must still ensure all researcher comply with their research training education requirements, submit any modification to the study prior to implementation, promptly report any unanticipated problems/adverse events/protocol deviations to the IRB, and submit a closure when the study is complete.

HUMAN SUBJECTS PROTECTION TRAINING AND CERTIFICATION

The IRB Office likely requires the study team to attain human subjects protection training and certification. Be sure to check with the IRB Office for the policy related to the institution's preferred certification method required for the IRB application. Depending on the type of research project the study team will be engaged in, some organizations specify the appropriate training course required, such as social-behavioral or biomedical research training. The Collaborative Institutional Training Initiative (CITI) Program[4] offers a commonly used certification program, so consult the IRB office to determine which types of certifications will be accepted to document investigator certification for HSR.

Research Projects Not Subject to IRB Oversight

Several forms of research are not subject to IRB oversight. Some forms, such as literary criticism or historical research, are specifically identified within the regulations[5]. The institution's IRB policies may include other types of research that can be determined as not meeting the federal definition of research and thereby not subject to IRB, including the following types of research.

QI/QA

Quality improvement activities (QI/QA) are often not designed to create new knowledge. However, they are used to create existing knowledge to focus on improving systems and organizational performance, thereby not being subject to IRB oversight (U.S. Department of Health and Human Services [HHS], n.d.). The line between quality improvement/ quality assurance initiatives and research is often blurred, and these can often overlap with research projects (see Table 1).

Table 1. Is it research or QI/QA?

	Research	QI/QA
Purpose	Test a hypothesis or study something new	Assess or improve a process or usual care (increase patient satisfaction, increase staff efficiency)
Benefits	Knowledge may/may not benefit current subjects	Knowledge directly benefits process/program/system
Risk/ Burden	Additional risks to produce measurable results	No additional risks, with the exception of possible privacy concerns
Methods	Data collection above and beyond normal healthcare practice	No experimental activities, data collection only to improve service/area/unit/program
Analysis	Determine statistically significant differences in comparison groups	Compare program/process/system to established set standards
Result	Answer a research question	Improves or creates a program/process/system that results in greater safety, efficiency, or satisfaction

Adapted from Reinhardt, A. & Ray, L. (2003). Differentiating quality improvement from research. *Applied Nursing Research, 16*(1), 2-8.

Classroom Research Projects

A classroom research project focuses on a specific topic within a single subject or course and is typically completed within a short duration. The educational focus of a classroom research project aims to develop knowledge related to the content course in relation to teaching research methodologies, critical thinking, and academic writing.

Capstone Projects

A capstone project is a multifaceted assignment that often requires students to integrate and synthesize knowledge from their coursework and apply it to real work scenarios to help bridge the gap between academic theory and practical application.

Not Human Subjects Research Determination

Non-human subjects research (NHSR) typically involves research protocols that do not include human subjects. However, some NHSR protocols may involve human subjects data that has been coded or deidentified for all private information or when the private information cannot be linked to the specific individual by the investigator either directly or indirectly through a coding system. Finally, NHSR may also be determined when the protocol does not involve a systematic investigation designed to develop or contribute to generalizable knowledge, thereby not meeting the federal dcfinition for research.

Informed Consent in Research Participation

The consent document is a written summary of the study. The informed consent process is a regulated process that ensures that research participants are adequately informed regarding risks and/or benefits from research participation. The consent process also allows participants to consider all options, participant responsibilities, and the right to withdraw. It is essential to offer adequate time for a participant to read the consent form and ask questions before agreeing to participate in the research. Thus, the information in the consent form must be available in lay terminology and a language understandable to the participant.

Research investigators are responsible for obtaining informed consent to ensure no human subjects will be involved in the research before obtaining legally effective informed consent. Investigators are responsible for obtaining an individual's signed permission to use or disclose the participant's protected health information (PHI) for research purposes by the Health Insurance Portability and Accountability Act of 1996 (HIPAA) Privacy Rule for research (45 CFR 160 and 164). PHI includes demographic data that relates to the individuals that can be used to identify the individual, including but not limited to name, address, birth date, and Social Security Number.

Waiver of Informed Consent

In the IRB application, it is important to distinguish who among the study team will be permitted to obtain consent and assent (for research with children) from the study participants. Whether it be the PI or other study team members, if the research involves a drug, device, or medical intervention, the IRB may have additional requirements for the consent form and procedure to obtain consent, so again, it is essential to review the IRB office's policies and procedures prior to preparing the IRB application.

IRB Application Preparation Reminders

When reviewing a new application, the IRB will review for consistency between the IRB application and all supporting documents (see Figure 2). It can be helpful for new investigators to follow the protocol as an outline or checklist to ensure all aspects are covered in detail in the application. Oftentimes, researchers have colleagues at other institutions who would like to assist in protocols. The IRB will ask investigators to have the external colleague check with the colleague's institution to determine engagement. The reviewing IRB cannot determine on the external colleague's behalf. If the research is federally funded, then a reliance agreement may be required, so determining engagement is very important in addition to following the institution's IRB policies.

Figure 2. IRB application submission checklist

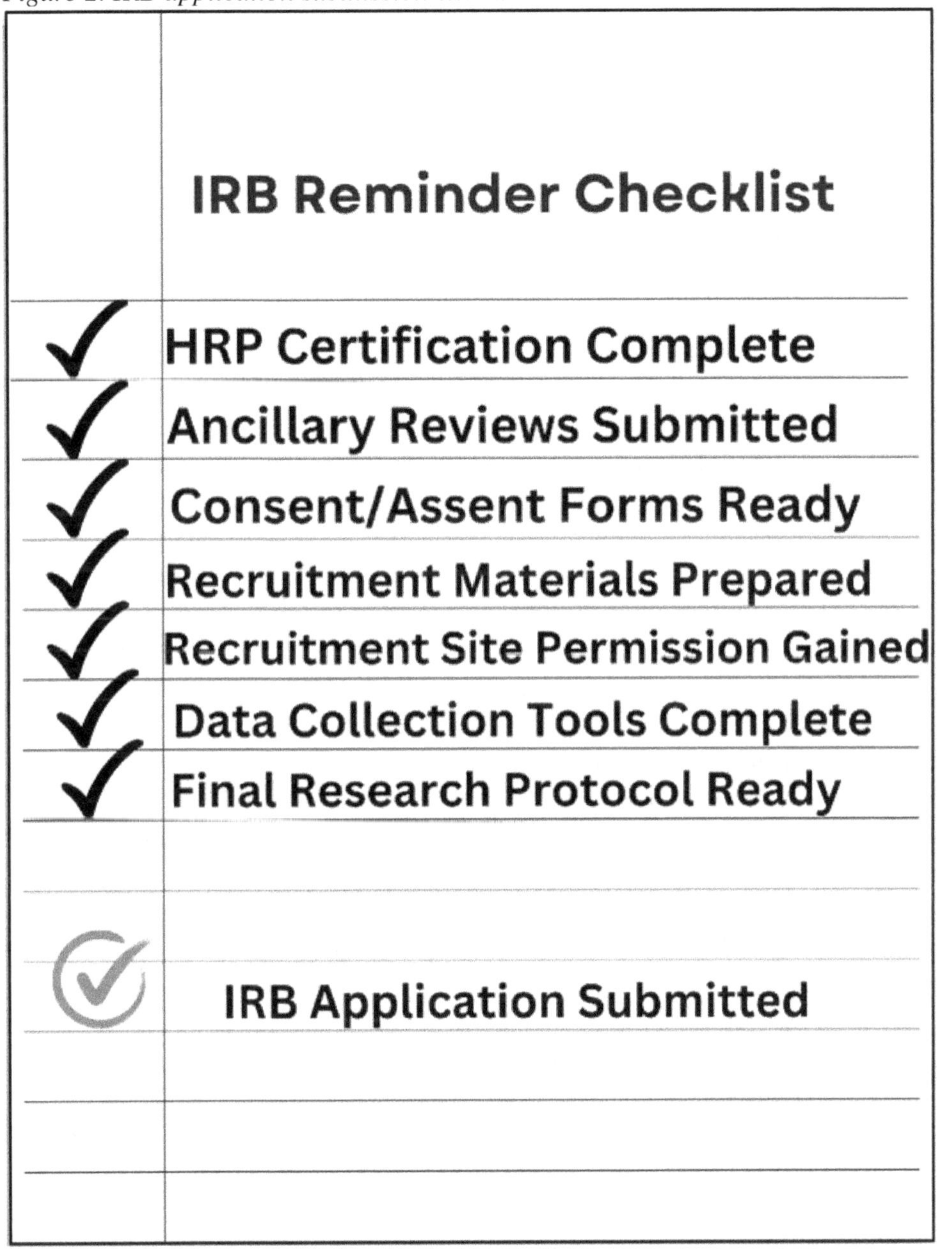

Suppose the study requires reviews ancillary to the IRB review (e.g., radiation safety, investigational pharmacy, conflict of interest). In that case, it is essential that the review process with those groups has been started. Some IRBs hold final approval until those committees approve. Regardless of when approval is required, it is always better to start the process with ancillary reviews sooner than later.

When completing the application, lay language should always be used. This lets anyone pick up the submission and understand the objectives and goals. It allows reviewers to understand what is being asked and expected of participants. When this is clear, it makes it easier for all involved in the review process. While completing this portion of the application, stop to think about the safest way to perform the research. Not all research is minimal risk, but there are always ways to minimize the risk of harm to subjects and others. Are there alternative methods that can be used even if the research is novel? Has a peer or mentor reviewed the protocol before submitting it to the IRB?

When adding recruitment methods, be clear about how participants will be recruited (U.S. Food and Drug Administration [FDA], 1998). Many investigators start with the medical record, and that is fine, but a request for a Waiver of HIPAA authorization will be needed. This allows access to medical records lawfully (U.S. Department of Health and Human Services [HHS], 2017). Remember, just because one can access the medical record does not mean the data does not need protection for privacy and confidentiality. While fundamentally, the request for the waiver may be straightforward, but the access justification must be reasonable and necessary. If an investigator intends to call potential participants, the investigator must have some form of relationship with the potential participant (e.g., patients in the investigator's department). All recruitment methods (e.g. flyers, phone scripts) should be submitted with the application.

How the data is to be handled and stored requires detail and specificity. Investigators must follow the institution's policies and best practices. Proper storage of data prevents breaches and further protections for study participants. Typically, records must be securely stored for at least three years after the research is completed and seven years if HIPAA applies. Each institution might require a longer time for record retention, so it is essential to review the IRB's document retention requirements. This protects the institution, and the investigator and study team should questions arise after publication.

Compensation includes time, travel, and inconvenience (Pandya & Desai, 2013). It should never be used to entice, unduly influence, or coerce participation in research. During a presentation on research participation compensation, the following anecdote was used:

> *"Would you have the tip of your pinky removed for $1,000? Think about it. How often do you use the tip of your pinky? Now imagine asking an unhoused person or a person with a substance abuse problem if they would participate in a research study removing the tip of their pinky for $1,000. Would this represent coercion or undue influence?"*

Bold print and flashing dollar signs are inappropriate when indicating compensation. Stating "free" exams or healthcare can also represent undue influence.

Consent documents need to match what has been stated in the application regarding subject participation. This seems to be an area where the two forms do not agree. If the application states that participants will have three visits, but the consent states four visits, then the discrepancy must be clarified. The documents need to be clear and fully inform participants of what is expected during the study. Institutions typically have a template with the required consent language. Ensure the most current template is being used and not an old copy borrowed from a colleague. Often, returned submissions can be avoided by simply using the most current template provided by the IRB office.

Risks are expected to be identified for participants within the research, and many research protocols may hold no direct benefits for the participants. They may simply offer overall benefits to society or the larger population. Often, investigators describe the means to mitigate the risks in a submission while concurrently identifying benefits that might not exist for participants. The IRB asks that risks be mitigated, but that does not mean the application will be denied if risks exist. It is essential to be honest about the level of risks and frequency. Thus, it is essential to clearly define the risks involved in participation and any benefits participants may receive.

CONCLUSION

In collaboration with investigators, the IRB staff exists to ensure the protection of human subjects in research and ensure that ethical and responsible research is conducted. IRB staff can be a valuable resource to investigators as they continue to refine their research skills under the federal regulations that apply to the research. Applying the fundamentals and best practices for human research protections can assist the IRB with efficiency in evaluating protocols submitted by new investigators.

REFERENCES

Carlson, G., & Hecht, S. (2024). Advocating for learners in telecollaboration exchanges: The role of ethical training and IRB approval. Japan Association for Language Teaching, *Computer-Assisted Language Learning Conference*. DOI: 10.37546/JALTSIG.CALL.PCP2023-03

Ciampa, K., & Wolfe, Z. (2019). Preparing for dissertation writing: Doctoral education students' perceptions. *Studies in Graduate and Postdoctoral Education,* 10(2), 86-108.

Clark, B., Lapid, M., & Wright, R. (2019). Institutional review boards: What clinician researchers need to know. *Mayo Clinic Proceedings Concise Review for Clinicians*, 515-525

Jach, E., Gloeckner, G., & Kohashi, C. (2020). Social and behavioral research with undocumented immigrants: Navigating an IRB committee. *Hispanic Journal of Behavioral Sciences*, 42(1), 3–17. DOI: 10.1177/0739986319899979

Lewis, C., & Throne, R. (2021). Autoethnography and other self-inquiry methods for practice-based doctoral research. In *Practice-based and practice-led research for dissertation development* (pp. 87-107). IGI Global.

Nucera, S., Tartarisco, G., Epasto, A., Smeriglio, D., Mazzeo, A., Pioggia, G., & Anastasi, A. (2018). Ubiquitous, wearable, mobile: Paradigm shifts in e-learning and diffusion of knowledge. In *Handbook of research on mobile devices and smart gadgets in K-12 education* (pp. 286–307). IGI Global. DOI: 10.4018/978-1-5225-2706-0.ch018

Office of Human Subjects Research Protections. (n.d.). *Understanding and submitting for an exempt determination.* National Institutes of Health.

Pandya, M., & Desai, C. (2013). Compensation in clinical research: The debate continues. *Perspectives in Clinical Research*, 4(1), 70–74. DOI: 10.4103/2229-3485.106394 PMID: 23533986

Reinhardt, A., & Ray, L. (2003). Differentiating quality improvement from research. *Applied Nursing Research*, 16(1), 2–8. DOI: 10.1053/apnr.2003.59000 PMID: 12624857

Reisig, M. D., Flippin, M., & Holtfreter, K. (2022). Toward the development of a perceived IRB violation scale. *Accountability in Research*, 29(5), 309–323. DOI: 10.1080/08989621.2021.1920408 PMID: 33877941

Ritchie, K. L. (2021). Using IRB protocols to teach ethical principles for research and everyday life: A high-impact practice. *The Journal of Scholarship of Teaching and Learning*, 21(1), 120–130. DOI: 10.14434/josotl.v21i1.30554

Rosenfeld, S. J., Shaler, G., & Hickey, R. (2023). The Harraseeket Conference–Revisiting systems for ethics oversight of research with human participants. *Research Ethics*, 19(3), 231–249. DOI: 10.1177/17470161231157053

Slovin, L. J., & Semenec, P. (2019). Thinking/writing within and outside the IRB box: Ethical disruptions of data in qualitative research. *Reconceptualizing Educational Research Methodology*, 10(1), 14–27. DOI: 10.7577/rerm.3241

Throne, R., & Hendon, M. (2024). Belmont 2.0: Ethical research expectations for PII in AI, ML, and data mining/scraping. In *Methodologies and ethics for social sciences research* (pp. 1-19). IGI Global.

Throne, R., Hendon, M., & Kozinski, J. (2023). Graduate student investigator: Best practices for human research protections within online graduate research. In Walters, K. (Ed.), *Dynamic curriculum development and design strategies for effective online learning in higher education*. IGI Global. DOI: 10.4018/978-1-6684-8646-7.ch001

U.S. Department of Health and Human Services. (2017). Research. https://www.hhs.gov/hipaa/for-professionals/special-topics/research/index.html

U.S. Department of Health and Human Services. (2021). Exemptions (2018 requirements). Office for Human Research Protections. https://www.hhs.gov/ohrp/regulations-and-policy/regulations/45-cfr-46/common-rule-subpart-a-46104/index.html

U.S. Department of Health and Human Services. (n.d.). Quality improvement activities FAQs. https://www.hhs.gov/ohrp/regulations-and-policy/guidance/faq/quality-improvement-activities/index.html

U.S. Food and Drug Administration. (1998, January). Recruiting study subjects. https://www.fda.gov/regulatory-information/search-fda-guidance-documents/recruiting-study-subjects

U.S. Food and Drug Administration. (2021, January 29). Institutional Review Boards frequently asked questions. U.S. Department of Health and Human Services.

ADDITIONAL READING

Bates, B. J., & Birch, B. (2013). Online research and obtaining human subjects/IRB approvals. In *Online instruments, data collection, and electronic measurements: Organizational advancements* (pp. 40–55). IGI Global. DOI: 10.4018/978-1-4666-2172-5.ch003

Durdella, N. (2022). Conducting research with human participants: An IRB guide for students and faculty. SAGE.

Martinez, J. (2023). Ethics and equity: Lessons learned from an institutional review board (IRB) chair of color. In *Cases on servant leadership and equity* (pp. 112-131). IGI Global.

Murphy, C. M., & Verden, C. (2011). A student's guide to navigating the IRB: How to successfully navigate a potentially overwhelming process. *Journal of the American Academy of Special Education Professionals*, 84, 95.

Quintana, A. J. (2021). Ensuring research integrity. In *Strategies and tactics for multidisciplinary writing* (pp. 192–201). IGI Global. DOI: 10.4018/978-1-7998-4477-8.ch015

U.S. Government Accountability Office. (2023). Institutional review boards: Actions needed to improve federal oversight and examine effectiveness. https://www.gao.gov/products/gao-23-104721

KEY TERMS AND DEFINITIONS

Human Research Protections Program (HRPP): A human research protection program (HRPP) is typically a collaboration between all individuals and groups involved in human subjects research, including IRBs and IRB staff, principal investigators, institutional leaders, faculty research supervisors, student researchers, and others among the research community. The HRPP typically offers training, education, and other guidance to ensure research with human subjects and human subjects' data is ethical, responsible, and scientifically sound (Throne et al., 2023).

Institutional Review Board (IRB): The Institutional Review Board (IRB) is an administrative board established within an organization or institution to ensure the protection of the rights and welfare of human research subjects recruited to participate in research activities conducted under the organization or institutional affiliation (Throne et al., 2023).

Ubiquitous Digital Technologies: Ubiquitous digital technologies refer to all everyday objects embedding intelligent microprocessors able to create architectures that underpin ubiquitous computing, such as networking, artificial intelligence, and wireless computing. Ubiquitous technologies employ sensitive human-computer interaction without spatial or temporal constraints (Nucera et al., 2018; Throne et al., 2023).

ENDNOTES

1 Kammerer, N. (2013). Advice on preparing IRB applications for qualitative research. Heller School, Brandeis University. https://www.brandeis.edu/anthropology/pdfs/advice-about-irb-applications.pdf

2 The Common Rule [45 CFR 46] sets forth the basic policy requirements for the composition and function of the IRB, criteria for IRB approval, informed consent requirements, and definitions, which multiple federal agencies have adopted to protect human subjects in research. In addition to the basic federal policy, additional protections for vulnerable populations are regulated. See also U.S. Department of Health and Human Services, Office for Human Research Protections

3 See 45 CFR 46.102(l)

4 The Collaborative Institutional Training Initiative (CITI Program) offers education leading to certification for biomedical and social-behavioral researchers involved in human subjects research (HSR) to ensure ethical and responsible protections of participants involved in research. HRPPs and IRBs widely use the program to document the researcher's HSR certification. See https://about.citiprogram.org/

5 See 45 CFR 46.102(l)(1)

Chapter 2
Facilitating Compliance Through Capacity:
Empowering Principal Investigators With a Comprehensive IRB Education Program

Chad Becker
https://orcid.org/0009-0007-6270-2728
National University, USA

Brianne Mongeon
National University, USA

ABSTRACT

This chapter presents how a comprehensive education program for Principal Investigators (PI) leads to an efficient IRB process providing in-depth guidance on data ethics, procedures, and curriculum, reduction in PI time investment, and elevating the quality of research protocols and applications through IRB education and training. The authors emphasize inclusivity and adaptability, catering to the needs of diverse stakeholders ranging from novice doctoral students to seasoned researchers and IRB professionals. The education program offers a range of services including 1:1 office hours, tailored learning sessions, individualized feedback, and live webinars to offer real-time engagement and interactive learning opportunities. Resources include a dynamic IRB website offering guidance, templates, FAQs, and other PI support materials. Regular newsletters provide updates, best practices, and upcoming events. By providing comprehensive training on human research ethics, regulatory requirements, and best practices, the program equips

DOI: 10.4018/979-8-3693-3848-3.ch002

PIs to conduct studies responsibly and ethically.

INTRODUCTION

In the landscape of academic research, ensuring the ethical treatment of human subjects is paramount. Institutional Review Boards (IRBs) play a pivotal role in safeguarding the rights and welfare of participants involved in research studies. However, navigating the IRB process can be daunting for Principal Investigators (PIs), both veteran researchers and those new to conducting human subjects' research. In this chapter, the authors delve into the significance of comprehensive IRB education programs tailored to meet the diverse needs of PIs, both students and non-students, exploring how such programs streamline the IRB process and enhance the quality of research protocols and applications. This chapter contains fewer references because it discusses an innovative IRB education program that is unique and has not been extensively researched. Since similar programs do not exist in many other institutions, there is a limited body of literature to draw from, making this content primarily focused on original insights and program development.

Importance of IRB Education Services

Understanding the intricacies of IRB regulations and ethical guidelines is essential for researchers to conduct studies responsibly. Without adequate education and training, PIs may struggle to develop protocols that adhere to these standards, potentially compromising participant safety and data integrity. Therefore, establishing robust IRB education services is imperative to equip PIs with the knowledge and skills to navigate human subjects' research's ethical complexities.

IRB and human research protections education programs play a critical role in fostering a culture of ethical research conduct by instilling a deep understanding of data ethics, procedural requirements, and best practices. This empowerment enables PIs to uphold the highest standards of integrity in their research endeavors. Human Research Protection Programs (HRPPs) are essential for ensuring the rights and welfare of human subjects, extending beyond the specific responsibilities of IRBs (Coley & Matuck, 2022). While IRBs primarily focus on reviewing and approving research protocols, HRPPs encompass all systems involved in human subjects research to safeguard participants. This comprehensive approach allows institutions to integrate various components, such as compliance offices, legal departments, and IRB staff, into a cohesive framework.

Education is a vital aspect of HRPPs, as it equips researchers with the necessary knowledge of ethical standards, regulations, and policies essential for conducting compliant research. An ongoing education program is indispensable, ensuring that researchers stay updated on evolving mandates and best practices. Written procedures further enhance the effectiveness of these educational efforts by providing clear guidelines for consistency and compliance (Barnard & Feige, 2022). Institutions must also define roles related to monitoring, auditing, and reporting to address any issues swiftly. By emphasizing and investing in continuous education about ethical standards and the IRB review process, HRPPs empower researchers to navigate the complexities of human subjects research effectively and responsibly.

Overview of Chapter Content

This chapter provides a thorough analysis of an innovative IRB education program tailored to assist PIs at every stage of their research journey. It underscores a multifaceted strategy, incorporating personalized guidance, customized learning sessions, live webinars, and diverse resources dedicated to demystifying the IRB process (Hicks et al., 2021; Smale, 2010). Through a synthesis of diverse experiences and insights, the chapter highlights the program's effectiveness in addressing the complex needs of researchers navigating the ethical review landscape. Additionally, the authors showcase the program's role in promoting understanding, compliance, and ethical conduct among PIs.[1]

The chapter begins by emphasizing the diverse experiences of PIs, highlighting the program's inclusive approach catering to stakeholders at various career stages. From novice doctoral students to seasoned researchers and IRB professionals, the program addresses a wide spectrum of needs. Through its flexible educational initiatives, individuals receive tailored support to navigate the IRB process with confidence. This comprehensive approach fosters an environment where researchers of all backgrounds can engage effectively with ethical review processes.

Next, through a detailed examination of the program's components and outcomes, the authors discuss its transformative impact on reducing the time investment required by PIs while simultaneously enhancing the quality of research protocols and applications reviewed by IRBs. The chapter explores the implications of such initiatives for promoting a culture of ethical research conduct and safeguarding the welfare of human subjects involved in research studies.

In the subsequent sections readers delve into each aspect of the IRB education program through explanations of its methodologies, benefits, and implications for research integrity and participant protection. By shedding light on the importance of tailored services, continuous improvement strategies, and adherence to best practices, the authors provide valuable insights for IRB/HRPP administrators, researchers, and stakeholders invested in promoting ethical research conduct. Through collaborative efforts and a commitment to excellence, the endeavor is to cultivate a research environment grounded in respect, trust, and protection for all participants involved.

UNDERSTANDING THE NEEDS OF RESEARCHERS

Understanding the diverse needs of their research community is crucial for HRPPs and IRBs to effectively fulfill their roles as guardians of ethical research conduct. By recognizing the varying skill levels, disciplinary backgrounds, and experiences of researchers, these oversight bodies can tailor their guidance and support to meet the specific needs of individuals at every stage of their research journey. This approach fosters a culture of inclusivity and support, empowering researchers to navigate the IRB process with confidence and integrity. Additionally, by addressing common challenges and misconceptions prevalent among researchers, HRPPs and IRBs can enhance compliance, streamline review processes, and ultimately uphold the ethical principles fundamental to human subjects research.

Diverse Skill Levels and Experiences

Researchers encompass a spectrum of skill levels and experiences, ranging from novice students embarking on their first research project to seasoned professionals with years of academic inquiry under their belts. Researchers bring unique sets of perspectives, challenges, and competencies to the research landscape. Understanding and accommodating this diversity is paramount in designing effective IRB education services.

Novice researchers, often graduate students or early-career academics may grapple with foundational concepts of research ethics and regulatory requirements and their applicability to the protection of human subjects in research. They may lack familiarity with the regulatory framework of IRBs, IRB protocols, and procedural nuances, leading to uncertainties and apprehensions in navigating the IRB process. These researchers might struggle with understanding the ethical principles of respect for people, beneficence, and justice, and how these principles translate into practical requirements for informed consent, risk minimization, and equitable subject selection.

In contrast, experienced researchers may possess a solid understanding of ethical principles but require updates on evolving regulations or guidance on complex study designs. These seasoned professionals might encounter new ethical challenges as they engage in innovative research involving novel methodologies, technologies, or vulnerable populations. Their needs are more likely centered on staying current with changes in federal regulations, institutional policies, and best practices in research ethics. Moreover, the disciplinary background of researchers significantly influences their educational needs. For instance, researchers in the social sciences may encounter different ethical dilemmas compared to those in biomedical fields.

Disciplinary Differences in Ethical Considerations

Researchers in social sciences often engage with participants in ways that pose different types of risk and ethical considerations than medical research (Hammersley & Traianou, 2016). Non-experimental social research studies might involve sensitive topics such as trauma, abuse, mental health, or social stigma. The potential for psychological harm or distress is higher, requiring social scientists to be adept at creating safe environments for participants, ensuring confidentiality, and providing appropriate referrals for support when necessary. Ethical considerations in social sciences often include managing the emotional well-being of participants, addressing power dynamics between researchers and subjects, and navigating issues of consent in contexts where complete disclosure might influence participant behavior and data integrity.

However, biomedical field researchers often face ethical issues related to physical risks. Biomedical research might involve invasive procedures, the administration of drugs or other interventions, and interactions with participants' bodies. This type of research requires stringent adherence to protocols that minimize physical harm, ensure informed consent about potential risks and benefits, and provide thorough monitoring for adverse events. Biomedical researchers must be skilled in assessing and mitigating physical risks, understanding regulatory requirements for clinical trials, and maintaining exacting standards of safety and care.

Both disciplines, while distinct in the risks they manage, share a common need for robust ethical training tailored to their specific contexts. Effective IRB education must address these diverse needs, providing social scientists with strategies for minimizing psychological risks and biomedical researchers with tools for safeguarding against physical harm. Additionally, interdisciplinary research brings its own set of challenges, requiring a comprehensive understanding of ethical principles that cut across different fields.

By recognizing and addressing the varied educational needs of researchers across different skill levels and disciplines, IRB education services can better equip all researchers to conduct ethically sound and compliant research. This nuanced approach ensures that both novice and experienced researchers are prepared to uphold the highest standards of research ethics, contributing to the protection and respect of human subjects across all fields of study.

Identifying Common Challenges Faced By Researchers

Across skill levels and disciplines, researchers encounter familiar challenges in navigating the IRB process. These challenges often stem from misconceptions about regulatory requirements, ambiguity in protocol development, or logistical hurdles in obtaining approvals. Identifying and addressing these challenges is instrumental in enhancing the efficacy of IRB education services.

One common challenge faced by researchers is the misconception that IRB approval is merely a bureaucratic hurdle rather than a fundamental ethical safeguard (Zywicki, 2007). This misconception can lead to rushed protocol development, inadequate consideration of ethical implications, and delays in the review process. Education programs must emphasize the ethical imperatives underlying IRB regulations, fostering a deeper appreciation for the principles of respect, beneficence, and justice (National Commission for the Protection of Human Subjects of Biomedical and Behavioral Research, 1978). A key component of a successful education program is making the content and the availability of services both broad enough to bring researchers to the services, yet specific enough to meet their needs.

Additionally, researchers may struggle with articulating their research protocols in a manner that aligns with IRB expectations. Developing clear, concise, and ethically sound protocols requires not only knowledge of regulatory requirements but also effective communication skills. Education services should provide guidance on articulating research aims, identifying potential risks, and crafting informed consent materials that are comprehensible to participants from diverse backgrounds.

Logistical challenges, such as navigating online submission platforms or understanding reviewer feedback can also impede the IRB process. Researchers may benefit from practical guidance on utilizing IRB systems efficiently, interpreting reviewer comments constructively, and addressing common pitfalls in protocol development.

Tailoring Education Services to Address Varied Needs

Recognizing the diverse needs of researchers necessitates a tailored approach to IRB education services, as one-size-fits-all solutions are unlikely to resonate with the heterogeneous research community. Educational initiatives should adopt a flexible, multifaceted approach that accommodates varying skill levels, disciplinary backgrounds, and learning preferences. Personalized consultations offer a valuable opportunity for researchers to receive individualized guidance and support tailored to their specific needs, whether addressing conceptual queries, reviewing draft protocols, or clarifying regulatory requirements. Crafting an education program that is broad enough to meet the needs of the diverse stakeholder community, while also being highly specialized, requires careful consideration of the varying levels of expertise, disciplinary backgrounds, and learning preferences present among researchers.

Customized learning sessions allow for in-depth exploration of topics relevant to researchers' areas of interest or expertise, enhancing the relevance and applicability of IRB education content. By catering to specific disciplinary contexts or methodological approaches, these sessions foster deeper engagement and understanding among participants. Moreover, asynchronous resources, such as online tutorials, instructional videos, and interactive modules, offer flexibility and accessibility to researchers with diverse schedules and learning preferences, providing on-demand support throughout the IRB process.

Incorporating feedback mechanisms into education programs enables continuous improvement and responsiveness to researchers' evolving needs. Soliciting input regarding the effectiveness of educational interventions, areas for improvement, and emerging challenges informs iterative enhancements to program content and delivery methods. By embracing a tailored approach to IRB education services, institutions empower researchers at all career stages to navigate the ethical complexities of human subjects research confidently and responsibly, fostering a culture of research integrity and enhancing the quality of protocols reviewed by IRBs.

CUSTOMIZED GROUP SESSIONS WITH DYNAMIC FEEDBACK

A comprehensive IRB education program for PIs includes group sessions tailored to each learner. The process is most efficient when it is self-schedulable with varied offerings for learners comprising a wide geographic demographic. Group sessions include 20-25-minute content presentations by IRB education staff members/group leaders. The group session format allows IRBs to present varied content such as eligibility criteria, recruitment materials, and content materials. To facilitate movement through the IRB application process, sessions are customized for each learner by offering oral and written feedback regarding pre-submitted supplemental application materials. The program can include expanded materials associated with the group sessions to include case studies and interactive exercises both during the sessions and available as related asynchronous learning materials.

Structure and Format of Group Sessions

Customized group sessions are regularly occurring and self-schedulable (scheduling system used to manage registrations for group sessions and website links to system making it easy for PIs to access the sessions). Upon registration and up to 24 hours before the start of the session, PIs submit supplemental application components (eligibility criteria, recruitment material, and consent/assent material) for group leader review. Sessions are limited to 6-8 registrants to help keep the learning environment student focused. Beginning with 20–25-minute content presentations, each session offers foundational information and materials (templates, guidance documents, and best practices). After content presentations, if the registrant provides material for review (not required) and if they agree to discuss their material in the presence of the other session registrants, feedback is provided verbally, and discussion ensues. Written feedback is provided through a follow-up email with an attachment in the form of right-hand margin comments and track changes. PIs can reschedule and resubmit to any group session as they work their way through the process and find that they need more guidance. Group sessions are not required of PIs but prove useful for compiling robust applications for initial submission. IRB analysts reviewing applications often suggest attendance at specific group sessions based upon the initial review of the application, if a PI shows a lack of understanding of process or material construction.

Group Session Topics: Eligibility Criteria, Recruitment Strategies, Informed Consent/Assent Process

Topics covered in customized group sessions include creating appropriate eligibility criteria for potential participants in human subjects research. Inclusion and exclusion criteria are described and modelled for learners as they work to create this key aspect of any application submission. Eligibility criteria constitutes an important component of both recruitment and consent materials, and many PIs struggle with articulating criteria that both represents potential participants that can provide the information needed for the study as well as ascribes to the best practices of ethical research: typically this relates to who the PI can feasibly access as participants with the level of permission they have but can also relate to topic selection as it relates to the risk level for participants (who the PI is seeking to study and why).

Group sessions are also available for understanding recruitment strategies and creating materials including various templates for different recruitment mediums. The content presentations also attend to recruitment processes that differ depending on delimitations of the population, characteristics of the study setting, and regulations governing access to research subjects. In the recruitment-based sessions, PIs also learn about site permission and the need for appropriate authorizations needed to access private contact information for human research study.

A necessity for a comprehensive education program is training regarding informed consent as a process and the materials involved. The content presentation for the consent session includes procedures for acquiring consent that differ based upon the level of risk to participants. Additionally, procedures differ depending upon data collection techniques as well as the age of participant, and group sessions provide the platform for PIs to make sense of the complicated process and variety of materials involved in attaining informed consent. Several templates are discussed for consent/assent processes, and part of each group session contains a website tour informing attendees of locations of all templates.

In a program like the one described in this chapter, there is room for an informational group session about the application platform and structure. It is a larger group than the group sessions with document review due to the absence of material submission and ensuing oral feedback to PIs. With a mostly presentational platform for this session, and the lack of feedback, the maximum number of attendees for a 1-hour session increases to 15: this allows for a detailed presentation with example application pages as well as a live question and answer session to attend to specific questions attendees may have.

Incorporating Dynamic Feedback Mechanisms

Registrants for customized group sessions about eligibility criteria, recruitment materials, and consent materials can upload related documents for review prior to the session. This format allows the group leader to prepare verbal and written feedback for each participant for the session meeting. A verbal review of feedback during the group session provides all PIs the opportunity to ask clarification questions and seek further advice beyond the content presentation. A follow-up email including written feedback is sent to each participant that submits material for review (the uploading of materials is not required to attend, which is helpful for PIs just beginning the process that want a primer before constructing their supplemental application materials). The format allows the group session to meet the specific needs of all PIs from those at an introductory stage to those fully engaged in the application submission process.

Case Studies and Interactive Exercises

As HRPP/IRBs work to further diversify pedagogical approaches to reach as many PI learners as possible, the intention is to incorporate case study learning as well as interactive exercises as both synchronous (part of group sessions) and asynchronous (available as extension learning from the website) techniques of content delivery. Case studies are a powerful tool in engaging learners (Hancock et al., 2021) within the IRB education framework, as they provide real-world scenarios that vividly illustrate complex ethical dilemmas and regulatory challenges. By analyzing specific cases, researchers can better understand the application of abstract ethical principles and regulatory requirements to concrete situations. This active learning approach promotes critical thinking, encourages discussion, and helps learners identify best practices and potential pitfalls in research involving human subjects. Incorporating case studies into IRB education allows both novice and experienced researchers to explore diverse perspectives, reflect on their decision-making processes, and gain practical insights into navigating the IRB process effectively.

QUARTERLY WEBINARS: ENGAGING RESEARCHERS AT SCALE

A comprehensive IRB education program for PIs provides opportunities to learn through various pedagogies including live interaction. While asynchronous resources and customized group sessions for learning about and constructing key components of an IRB application are important component of the program, live webinars provide another means of reaching PI learners, especially those that like the community feel of large, online group settings. The hour-long webinars include a content presentation and offer a chat for interaction between attendees. The structure of the webinars also includes a Q&A feature that allows for live answers from additional IRB staff as well as the means to later provide to attendees the full list of Qs and As. The webinars include a series of rotating topics invaluable to PIs including introductions to the IRB process, materials and procedures, and the application format. The webinars are recorded for future asynchronous use and linked from the website. The program also incorporates guest speakers and expert panels into the ongoing live webinar format.

Advantages of Webinars for Education Delivery

Live, synchronous webinars offer timely access to crucial aspects of IRB application submissions. While a comprehensive IRB education program for PIs contains vast resources available asynchronously, it is important to engage PIs in a more traditional face-to-face manner (Fischer et al., 2020). The group sessions and office hours provide a range of synchronous learning experiences for PIs, but a live webinar adds a layer of community exposure that not only helps the HRPP make relevant improvements to offerings (by gauging attendees' reactions to the topics) but also provides a large group setting for PIs to understand that their questions are valid, and that there are hundreds of other PIs working through the same issues/problems at any given moment. Attendance for live webinars averages between 100 and 150 PIs.

Rotating Topics to Cover Wide Range of Issues

Considering the significant percentage of novice researcher PIs that utilize the IRB education offerings, and the general pace at which they move through the process, it is important to select relevant topics that are useful for a wide range of PIs. This involves choosing broad curricular topics (introductory in nature) and repeating them often enough to find a balance between reaching as many people as possible but also offering educational opportunities to a constantly changing demographic, as many of the PIs move through their programs.

Q&A Sessions for Participant Engagement

Q&A sessions during IRB webinars significantly enhance engagement by providing researchers with the opportunity to clarify uncertainties and deepen their understanding of the IRB process. These interactive segments allow participants to pose specific questions related to their unique research projects, receiving tailored advice from experts. This direct engagement helps demystify complex regulatory requirements and ethical considerations, making the IRB process more accessible and less intimidating. Moreover, real-time feedback and dialogue foster a sense of community among researchers, encouraging the exchange of ideas and experiences that can lead to innovative solutions and best practices.

Additionally, Q&A sessions contribute to a dynamic learning environment where researchers can learn from their peers' inquiries and the responses provided. This shared learning experience ensures that participants are exposed to a wide range of issues and perspectives, broadening their understanding beyond their immediate concerns. The interactive nature of Q&A sessions keeps researchers actively involved, maintaining their attention and interest throughout the webinar. By addressing both common and unique challenges faced by researchers, these sessions enhance the overall effectiveness of IRB education, ensuring that participants are better prepared to navigate the complexities of conducting ethical research involving human subjects.

Guest Speakers and Expert Panels

Utilizing HRPP administrators and staff, as well as IRB committee members, a comprehensive IRB education program includes opportunities for PIs to engage with a variety of sources providing context and expertise about both objective regulatory requirements and techniques for navigating the institution's specific IRB process. Spaced between live webinars in the yearly event schedule, guest speakers and expert panels provide an interactivity and diversity of perspective providing more pathways to learning for demographically broad PI stakeholders.

INDIVIDUALIZED IRB OFFICE HOURS

A comprehensive IRB education program for PIs includes individualized office hours to provide the personalized help often needed by novice researchers. Unlike one-on-one consultations for single IRB reviews, which focus on coordinating and navigating the requirements across multiple sites (Burr et al., 2019; Johnson et al., 2022; Klitzman et al., 2017), this program emphasizes broader IRB education. Sessions are easily scheduled through the website and available regularly, with learners able to upload topics or questions in advance for the host to prepare. These consultations address specific concerns and challenges, leveraging the expertise of seasoned researchers to enhance the PI's understanding of IRB processes, improve submission quality, and increase likelihood of securing IRB approval. This dedicated space for personalized assistance ensures researchers receive the targeted support necessary to navigate complex regulatory requirements successfully.

Personalized Assistance for Principal Investigators

With the wide range of resources available through a website and other asynchronous methods of delivery, and in addition to the synchronous, customized group sessions, an education program like the one described in this chapter can also include 1:1 personalized assistance for PIs. Emphasis is put on directing PIs to the aforementioned resources and services, but individualized help is a regular necessity considering the complexity that accompanies IRB processes. This can be especially true with novice PIs as many are navigating the application process for the first time. Whether it be to Deans, Department Chairs, Program Directors, Office Supervisors, or affiliated departments across the institution, consistent messaging is given about how to approach the IRB process and materials. This curriculum has been carefully considered to guide PIs through a course of action with intentionally different forms of content delivery facilitating diverse learning environments serving novice researchers. Often, being able to speak with a staff member about study-specific questions or issues is the very type of assistance that can lead a PI to fully grasp the requirements of their study application, clarify complex circumstances that arise frequently within research, or even contribute to an expedited submission of an application for approval.

Benefits of One-On-One Consultations

One-on-one consultation services for an IRB education program offer numerous benefits, significantly enhancing the learning experience for researchers. Personalized consultations provide tailored guidance and support, ensuring that individual questions and concerns are addressed in-depth. This individualized approach helps researchers better understand complex ethical and regulatory requirements, leading to more thorough and compliant research proposals. Additionally, one-on-one sessions foster a more engaging and interactive learning environment, encouraging researchers to actively participate and seek clarification without the hesitation that can occur in larger groups. These consultations can also identify specific knowledge gaps and areas for improvement, allowing for targeted education and skill development. Overall, one-on-one IRB education consultations enhance the quality and compliance of research, fostering a more ethical and efficient research environment.

COMPREHENSIVE IRB WEBSITE: A HUB FOR RESOURCES AND INFORMATION

A comprehensive IRB education program for PIs includes a website offering a wide variety of resources accessible through a range of modalities; understanding the demographics that the program is directed toward to guide creation of user-friendly interface (different types of resources including videos, documents, infographics, example application components; etc...); centralization of materials includes as benefit specific locations to send researchers for help by admins/analysts; interactive nature of tools; regular updates and maintenance to align process and policy to advancements in functioning of HRPP/IRB.

Creating a User-Friendly Interface

An IRB's website must prioritize a user-friendly interface to cater effectively to both novice and experienced users. An intuitively designed interface is essential, incorporating clear menus, accessible navigation tools, and a cohesive layout that seamlessly directs users to the resources they need. Implementing responsive design principles is crucial, ensuring the website's accessibility and functionality across various devices, including desktops, tablets, and smartphones. To achieve this, the website should employ a clean, modern design that minimizes clutter and highlights essential information, enabling users to find what they need quickly and efficiently.

It is also important to seek feedback from the end-users and research community to identify and address potential pitfalls and areas for improvement. This iterative process should involve diverse user groups, reflecting the spectrum of skill levels and experiences among researchers. Feedback from these groups can inform refinements that enhance usability, such as simplifying complex processes, and ensuring that critical resources, like IRB protocols and submission guidelines, are easily accessible. By continuously refining the user interface based on real-world feedback, the website can better meet the diverse needs of its audience, supporting researchers in navigating the IRB process with confidence and ease.

Centralized Access to Educational Materials

Centralized access to educational materials on an IRB website offers significant benefits for researchers. By consolidating resources in one accessible location, the website becomes a valuable repository of guidance and support, aiding researchers in navigating the intricacies of the IRB process. This centralized approach simplifies the research journey, providing PIs and other stakeholders with easy access to essential information, such as IRB protocols, submission guidelines, and best practice guides. Researchers can efficiently locate and utilize these resources, ensuring they have the necessary knowledge and tools to conduct ethically sound and compliant studies.

In addition to supporting researchers, centralized access to educational materials is a crucial internal tool for the IRB. The website enhances efficiency and effectiveness in supporting research endeavors by streamlining administrative processes and consolidating resources. IRB staff and analysts can easily reference and share educational materials with researchers, providing consistent guidance and support throughout the review process. This internal tool ensures that all team members are working with the same up-to-date information, facilitating smoother communication and collaboration. Furthermore, the centralized repository enables the IRB to maintain compliance with evolving policies and regulations, as updates can be promptly disseminated to both internal stakeholders and external researchers.

Interactive Tools and Templates

Integrating interactive tools and templates into the IRB website can significantly enhance user engagement and streamline processes. These tools might include online application forms, consent form templates, and project tracking systems that facilitate efficient submission and review of research proposals. By offering customizable templates and automated tools, the website not only saves time but also ensures that submissions meet regulatory requirements, reducing the likelihood of errors and the need for revisions.

Additionally, incorporating a robust FAQ section and detailed instructional videos can further support users navigating the IRB process. These resources can address frequent questions and provide step-by-step guidance on completing forms and understanding regulatory requirements. A well-organized document repository containing sample IRB submissions, best practice guides, and policy updates can serve as a valuable reference for researchers at all experience levels. Regularly updating these resources ensures that users have access to the latest information and guidance, fostering a more informed and confident research community. By combining interactive tools with comprehensive educational materials, the IRB website can effectively streamline processes, enhance user engagement, and support researchers in conducting ethically sound and compliant studies.

Regular Updates and Maintenance

Regular updates and maintenance are essential to keep the IRB website relevant and effective. This involves routinely updating content to reflect the latest regulatory changes, guidelines, and best practices in research ethics and compliance. Scheduled maintenance ensures that all links, tools, and functionalities remain operational. Gathering user feedback through surveys or feedback forms can provide valuable insights into areas needing improvement. By consistently refreshing the content and technology, the IRB website remains a reliable and up-to-date resource for all stakeholders involved in the research approval process.

LEVERAGING ASYNCHRONOUS LEARNING: YOUTUBE VIDEOS

In the digital age, asynchronous learning has emerged as a powerful educational tool, offering flexibility and accessibility to learners across diverse backgrounds and schedules. Among the myriad mediums available, YouTube videos stand out as a dynamic platform for delivering educational content. YouTube as a medium for asynchronous learning within the context of IRB education can serve as a dynamic tool for the research community to interact with the IRB. By leveraging the advantages of asynchronous learning, developing educational videos on key topics, and enhancing accessibility and flexibility for researchers, institutions can harness the full potential of YouTube as a valuable resource in IRB education.

Advantages of Asynchronous Learning

Asynchronous learning, especially through platforms like YouTube, presents a change in basic assumptions of IRB training and education. Unlike synchronous methods that demand rigid adherence to preset schedules, asynchronous learning offers researchers the freedom to engage with educational content on their own terms. This flexibility is paramount in accommodating the varied schedules of researchers, who may be navigating complex project timelines and professional obligations. By removing the constraints of time and location, asynchronous learning empowers researchers to take charge of their educational journey, promoting autonomy and self-directed learning.

One of the key benefits of asynchronous learning, particularly through YouTube videos, is the ability for researchers to access and revisit material as needed. Unlike traditional classroom settings where information is delivered once, asynchronous platforms enable users to pause, rewind, and replay content at their own pace. This feature is invaluable for comprehending complex concepts or reviewing specific topics in greater detail. Additionally, the accessibility of YouTube allows researchers to engage with educational videos anytime, anywhere, whether they are in the lab, at home, or in the field. This accessibility fosters continuous learning and skill development, empowering researchers to stay informed and updated in the rapidly evolving field of research ethics.

Developing Educational Videos on Key Topics

Developing educational videos on key IRB-related topics is instrumental in equipping researchers with the knowledge and skills necessary to navigate the complex landscape of research ethics. These videos should address a diverse array of subjects, ranging from the foundational principles of IRB processes to the intricacies of ethical considerations and regulatory requirements. For instance, researchers often encounter challenges in effectively obtaining informed consent from participants or recruiting subjects in an ethical and equitable manner. By creating targeted videos on these specific topics, IRBs can provide practical guidance and best practices to researchers, ensuring they understand their responsibilities and obligations in these critical areas.

Each educational video should strike a balance between conciseness and comprehensiveness, delivering information in a clear and accessible manner. To enhance engagement and retention, IRBs should employ plain language and incorporate visually appealing graphics or animations to illustrate key concepts. Moreover, leveraging the expertise within their teams and collaborating with subject matter experts ensures the accuracy and relevance of the content. By incorporating real-world examples and case studies into the videos, IRBs can contextualize abstract concepts and demonstrate their application in researchers' day-to-day activities.

Accessibility and Flexibility for Researchers

YouTube's platform offers unparalleled accessibility and flexibility for disseminating IRB educational content. Videos can be accessed anytime and anywhere, provided there is an internet connection, making it easier for researchers across the globe to benefit from the training. YouTube's features, such as subtitles, transcripts, and playback speed control, further enhance accessibility, catering to diverse learning needs and preferences. For instance, researchers who are non-native English speakers or those with hearing impairments can utilize subtitles, while busy professionals might prefer adjusting the playback speed to optimize their viewing time.

Practical Necessities for Program Implementation

To implement a comprehensive educational program like the one described in this chapter, there are several practical components that are required. We've discussed online services that require a website hosting service, which could be internal to the institution or externally hosted. Further, the group sessions require a scheduling software system that can accommodate scheduling, send reminder notifications, contain the capacity to upload documents for group sessions, send follow-up notifications, and track users for quality assessment and quality improvement purposes. In addition, for meetings that the educational program offers, an online meeting platform is needed to hold the sessions. For some asynchronous help materials (videos), an online video hosting service is needed.

Having previously referred to the makeup of the education program staff, more information might assist in understanding, as well as replicating, such a program. The institutional unit that constitutes the HRPP/IRB offices at the home institution for this educational program contains 2 director-level positions, 2 analyst/reviewer-level positions, and 2 educational analyst-level positions. While all 6 staff members contribute across the spectrum of HRPP/IRB responsibilities, the 2 educational analyst-level positions constitute the staff of the IRB's comprehensive educational program.

The two educational analyst-level positions are filled by people with atypical backgrounds for IRB staff members with one holding a terminal degree in the discipline of curriculum and instruction, and one holding a terminal degree in the discipline of instructional design (note: that both are within the larger discipline of education, providing solid grounding for not only the substance of the educational curriculum but also the best practices for delivery of that content). Additionally, both educational analyst-level positions were filled by entry level IRB staff with limited training and experience. Ensuing professional development has been key to ensuring sound human subjects research protection that guides the program. And, the education-based backgrounds of the educators facilitated the rapid and substantive construction of the education program. The full IRB board contains a chair that is drawn from the faculty of the institution and is constituted mostly by the other faculty. These members are involved at the full-board review level and engage in professional development, but they are not involved in the education program. Additionally, while the contributions of the director-level and analyst/reviewer-level staff should be credited in the overall design and updating of the education program, two full-time staff provide all that is described in this chapter.

EVALUATING EFFECTIVENESS AND CONTINUOUS IMPROVEMENT

In the realm of IRB education, ensuring the effectiveness of educational services is paramount. This section delves into the critical process of evaluating the impact of education services on researcher competence and fostering continuous improvement. By establishing a feedback loop for education services, assessing the impact on researcher competence, implementing an iterative process for service enhancement, and incorporating best practices and emerging trends, IRBs can enhance their educational offerings and better support researchers in conducting ethically sound and compliant research.

Feedback Loop for Education Services

A robust feedback loop is essential within the IRB framework, facilitating the continuous improvement of educational content and materials. This loop encompasses various avenues, including structured surveys, informal feedback during office hours or group sessions, and proactive engagement with stakeholders. IRB staff will actively seek input from users such as researchers, faculty, and fellow IRB members, regarding their experiences with educational content, materials, and resources.

A feedback loop's effectiveness lies in providing real-time insights into the strengths and weaknesses of IRB education initiatives. By gathering feedback on the clarity, relevance, and utility of educational content, IRBs can identify areas for improvement and make data-driven decisions to enhance the overall quality of their educational offerings. For example, feedback may reveal that certain topics are not adequately covered or that specific resources are difficult to access, prompting adjustments to content development or website design.

Furthermore, a feedback loop enables IRBs to stay nimble and responsive in their educational efforts. As the research landscape evolves and regulatory requirements change, IRBs must adapt their educational content to address emerging challenges and trends effectively. By continuously gathering feedback and incorporating suggestions for improvement, IRBs can ensure that their educational materials remain relevant and up to date. This iterative process of refinement allows IRBs to maintain agility in their educational strategies, effectively meeting the evolving needs of researchers and fostering a culture of continuous learning and improvement within the research community.

Incorporating Best Practices and Emerging Trends

To ensure the continued relevance and effectiveness of IRB education efforts, it is essential to incorporate best practices and stay abreast of emerging trends in research ethics and compliance. IRBs should actively monitor developments in regulatory guidelines, ethical frameworks, and educational methodologies to inform their educational initiatives. For example, most recently, changing laws around women's health had significant impacts on the way IRBs can review certain research. By staying informed about evolving regulatory landscapes, IRBs can enhance the quality and comprehensiveness of their educational content.

Moreover, IRBs must remain vigilant in identifying and adapting to emerging trends in research ethics and compliance. This includes staying informed about innovative technologies, methodologies, and research practices that may impact the ethical conduct of research. For example, advancements in data analytics and artificial intelligence raise new ethical considerations regarding privacy, consent, and data protection. By proactively addressing these emerging trends in their educational materials, IRBs can help researchers navigate these complexities and uphold ethical standards in their research endeavors.

Delimitations

The delimitation of this IRB education program lies in its clear focus on social, behavioral, and educational research domains. While it provides robust support tailored to the specific needs of researchers in these fields, questions arise regarding its scalability across institutions of varying sizes. The program's effectiveness in larger institutions with diverse research portfolios and varying levels of IRB engagement may require adaptation and additional resources to ensure comprehensive coverage and support. Evaluating its scalability involves considering the program's capacity to accommodate differing institutional infrastructures and research scopes, aiming to maintain equitable access to effective IRB education and support services across diverse academic settings.

Additionally, the program's focus on novice researchers, particularly graduate students, introduces a potential student bias. While the emphasis on early-career researchers is valuable, the program could benefit all researchers, regardless of experience. This descriptive evaluation of the IRB education program highlights its strengths and limitations, noting that more research is needed to explore its broader applicability and effectiveness. Such research would provide deeper insights into optimizing IRB education programs to serve a wider range of researchers and institutional contexts effectively.

Challenges to Implementing a Comprehensive IRB Educational Program

There are several challenges to implementing a comprehensive IRB educational program. Since many IRBs do not include educational elements in their framework, arguing for and procuring "buy-in" from an institution can be challenging. Some administrators may perceive such initiatives as unnecessary expenditures, viewing them as an added burden rather than a valuable investment in research quality and compliance., but preliminary findings from quality assessment and quality improvement efforts have shown encouraging evidence for decreasing the time that PIs spend in the IRB application and approval process. This reduction in time is especially crucial when dissertation-level PIs make up a sizable portion of IRB submissions, as inefficiencies can lead to delays that affect their time-to-degree and overall satisfaction with their academic programs. Expediency in the application and approval process not only enhances individual researcher experiences but also has broader implications for institutional productivity and research output, ultimately impacting the efficacy of the entire academic program.

Another challenge to creating and facilitating a comprehensive IRB educational program is finding the right staff members. The authors believe that candidates must not only be personable and pedagogically skilled but also possess the capacity for the specific demands of the position. Building trust is essential for the success of a program like the one described here, as it fosters an environment where researchers feel comfortable seeking guidance and engaging openly with the IRB. Trust is particularly critical in navigating complex regulatory requirements and ethical considerations, where clear communication can make all the difference. This necessitates a careful approach to the hiring process in the IRB. The nature of the job descriptions often emphasizes the importance of soft skills, which can sometimes be overlooked in favor of technical expertise (Burgett-Moreno et al., 2024). Additionally, the decision not to require a minimum of two years of IRB experience, as noted earlier in the chapter, calls for thorough consideration of the applicants' sophistication and interpersonal capabilities. Therefore, it is crucial to be deliberate in the interview process, employing tailored questions and scenario-based assessments that reveal how candidates manage real-world challenges and build rapport with diverse stakeholders. Ultimately, selecting the best candidates to produce and deliver a comprehensive IRB educational program hinges on this nuanced understanding of both skills and relational dynamics, ensuring that the program can effectively support researchers in their work while fostering a collaborative culture.

Building a comprehensive education program requires including many components and often the need to begin "from scratch" due to limited resources for programs like this. The hope is that this chapter helps to bring into focus what is needed to create one of these types of programs. Doing so can be difficult simply in its construction and might be overwhelming. We believe that there does need to be dedicated staff (as we have shown: two full-time staff) to be able to deliver, grow, and update the program to keep it relevant. We have found the best approach is grounded in pedagogical best-practices and federal regulatory framework leading to a coordinated, methodical, and thoughtful approach.

Importantly, effective training must encompass two key areas: understanding the reasons and procedures for IRB protection, and learning how to relate effectively with the IRB, including what is needed from the IRB's perspective. As highlighted by Carline et al. (2007), this dual focus is critical; one respondent noted the necessity of understanding "the hot buttons for the IRB and what they thought about." This insight is essential for researchers to navigate the IRB process more effectively.

To achieve these educational goals, the program should incorporate diverse instructional strategies, such as interactive workshops and case studies that reflect real-world challenges faced by researchers. Such methods not only enhance engagement but also help demystify the IRB process, fostering a deeper understanding of both regulatory requirements and the IRB's operational mindset. By promoting open dialogue and encouraging researchers to engage with the IRB, institutions can foster a more productive relationship that benefits the research community.

CONCLUSION

Throughout this chapter, a descriptive analysis was provided about the integral role that a comprehensive IRB education program plays in supporting PIs and enhancing the quality of research protocols. The importance of tailored educational services that cater to the diverse needs of PIs, including novice doctoral students and seasoned researchers, was emphasized. Key components of the program, such as individualized office hours, customized group sessions, webinars, and a comprehensive website, were discussed in detail. These elements collectively reduce the time investment required by PIs and improve the quality of IRB submissions by equipping researchers with the necessary knowledge and skills to navigate ethical and regulatory requirements effectively.

Furthermore, the authors highlighted the program's focus on social, behavioral, and educational research, noting the potential limitations in its applicability to biomedical fields. The discussion also addressed the scalability of the program for institutions of varying sizes, suggesting that adaptations and additional resources may be necessary for larger or more diverse research settings. Additionally, the potential bias towards novice researchers was acknowledged, while recognizing that the program's benefits could extend to all researchers. This chapter underscores the need for more data-driven research to fully evaluate the effectiveness of IRB education programs and optimize them for broader application.

Future Directions in IRB Education Services

The future of IRB education services lies in the development of more adaptive, scalable, and comprehensive programs that address the evolving needs of researchers across various disciplines. One critical direction is the integration of advanced technologies such as AI-driven personalized learning platforms, which can offer tailored educational content and real-time feedback to researchers based on their specific needs and progress. Expanding the focus beyond social, behavioral, and educational research to include more extensive support for biomedical and clinical research is also essential. This involves creating specialized modules that address the unique ethical and regulatory challenges in these fields. Additionally, fostering a culture of continuous improvement and learning within institutions is vital. This can be achieved by implementing regular training updates, interactive workshops, and collaborative forums where researchers can share experiences and best practices. Emphasizing the importance of lifelong learning, these initiatives should support both novice and experienced researchers, ensuring they stay abreast of the latest IRB requirements and ethical standards. Moreover, enhancing accessibility through online platforms and asynchronous learning materials will ensure that IRB education is available to researchers regardless of their location or institutional resources. Overall, these future directions aim to create a more dynamic and inclusive IRB education framework that supports ongoing researcher development and promotes exacting standards of ethical research.

Importance of Ongoing Support for Researcher Development

Continued support for researchers with IRB education is crucial for fostering an environment of rigorous research ethics and high standards. This support ensures researchers are well-versed in ethical principles, regulatory requirements, and best practices for protecting human subjects, thereby preventing ethical breaches that could compromise research integrity. Institutions benefit from ongoing IRB education by reducing the risk of non-compliance incidents, which can lead to costly legal consequences and damage to institutional reputation. Furthermore, regular IRB training helps researchers design and conduct studies that are ethically sound which can enhance the overall quality and credibility of the research output.

For researchers, consistent IRB education provides a clear understanding of how to navigate complex ethical dilemmas, ensuring that they can make informed decisions that prioritize participant welfare. This knowledge fosters a culture of ethical mindfulness, where researchers are continuously aware of their responsibilities towards participants and the broader community. Additionally, well-supported IRB processes streamline the approval and monitoring of research projects, reducing delays and facilitating smoother progress. Both institutions and researchers benefit from a robust IRB education framework, as it leads to more ethical, reliable, and respected research outcomes, promoting public trust and advancing scientific knowledge.

REFERENCES

Barnard, D., & Feige, M. M. (2022). Documentation, policies, and procedures. In Bankert, E. A., Gordon, B. G., Hurley, E. A., & Shriver, S. P. (Eds.), *Institutional review board: Management and function* (3rd ed., pp. 45–54). Jones & Bartlett Learning.

Burgett-Moreno, M., Cartmell-McGlyn, A., & Mongeon, B. (2024, May 21–23). HRPP/IRB staffing: Onboarding, education, and empowerment [Conference presentation]. AAHRPP Annual Conference, San Diego, CA, United States. Burr, J. S., Johnson, A. R., Vasenina, V., Bisping, S., Coleman, R. W., Botkin, J. R., & Dean, J.

Burr, J. S., Johnson, A. R., Vasenina, V., Bisping, S., Coleman, R. W., Botkin, J. R., & Dean, J. M. (2019). Implementing a central IRB model in a multicenter research network. *Ethics & Human Research*, 41(3), 23–28.

Carline, J. D., O'Sullivan, P. S., Gruppen, L. D., & Richardson-Nassif, K. (2007). Crafting successful relationships with the IRB. *Academic Medicine*, 82(10, Suppl), S57–S60. DOI: 10.1097/ACM.0b013e31813e66d4 PMID: 17895692

Coley, C. H., & Matuk, J. (2022). Overview of the human research protection program. In Bankert, E. A., Gordon, B. G., Hurley, E. A., & Shriver, S. P. (Eds.), *Institutional review board: Management and function* (3rd ed., pp. 29–39). Jones & Bartlett Learning.

Fischer, C., Xu, D., Rodriguez, F., Denaro, K., & Warschauer, M. (2020). Effects of course modality in summer session: Enrollment patterns and student performance in face-to-face and online classes. *The Internet and Higher Education*, 45, 100710. DOI: 10.1016/j.iheduc.2019.100710

Hammersley, M., & Traianou, A. (2016). Ethics in qualitative research: Controversies and contexts. SAGE.

Hancock, D. R., Algozzine, B., & Lim, J. H. (2021). *Doing case study research: A practical guide for beginning researchers* (4th ed.). Teachers College Press.

Hicks, R. W., Hines, K., & Henson, B. (2021). Demystifying the Institutional Review Board. *AORN Journal*, 114(4), 309–318. DOI: 10.1002/aorn.13498 PMID: 34586652

Johnson, A. R., Pautler, M., Burr, J. S., Abdelsamad, N., VanBuren, J. M., Rigtrup, L. M., Dean, J. M., & Rothwell, E. (2022). Using single IRB consultations to meet the educational needs of investigative teams. *Contemporary Clinical Trials Communications*, 29, 100971. DOI: 10.1016/j.conctc.2022.100971 PMID: 36033361

Klitzman, R., Pivovarova, E., & Lidz, C. W. (2017). Single IRBs in multisite trials: Questions posed by the new NIH policy. *Journal of the American Medical Association*, 317(20), 2061–2062. DOI: 10.1001/jama.2017.4624 PMID: 28445582

National Commission for the Protection of Human Subjects of Biomedical and Behavioral Research. (1978). *The Belmont Report: Ethical principles and guidelines for the protection of human subjects of research.* https://www.hhs.gov/ohrp/sites/default/files/ the-belmont-report-508c_FINAL.pdf

Smale, M. A. (2010). Demystifying the IRB: Human subjects research in academic libraries. *portal. Portal (Baltimore, Md.)*, 10(3), 309–321. DOI: 10.1353/pla.0.0114

Zywicki, T. J. (2007). Institutional Review Boards as academic bureaucracies: An economic and experiential analysis. *Northwestern University Law Review*, 101(2), 861–895.

ADDITIONAL READING

Abbott, L., & Grady, C. (2011). A systematic review of the empirical literature evaluating IRBs: What we know and what we still need to learn. *Journal of Empirical Research on Human Research Ethics; JERHRE*, 6(1), 3–20. DOI: 10.1525/jer.2011.6.1.3 PMID: 21460582

Babb, S. (2020). *Regulating human research: IRBs from peer review to compliance bureaucracy*. Stanford University Press. DOI: 10.1515/9781503611238

Bledsoe, C. H., Sherin, B., Galinsky, A. G., Headley, N. M., Heimer, C. A., Kjeldgaard, E., Lindgren, J., Miller, J. D., Roloff, M. E., & Uttal, D. H. (2007). Regulating creativity: Research and survival in the IRB iron cage. *Northwestern University Law Review*, 101(2), 593–641.

Burris, S., & Moss, K. (2006). U. S. health researchers review their ethics review boards: A qualitative study. *Journal of Empirical Research on Human Research Ethics; JERHRE*, 1(2), 39–58. DOI: 10.1525/jer.2006.1.2.39 PMID: 19385877

Dingwall, R. (2012). How did we ever get into this mess? The rise of ethical regulation in the social sciences. *Studies in Qualitative Methodology*, 12, 3–26. DOI: 10.1108/S1042-3192(2012)0000012004

DOI: 10.1108/S1042-3192(2012)0000012004

Emanuel, E. J., Grady, C. C., Crouch, R. A., Lie, R. K., Miller, F. G., & Wendler, D. D. (2008). *The Oxford textbook of clinical research ethics*. Oxford University Press. DOI: 10.1093/oso/9780195168655.001.0001

Feeley, M. M. (2007). Legality, social research, and the challenge of Institutional Review Boards. *Law & Society Review*, 41(4), 757–776. DOI: 10.1111/j.1540-5893.2007.00322.x

DOI: 10.1111/j.1540-5893.2007.00322.x

Grinnell, F., Sadler, J. Z., McNamara, V., Senetar, K., & Reisch, J. (2017). Confidence of IRB/ REC members in their assessments of human research risk: A study of IRB/REC decision making in action. *Journal of Empirical Research on Human Research Ethics; JERHRE*, 12(3), 140–149. DOI: 10.1177/1556264617710386 PMID: 28558484

KEY TERMS AND DEFINITIONS

PI Education: Principal Investigator Education refers to structured learning programs designed to enhance professional skills and knowledge about Institutional Review Board requirements.

Tailored Services: Customized IRB services specifically including self-scheduled, customized group sessions with varied content, pre-submitted material review, dynamic feedback, and additional case studies and interactive exercises, both in sessions and as asynchronous resources.

1:1 Consultations: One-on-one meetings between an IRB Specialist and a PI to provide personalized IRB advice, guidance, or support on their specific needs related to application submission.

Office Hours: Designated times when IRB Specialists are available to meet with student or faculty researchers to answer questions, provide support, and/or discuss concerns.

Dynamic Feedback: Real-time, interactive, and constructive responses provided by IRB Specialists to researchers based on their IRB application, materials, or research development ideas, aimed at successfully navigating the IRB process.

Application Process: The series of steps or procedures that a researcher must follow to apply for IRB approval, often including submitting forms, providing documentation, and possibly undergoing multiple revision cycles.

Comprehensive Website: A detailed and extensive online platform that provides a wide range of information, resources, and functionalities to users, covering all aspects of the IRB process.

Asynchronous Resources: Educational or informational materials that can be accessed and used by individuals at any time, without the need for real-time interaction or participation.

Continuous Improvement: An ongoing effort to enhance IRB resources, services, processes, and practices by regularly assessing performance and implementing changes to achieve better outcomes.

Best Practices: Widely accepted, optimal methods or techniques that are recognized as superior and effective within the IRB field, often based on research, experience, and evidence.

ENDNOTES

1 A portion of the concepts presented in this chapter were previously shared in a conference poster session at the PRIM&R Annual SBER Conference, December 3, 2023. https://primr23-sber23.eventscribe.net/fsPopup.asp?PosterID=632282&mode=posterInfo

Chapter 3
How Much Is Enough?
Informed Consent in Healthcare Minimal-Risk Research and Quality Improvement

Paula Garcia McAllister
https://orcid.org/0009-0000-0848-2591
Northern Arizona Healthcare, USA

ABSTRACT

This chapter provides key takeaways for investigators and clinical researchers in healthcare settings doing minimal-risk research and quality improvement. Following an in-depth overview of the current regulations for informed consent (IC) with a focus on the 2018 regulatory revisions, it then describes the current state of IC in quality improvement activities commonly employed in learning healthcare systems and the ethical challenges they present. This chapter does not address the complexities of IC in clinical trials or in research with greater than minimal risk because the IC processes in such activities are well established. Instead, the focus is on what the regulations state regarding the requirements of informed consent; when and how consent can be waived; how research and quality improvement activities differ from the viewpoint of IC; and what information potential research subjects need to make decisions about participating in minimal-risk research and quality improvement in healthcare settings.

DOI: 10.4018/979-8-3693-3848-3.ch003

INTRODUCTION

Informed Consent (IC) is a fundamental cornerstone of ethical practice in human subjects research. Enshrined in the Belmont Report (National Commission, 1979) under the ethical principle of *respect for persons*, informed consent must include information, comprehension, and voluntariness (Eyal & Magalhaes, 2022). Since the adoption of the Common Rule in the 1970s, achieving this bar ethically and practicably has proven to be one of the more challenging aspects of conducting research with human participants.

The implementation of informed consent as a process has focused mainly on the informed consent form (ICF), a detailed document that can run in length to 20 pages or more and can be cumbersome and overwhelming to prospective participants (Bankert, 2022; Eeckhout et al., 2023; Grady, 2022). The regulations under the United States Department of Health and Human Services Office of Human Research Protection (OHRP) allow for various alternatives including waiver or alteration of informed consent, waiver of documented consent, and broad consent for secondary research, all conditioned on specific criteria (Protection of Human Subjects, 2009, 2018). In clinical settings conducting more than minimal risk research, IC is not only governed by OHRP rules, but also by the Food and Drug Administration (FDA) and Health Insurance Portability and Accountability Act (HIPAA) regulations. The over-layering of regulatory bodies that oversee various types of research and research-like activities creates questions and confusion, especially when conducting activities that cannot be clearly designated as either research or quality improvement but instead consist of elements of both (Finkelstein et al., 2015; Hunt et al., 2021).

Investigators need to know which circumstances trigger the various sets of rules, how the rules should be implemented, and what steps can be taken to deliver informed consent in a practical and achievable manner while maintaining the ethical intent of *respect for persons*. This chapter describes the changing landscape of informed consent within the post-2018 Common Rule revisions (Protection of Human Subjects, 2018) and provides IC strategies when conducting minimal-risk research and quality improvement in healthcare settings.

INFORMED CONSENT IN RESEARCH

The regulations regarding IC are well established in 45 CFR Part 46 and, in 2018, they were expanded and updated to address current research methodologies. For research that is considered more than minimal risk, very specific basic elements plus a set of additional elements are required in IC (see 45 CFR 46.116 General Requirements for Informed Consent). However, if certain criteria are met, IC can be waived or altered. In addition, the documentation of consent, that is, the signing of the consent form by either the participant or their legally authorized representative (LAR), can also be waived or altered. Nevertheless, most clinical trial research involves the use of full-length consent forms that can be overwhelming to prospective participants and to research personnel tasked with obtaining consent.

Current criticisms of the informed consent process have focused on the perceived burdens experienced by researchers and participants alike, as well as on the ethical questions that arise when such hardships are placed on vulnerable participant populations (Anderson et al., 2017; Godskesen et al., 2023; Pietrzykowski & Smilowska, 2021; Zondag, et al., 2023). Indeed, a recent study by Eeckhout et al. (2023) identified barriers to IC such as patient level of education and health literacy, lengthiness of IC documents, and a feeling of intimidation among patients when reading consent forms with complex medical and legal jargon. Godskesen et al. (2023) interviewed clinical research nurses responsible for seeking consent from prospective patient participants in order to discover perceived issues and barriers. The clinical research nurses identified a misunderstanding of voluntariness, language barriers, and problems addressing the needs of diverse patients as ethical challenges when conducting IC.

Anderson et al. (2017) conducted focus groups with former research participants and with research staff responsible for obtaining IC from prospective participants. Their findings among the former research participants revealed that although participants viewed their research participation positively, they identified shortcomings with the informed consent process and content. The former research participants specifically pointed out that ICFs provided a great deal of information, much of which they saw as unneeded, and used complex medical jargon making the form hard to understand. Research staff observed the same shortcomings regarding the length and complexity of consent forms, and stated that they would like more tools for explaining the research to prospective participants in ways that helped them see the study in larger context.

In a systematic review of fourteen peer-reviewed research articles on clinical trial participants' comprehension of ICFs, Pietrzykowski & Smilowska (2021) found that subjects had varying levels of understanding of study details. Each of the articles included in the systematic review employed questionnaires to objectively evaluate participants' comprehension. They found that the highest levels of understanding among research participants was around voluntary participation, blinding of participant group assignment, and freedom to withdraw at any time. Participants were found to struggle with understanding the concept of placebo, randomization, safety, risks, and side effects; and, they had limited comprehension of research benefits. It is notable that the more technical the explanation of how the research was conducted was, the less information was comprehended by the research participants. The study authors argue that the shortcomings in ICF comprehension undermine the ethical foundations of the conduct of clinical trials, suggesting that the research community should question whether patients are truly exercising their autonomy in their decision-making process.

These studies show that IC can be cumbersome and misguided in clinical trials that require a full consent process. However, this dynamic changes in quality improvement initiatives where patient consent is complicated by large numbers of participants or by the use of retrospective research design.

IC in Quality Improvement (QI)

The past decade or so has seen an explosion in the analysis of in-context patient care data with the goal of continually improving care, specifically by focusing on quality improvement (QI), comparative effectiveness, and medical registries (Stiegler & Tung, 2017). With the implementation of the electronic medical record (EMR), the amount of patient data available has increased extensively in all areas. More than ever before, healthcare researchers can access detailed information on all aspects of a patient's care, including demographics, lab results, medications, diagnoses, treatment orders, insurance carriers, and so on—in short, every aspect of a patient's hospital stay from admission to discharge. These activities can raise the level of risk to patients and create the need for some degree of ethical oversight (Hall et al., 2020; Hunt et al., 2021; Nerenz, et al., 2003; Perneger, 2004).

QI has not been held to the same standards for ethical approval as research, and most QI projects do not require any type of consent from patients (Carrithers & Morain, 2022; Denison, et al., 2024; Johnson, et al., 2006). This relaxation of regulatory guidelines leads to the question: How can healthcare investigators make sure they are doing the right thing when carrying out quality improvement projects? At the heart of the matter, participants have a right to know that they are participating in research, whether minimal risk or not. Do they also have the right to know that their information is being used in analyses for QI purposes? If so, it can be argued that some form of informed consent should be included. The following review of the baseline federal requirements regarding patient consent may provide some guidance on how investigators can implement a level of consent that would be appropriate for the QI space.

CURRENT STATE OF THE REGULATIONS

The revised Common Rule that took effect in 2018 maintained most of the original regulations from the initial 45 CFR 46: The Protection of Human Subjects, published in 1974 as part of the National Research Act. Chapter 46.116 covers general requirements for informed consent and chapter 46.117 covers documentation of informed consent. These sections outline the rules of engagement for informing patients about the research and seeking their consent. The rules are categorized into three types: general requirements, basic elements, and additional elements. The general requirements stipulate obtaining legally effective IC prior to participation in research that falls under the federal regulations. Research, as a term, has a very specific meaning in the federal regulations as "a systematic investigation, including development, testing, and evaluation, designed to develop or contribute to generalizable knowledge" (45 CFR 46.102(l)). If a study does not meet this definition, then it does not fall under the regulatory requirements of the US Department of Health and Human Services Common Rule and the National Research Act.

The regulations in 45 CFR 46.116 begin with a list of baseline requirements that represent the minimum expectations when approaching prospective participants about a research study. The general requirements of IC include the following items and characteristics (45 CFR 46.116(a)):

- Prospective research participants or their legally authorized representatives (LAR) have sufficient opportunity to discuss and consider whether or not to participate and that the possibility of coercion or undue influence is minimized.
- The information is provided in a language understandable to the participant or LAR, which is often measured by reading level tests such as the Flesch-Kincaid Reading Ease score.
- The information is sufficient for a reasonable person to be able to make an informed decision about whether or not to participate, and the prospective participant has had an opportunity to discuss the information.
- Key information that is most likely to assist a prospective participant or LAR in understanding the reasons why one might or might not want to participate in the research is presented in sufficient detail and in a way that facilitates comprehension.
- No exculpatory language is present through which the prospective participant or LAR is made to waive or appear to waive any of the participant's legal rights.

In addition to the general requirements listed above, full informed consent materials that are not waived or altered must include the following basic elements (45 CFR 46.116(b)):

- A statement that the activity for which the investigator is seeking consent is research which includes a purpose for the research, expected duration of participation, a description of procedures, and whether any of the procedures are experimental

- Risks and benefits: A list of any reasonably foreseeable risks or discomforts and a description of any benefits that might be expected either to the individual participant or to others
- Information about any appropriate alternative procedures or treatments that might be beneficial to the participant
- Information about how confidentiality will be maintained, unless it is not maintained according to the research protocol
- For research that is more than minimal risk, an explanation as to whether any compensation or medical treatments would be available in the event of harm or injury, what those may be, and how to find out more information about them
- Contact information for answers to pertinent questions about the research and research subjects' rights, and whom to contact in the event of research-related injury or harm
- A statement that participation in the research at hand is voluntary and that refusal to participate is without penalty and will not deny services, treatments, or benefits to which the participant is otherwise entitled

The 2018 revised Common Rule added to the basic elements a provision specific to biomedical and hospital-based research that involves the collection of identifiable private information or identifiable biospecimens (45 CFR 46.116(b)(9)). Such research must include one of the following:

- A statement that identifiers might be removed from the identifiable private information or identifiable biospecimens and that, after such removal, the information or biospecimens could be used for future research or distributed to another investigator for future research without additional informed consent from the participant or their legally authorized representative, if this might be a possibility
- A statement that the participant's information or biospecimens collected as part of the research study, even if stripped of identifiers, will not be used or distributed for future research studies

These additional provisions were included to account for the possibility of data and samples, such as health information or biospecimens, being used in secondary research. Under CFR 45 Part 46.102, identifiable private information and identifiable biospecimens refers to information or biospecimens "for which the identity of the subject is or may readily be ascertained by the investigator or associated with the information or biospecimen" (45 CFR 46.102). Healthcare organizations also include the 18 HIPAA identifiers when determining whether data is identifiable (see Office of Civil Rights, 1996).

The regulations identify several additional elements to the IC information, if appropriate. These include statements regarding:

- Whether there is an increased risk in the event that the participant becomes pregnant while enrolled in the study
- That a participant's enrollment may be terminated at the discretion of the investigator
- Any additional costs that the participant might incur
- The consequences of a participant's withdrawal from the study and procedures for orderly termination of participation
- That new findings that may influence a subject's willingness to continue in the study will be shared with the participant
- The approximant number of participants to be enrolled in the study (45 CFR 46.116(c))

To address contemporary concerns, the revised Common Rule included the following provisions as additional elements: a statement that biospecimens, even if identifiers are removed, may be used for commercial profit that may or may not be shared with the participant; whether clinically relevant results will be disclosed to subjects, and under what conditions; and whether the research might include whole genome sequencing (in the case of biospecimens collection) (45 CFR 46.116(c)(7-9)). The additional provisions regarding the possibility of biospecimens being used for commercial profit and whole genome sequencing reflect the extent to which biomedical research has advanced since the inception of the Common Rule, and how these advances have the potential to increase possible risks to participating in research. By adding them to the elements of consent, OHRP has acknowledged the need to update regulations to meet the changing landscape of research.

Broad Consent

Also included in the revised Common Rule is a provision referred to as broad consent, which is meant to streamline the utilization of identifiable private information or identifiable biospecimens in secondary research. The regulations on broad consent outline how participants should be informed of the storage, maintenance, and potential secondary uses for their identifiable private information and identifiable biospecimens collected for research studies and for non-research purposes. The intent of broad consent was to ameliorate the burden of obtaining participant consent for every use and follow-up use of identifiable private information and identifiable biospecimens.

In addition to many of the basic and additional elements listed above, IC forms for broad consent have additional required elements outlined in 46.116(d), which include:

- A general description of the types of research that may be conducted in which the participants' data may be used
- A description of the identifiable private information or identifiable biospecimens that might be used in the research, whether it will be shared with other institutions, and the types of researchers and institutions that might conduct this research
- A description of the period of time that the identifiable private information or identifiable biospecimens may be stored, maintained, and used for research purposes
- A statement that participants or their LARs will not be provided details about specific future research studies in which their identifiable private information or identifiable biospecimens is used, unless otherwise arranged
- A statement that clinically relevant research results may not be disclosed to the participant, unless otherwise specified
- Contact information for questions regarding a participant's rights and storage and use of a participant's identifiable private information or identifiable biospecimens, and whom to contact in the event of research-related harm

One important caveat to the broad consent process arises in the event of a participant declining broad consent. That is, researchers need to determine what is to be done in the event that a participant does not want their identifiable private information or identifiable biospecimens to be used in secondary research and refuses the broad consent. In such cases, participants who refuse the broad consent would expect to have their identifiable private information or identifiable biospecimens removed from the possible pool of included samples. In a busy research environment, this has been seen as overly complicated and risky because of the possibility of a refusing participant's identifiable private information or identifiable biospecimens inadvertently being included in a secondary research study, thereby exposing the research institution to liability.

Recent posts on the use of broad consent in a human research protection discussion forum reveal that few institutions employ the broad consent option (personal communication on PRIM&R IRB Forum, January 25 2023). Instead, they simply de-identify the private information and biospecimens prior to making them available for secondary research without IC. This practice is acceptable because the additional elements that were added in the 2018 revision allow for private information and biospecimens to be used in secondary research as long as they are de-identified (45 CFR 46.116(b)(9)). The reasoning is that once the private information and biospecimens are de-identified, there is no human subject for the regulations to protect; therefore, consent is not needed, and investigators can do what they wish with the de-identified private information and de-identified biospecimens.

General Waiver or Alteration of Consent

The federal regulations outlined in 45 CFR 46 have always allowed for a general waiver or alteration of consent if certain conditions are met. A waiver refers to the elimination of any informed consent whatsoever, whereas an alteration refers to modifying the informed consent information that is provided to potential participants. For instance, an alteration of consent may involve not providing all of the information listed as basic or additional elements in the discussion above, or it might mean the absence of the signature on the informed consent form (see more below).

Investigators requesting waiver or alteration of IC must demonstrate to the IRB that the following five conditions are met. All five of the conditions for allowing waiver or alteration of IC must be presented and approved by the institutional review board (IRB). These conditions include the following:

- The research involves no more than minimal risk to the subjects (i.e., no more than the risk encountered in everyday life).
- The research could not practicably be carried out without the requested waiver or alteration.
- If the research involves identifiable private information or identifiable biospecimens, the research could not practically be carried out without using such information or biospecimens in an identifiable format.
- The waiver or alteration will not adversely affect the rights and welfare of the subjects.
- Whenever appropriate, the subjects or legally authorized representatives will be provided with additional pertinent information after participation.

Researchers who seek approval of waiver or alteration of consent must present their justification for the request to the IRB. It is the responsibility of the IRB members to consider whether the research meets the criteria listed above and thus qualifies for approval of the waiver or alteration.

In the case of alteration of IC, the general required elements must still be provided; however, there is leeway regarding what basic or additional elements are included. The rationale for maintaining the general requirements is that, if the research falls under the federal regulations, then it should follow the principles of the Belmont Report, and the rules and regulations passed under the National Research Act and described in 45 CFR 46. This requirement directly addresses the ethical principle that research participants have the right to know that they are participating in research and what that participation involves and to be given sufficient time to discuss and think about whether or not they want to participate.

In the case of broad consent, researchers should keep in mind that consent cannot be waived for a participant who refuses to sign a broad consent. The DHHS Protection of Human Subjects, Part 46.116(f)(1), states:

> *"If an individual was asked to provide broad consent for the storage, maintenance, and secondary research use of identifiable private information or identifiable biospecimens...and refused to consent, an IRB cannot waive consent for the storage, maintenance, or secondary research use of identifiable private information or identifiable biospecimens."*

In other words, the waiver of consent cannot be used as a work around for a failed broad consent.

Documentation of Consent

Investigators seeking informed consent from participants often equate obtaining IC with securing a signature from the participant or their LAR when, in fact, IC is meant to be a process that includes a conversation between the researcher and the prospective participant and their representatives. Indeed, the regulations in 46.117 quickly call out the need to document IC with a written form that is approved by the IRB and signed by the participant or their LAR. In addition, participants should be given a copy of the IC that they signed, whether electronically delivered or on paper.

However, the Common Rule has always allowed for some latitude in the documentation of a signed IC. An IC form does not have to be signed if it creates additional risk to the participant by linking them to the research, and the principal risk of harm is breach of confidentiality (45 CFR 46.117(c)(1)(i)). This avenue is often utilized in research studies where the participants undertake activities that are illegal or their identities are especially sensitive to law enforcement, immigration, or other authorities. Other instances where signature may be waived are when the research presents no more than minimal risk and does not involve procedures in which written consent is normally required or in cultures where signing forms is not the norm. Alternatively, voice recorded consent in research that involves recording participants, such as interviews, has also been accepted in lieu of written signatures. Of course, all of these alternative methods must be reviewed and approved by the IRB and are only allowable if the research is minimal risk.

The leeway provided by the waiver or alteration of documented consent gives investigators important guidance in ensuring the responsible conduct of research and ethical patient involvement in minimal risk research. An example of an alteration of consent is Pradeep et al. (2024) that looked at malnutrition and types of treatment among cancer patients in an outpatient clinical setting. This study involved providing a nutrition screening tool to patients who came into the oncology clinic. Patients whose screen showed poor nutrition were recommended consultation with the dietician. Data from the electronic medical record was collected on all of the screened patients to analyze which cancer patients had a greater propensity for being malnourished. Information on this patient group was then used to draw inferences about how clinical treatment of different types of cancer could be associated with malnutrition. Although this exploratory research project was considered minimal risk because malnutrition screening and dietician consultation are standard care and no other patient interaction took place, the IRB recommended a modified one-page IC form that informed patients that their data would be used and provided an opportunity to opt out of the study if they so wished. This example shows how IC can be implemented in an altered format without overly burdening patients and clinicians with a lengthy IC form and process, while also informing patients about the study and offering them an option to not participate.

Is QI Exempt from IC? Is it Research?

The proceeding discussion focused on the inarguable centrality of informed consent in activities that are identified as research. Quality improvement activities, on the other hand, have long been seen as different from research based on several distinguishing markers such as the purpose of the activity, the methods used, the nature of the risk faced by patients, and the roles or job positions of the investigators. Since the early 2000's, greater attention to the quality of care in hospitals and medical centers has prompted questions about whether QI activities amount to research. And, if so, do they necessitate IRB review? Traditionally, the goal of publication or presentation of results has been a distinguishing feature that could be applied when identifying a project as research rather than QI. However, as QI grows as an academic discipline in and of itself, thus expanding the dissemination of findings from all types of healthcare centers, a closer scrutiny of all aspects of QI activities should be considered when determining whether an activity is research or strictly QI (Hall, et al., 2020; Johnson, et al., 2006; Nerenz, et al., 2003; Stiegler & Tung, 2017).

In an early discussion on this topic, Nerenz et al. (2003) argued that the determination of whether an activity is research or QI rests on the fundamental objective or intent of the investigators. If the fundamental objective is to contribute to generalizable knowledge, a long-held characteristic included in the definition of research (45 CFR 46.102(l)), then the activity is deemed to be research. If the fundamental objective is to improve a local process, meaning a process that exists within a specific hospital or medical center, then the activity is considered to be QI. However, not all activities can be neatly classified as contributing to generalizable knowledge or intended to improve a local process. The authors elaborate on the criteria for generalizability to include sampling methods, selection of independent and dependent variables, the types of statistics used in analysis, and the dissemination of results in academic papers and presentations. Nerenz et al. (2003) acknowledge the potential for QI that has been deemed to not be research to still present some risk to patients and therefore they recommend some type of review where QI activities are evaluated on a scale of 1 to 5 based on potential risks to patients weighed against potential benefits.

Also looking at alternative forms of review, Perneger (2004) proposed a "management ethics committee" (p. 343) whose duties would include review of quality improvement projects as well as management decisions that could affect the quality of care. Predating the time of widespread electronic medical records that greatly expanded the scope and nature of QI activities, the QI initiatives of concern in Perneger's article focused on management or policy decisions such as decreases in personnel who deliver direct patient care in order to save money, or problems with paper-based patient records. Despite this outdated perspective on the full range of ethical issues that appear in today's QI space, Perneger acknowledges the impracticality of seeking informed consent from patients included in QI initiatives. He calls for a "community informed consent" (p. 344) that is obtained from a panel of former patients and other community representatives as a means of addressing the spirit of *respect for persons* and the Belmont Report. By adding review requirements such as a management ethics committee and a community informed consent, investigators and healthcare leaders would be able to show careful consideration of the ethical concerns that are inherent in QI initiatives. Granted, this would cause QI projects to be subject to more administrative red tape than usual, but, as Pereneger argues, QI projects would be more formalized and better planned because of these additional reviews.

Johnson et al. (2006) presented this question of QI or research to academic medical centers in the United States to find out how their IRBs address QI activities. They developed a survey asking respondents if they had policies on distinguishing QI from research, and to identify the criteria used to guide their decisions in making this distinction. The 34 institutions that responded to their survey identified elements that classified an activity as research to include: randomization of study subjects, the use of interventions outside of the standard of care, additional risk or collection of data such as private health information (PHI) outside of standard practice, and the intention to publish or present findings. Activities more likely to be classified as QI impacted a specific organizational environment or healthcare facility rather than contributing to the broader scientific community. Also, if the project only monitored existing practices or processes within the organization and involved data accessed by the investigators as part of their routine and usual responsibilities, the project was more likely to be classified as QI rather than research. Some respondents also identified the need for informed consent as an element that would classify an activity as research. However, this reflects a degree of circular logic—if the activity requires informed consent, it is research; and if it is research, it may require informed consent. Perhaps this feature should not be a criterion for research versus QI classification. As a deliverable for their survey findings, Johnson et al. (2006) present an algorithm for IRBs to use as a guide to distinguish QI from research. Criteria in the algorithm include the intent to disseminate outside of the institution, the potential to contribute to generalizable knowledge, randomization of subjects, the presence of an intervention, an increase in risk to patients outside of standard care, an increase in patient information beyond standard care, sending PHI outside of the institution, and the potential infringement on privacy or confidentiality. They recommended that activities that involve any one of these criteria should be reviewed by the IRB as research.

The controversy surrounding how to classify QI is still actively debated, particularly within academic medical centers. At one large teaching hospital, the leadership created a Quality Improvement Review Board (QIRB) comprised of members from the institution's compliance office, the research-oriented IRB, and faculty from nursing, pharmacy, medicine, and biomedical sciences (Denison, et al., 2024). The authors very clearly state that research and QI are fundamentally distinct; however, those distinctions were not readily recognized by many within their organization. Therefore, they developed a mechanism by which projects are first identified as either human subjects research or QI. Projects that are determined to be QI go through an application process much like an IRB process but designed to evaluate the characteristics and activities that are more prevalent in QI. The QIRB applies a decision tree that distinguishes between QI and research in which studies exhibiting elements traditionally associated with research, such as contributing to generalizable knowledge, the randomization of subjects, and the presence of a control group, are referred to the research IRB. Features of studies referred to the QIRB include a goal to improve upon a performance gap at the local site; the use of evidence-based practices; the inclusion of all patients, not just a sample group; the monitoring of data over time; and the continuous implementation of findings to improve clinical outcomes. Since the implementation of the QIRB, Denison et al. (2024) report several benefits, including an easing of the burden on the IRB staff; an elevation of QI as a scholarly activity; increased protection for the institution, faculty, and participants; and greater rigor of QI methods.

A LOOK AT LEARNING HEALTHCARE SYSTEMS

With the passage of the Affordable Care Act (ACA) in 2010, which tied Medicare reimbursement to quality of care, hospitals and hospital systems have taken a more proactive stance toward improving patient care outcomes. The learning healthcare system (LHS) model is a way to approach this objective by applying patient data into self-reflective evaluation of care in order to develop improved care process models. LHS is described as the continuous development of knowledge informed by the vast computational power of the electronic medical record (EMR) and applied to the improvement of patient care (Institute of Medicine, 2013). The LHS approach is one way that healthcare has directed greater attention to improving health outcomes because it dives deeply into specific targeted patient data, abstracted from the EMR and other computerized data sources, to conduct multiple real-time analyses that are used to inform individual healthcare centers about their specific quality of care. In the last decade or so, this has been transformational due to the massive size and scope of data that can by analyzed, parsed, compared, and reported in ways that have never been possible before. LHS methodology allows hospitals and medical centers to put their own patient data to use by pointing out gaps in care and safety and identify areas for improvement, thus contributing to informed quality of care within their own health systems (Kass et al., 2013).

LHS practice that involves the widespread collection and use of patient data gives rise to questions about how patient populations whose private health information is collected, stored, and analyzed for LHS purposes are informed about and provided an opportunity for consent regarding the use of their data. The answer is that, in most cases, they are not provided with an opportunity to consent, nor is it always appropriate to do so (Denison, et al., 2024; Faden et al., 2014). Several arguments can be made for excusing minimal-risk LHS activities from seeking IC. First, there is the feasibility of trying to obtain consent from the high numbers, sometimes in the tens of thousands or more, of patients whose private health information and biospecimens may be used in these activities. Investigators can argue the impracticability of seeking consent for this reason. Second, although information extracted from the EMR is identifiable upon collection, it is almost always de-identified when presented in data reports (Carrithers & Morain, 2022) and, therefore, LHS activities are not considered to be human subjects research that is overseen by the IRB. Another reason for not seeking consent is that the information collected is usually retrospective, mostly minimal risk, and arguably not generalizable. Such activities have historically been classified by the regulations as not human subjects research and, therefore, do not necessitate IC.

However, given the increasingly sophisticated ways of extracting data from electronic sources, QI can delve into patient characteristics that have historically required additional safeguards according to the regulations. For example, a QI inquiry looking at patient outcomes can be refined to categorize patients along racial, ethnic, socio-economic, and other demographic lines. Investigators can easily extract large data sets that include vulnerable populations such as Native Americans, low-income individuals, or people with histories of substance abuse without seeking the safeguards that an IRB would normally require. Healthcare systems should equip themselves with policies and guidelines for the role of IC in QI and other investigations in order to maintain protection of vulnerable populations and sensitive patient groups.

QI and LHS activities do not typically require the same level of ethical approval that has been required of research studies (Fiscella, et al., 2015; Hunt et al., 2021). To make up for this gap, several studies published in *The Hastings Center Report* focus on the ethical oversight of data collection in QI and LHS activities currently taking place in learning heath care systems (see *Hastings Center Report* volume 43, issue 1 and *The Hastings Center Report* volume 52, issue 3). Faden et al. (2013) and Faden and Kass (2022) presented an ethics framework addressing IC by embedding ethical responsibilities within the policies and practices put in place by the healthcare organization and carried out by researchers, clinicians, healthcare system administrators, and even payers, purchasers, and patients. They describe a framework consisting of seven moral obligations, which include

- Respecting the rights and dignities of patients,
- Respecting clinical judgements of clinicians,
- Providing optimal clinical care to each patient,
- Avoiding nonclinical risks and burdens on patients,
- Addressing health inequalities,
- Conducting continuous learning activities that improve the quality of clinical care and healthcare systems, and
- Contributing to the common purpose of improving the quality and value of clinical care and healthcare systems. (Faden et al., 2013, p. 18)

The implementation of this framework assumes a healthcare organization is ethically robust and has the capacity and resources to enact a high level of patient engagement, transparency, and accountability. Ideally, healthcare systems can achieve all seven of these obligations, but given the fast-paced nature of technological changes and after the devastation that COVID wreaked on many healthcare systems, most hospitals would be hard-pressed to meet all of the principles outlined in this framework. Instead, the value of this framework is to consider the seven moral obligations as goals toward which we should strive. Many hospitals and healthcare systems approach at least the first four of these obligations (Morain, et al., 2022). In addition, hospitals have employed increased patient involvement with the implementation of patient advisory committees and similar patient engagement activities that have the potential to help shape a LHS research agenda (see Bombard et al., 2018 for a fuller description of patient engagement models).

Variable practices regarding the implementation of IC in LHS activities can have real-world consequences affecting patient care. In a research study focusing on an IC process within a LHS model, Zondag et al. (2023) reported on a cardiovascular (CV) risk management program that allowed patients to opt out of participating in the analysis yet collected and maintained their health information as part of their care. The study compared patients who consented, patients who did not consent, and patients who did not respond to the consent request along various health risks and demographic factors. This IC study was secondary to the overall CV LHS study and was waived for consent by the healthcare organization's IRB. Zondag et al.'s (2023) findings showed that in asking consent, factors such as age, education level, CV risk factors, and clinical referral area (e.g., geriatrics, nephrology) impacted whether a patient consented, declined consent, or simply did not respond. Their analysis demonstrated participation bias coming into play, thus potentially undermining the validity and generalizability of results. The researchers faulted the IC process adopted by the CV LHS program with causing a misrepresentation of the target population and hampering the generalizability of findings. Their regression model showed that certain types of patients were more likely to either not consent or not respond and resulted in diminishing the representation of those patients in the CV LHS study. The researchers noted that those patients not included were at risk of a decrease in quality of care, risk factor management, and appropriate treatment strategy because they did not take advantage of the feedback loop that is part of the LHS model. Zondag et al. (2023) argue for a reassessment of the use of a traditional IC process in LHS models that employ routine clinical data, which is looked at more closely in the next section.

Other Approaches to Informed Consent

Since the inception of the LHS model and large-scale quality improvement, ethicists and researchers alike have been proposing alternative approaches to the traditional IC format with the goal of making it useful and relevant to prospective participants and the purposes of the activity at hand. A recent training video from the Office of Human Research Protection (OHRP, 2024) provides guidance on how to make consent forms more meaningful to potential research participants through a participant-centered approach. This approach promotes adopting a communication style that is comprehensible to potential participants and has their perspectives in mind. OHRP acknowledges that some consent forms include too many unnecessary details written in overly complex language that is more appropriate for an audience of scientific reviewers than prospective participants. In order to engage this approach successfully, investigators must have intimate knowledge of the population involved in order to explain the study so that it is relevant to them within their given context. When taking such an approach, it may not be critical to include all of the required items listed in the elements of consent in 45 CFR 46.116. The video training recommends focusing consent forms on the information that potential participants need in order to make an informed decision, presenting the information in language that is clear and easy to understand, and using strategies to communicate effectively to the target audience. By providing such guidance, OHRP acknowledges the patient-friendly shortcomings of traditional IC and demonstrates its support for flexibility in IC for the sake of improved patient comprehension and engagement in research.

Eeckhout et al. (2023) recommended making the IC process more participant-centered in their qualitative analysis of semi-structured interviews with research staff at a university research hospital. The staff identified barriers in communication as the main reason for hindering IC and provided strategies to improve the IC process. Their improvements included adapting consent documents to improve readability, providing education to the general patient population so that they can make better decisions for themselves when they are approached about a study, and training research staff on how to be more effective communicators among diverse patient groups. Current practices of constructing lengthy and technically worded consent documents that are often standardized across multiple sites, such as in the case of multi-site research projects, need to be re-considered if researchers want to respond to the call for inclusive research participant recruitment. Adapting consent documents to meet prospective participants' comprehension needs may be a challenge, but it is something that IRBs and research support staff should implement whenever possible.

In a call for ethical oversight of QI to include IC, Fiscella et al. (2015) describe the challenges of balancing potential risks to participants with the imperative of improving quality of care. They recommend expanding the institution's oversight committee already in place, such as the local IRB, to include patients who may be impacted by the QI activities and ethics experts experienced with weighing risks and benefits of clinical care such as those who sit on the clinical ethics committee, if one is in place. Such an expanded oversight committee would provide guidance as to whether IC was needed in particular study and the best way to approach it. They also identified the possibility of developing a universal consent that would be presented to patients upon enrollment in care with the option of removing their data if they prefer. This option would be mechanized by an automatic email or text message that alerts patients to upcoming studies in which their health data may be included and allows them to choose to opt-in or opt-out. Such a complex system may be difficult to implement given the changing, fast-paced nature of healthcare, but it may have possible potential in the future.

Bhupathi and Ravi (2017) presented an alternative approach to IC that involves a question/answer format and complies with the International Declaration of Helsinki guidelines (World Medical Association, 2018). The authors present this conversational format as an alternative to lengthy multi-page written consent forms for use in clinical trials; however, it could also be quite useful in more minimal risk studies and quality improvement. Markman et al. (2023) recommended engaging members of the community in an informed consent training program to improve researchers' communication skills and train them to think in the moment about real questions that potential participants might have about taking part in research. This type of contemporary community outreach comes close to responding to the ethical obligations in Faden et al.'s (2013) common purpose framework. Within such a framework, everyone involved is "obligated to work toward the goal of continuously participating in learning activities" that improve care (Grady & Wendler, 2013, p. 33).

Morain et al. (2022) described the approach of embedding clinical trials into clinical care as pragmatic clinical trials (PCTs) in a learning healthcare system. In PCTs, interventions are evaluated under real-word conditions in order to inform care delivery and policy development. They recognize the ethical and regulatory challenges with such activities and propose eight dimensions in which healthcare systems can demonstrate *respect for persons* within the context of the trial at hand. The eight dimensions are: engaging patients and communities in research design and execution, promoting transparency and open communication, maximizing agency, minimizing burdens and promoting accessibility, protecting privacy and confidentiality, valuing interpersonal interactions, providing compensation, and maximizing social value. Granted, several of these elements, such as protecting privacy and confidentiality and minimizing burdens, should always be part of any hospital-based research, QI, or LHS activity. The authors argue that by implementing each of these dimensions, written informed consent becomes less necessary because the Belmont principles on which it is based—*respect for persons* and autonomy—are inherently addressed in the practice of these eight dimensions.

Considering what patients would prefer, Jones et al. (2020) collected first-hand insight from cancer patients on their preferences regarding IC in LHS activities. Using a qualitative democratic deliberation method, the researchers asked 217 cancer patients to indicate their preferences from among four policy choices ranging from no notification of personal health information use to notification with an option to have personal health information removed. The majority of patients preferred the opt-out policy, stating that it balanced their desire for individual control over their own data with the option to contribute to the common good. While patients understood that their personal health information had the potential to inform research and add to the body of knowledge that could assist future patients, they also reserved the right to have their privacy protected. Several possible direct harms identified by this group of patients included: the stigma of having their medical condition be disclosed, financial hardship due to health insurance denying coverage, being harassed by marketers and the like, exploitation from pharmaceutical companies, and other unknown potential harms. The healthcare industry should look carefully at the possible dangers faced by patients, whether real or perceived, and consider how these dangers can either be mitigated or resolved through education or other means.

The discussion of new approaches to IC is rich in suggestions for improvement and philosophical considerations that can guide investigators in various types of healthcare minimal risk research and QI, including LHS environments. These examples point towards practical solutions while still working within the ethical conduct of research.

CONCLUSION

This chapter revisited the Common Rule regulations originally enacted in the early 1970s and revised in 2018 in order to come back to center on why we conduct IC and what IC should entail. It explored how the changing landscape of healthcare and the commitment to improving patient care have complicated IC calling into question what is ethical and what is practical when it comes to IC. Certainly, when thousands of patients' data is involved and de-identification can be promised, it makes sense not to seek IC. However, investigators should keep in mind that high-powered analytics can reveal a lot about certain groups of patients, some of which may be sensitive or vulnerable groups. Investigators should therefore consider finding ways to inform patient groups communally, involve patient representatives in planning QI and LHS initiatives, and seek their counsel in deciding which activities should include a consent process and which should not. By incorporating robust patient engagement strategies in minimal risk research, QI, and LHS activities, healthcare organizations can be transparent and accountable while creating a true patient-provider partnership in healthcare.

When asking the question of whether to obtain consent from patients in minimal risk research, QI, and LHS activities, there are no simple answers. Although each of these areas has the goal of improving patient care, long-held beliefs about research participant autonomy and *respect for people* should not be overlooked when it comes to the question of consent in these contexts. This chapter provided several alternatives and frameworks that investigators can employ in order to maintain the ethical intentions of the research regulations. It is not overly burdensome to make patients aware of the wide range of uses for their data either through a blanket statement upon admission to the hospital or by public service announcements played on monitors throughout the hospital or medical center. More information and official guidance are needed on how healthcare facilities, large and small, can implement such awareness campaigns.

Another strategy for healthcare facilities grappling with the balance between IC and large-scale data analysis for QI or LHS activities is to find the resources to provide patients with education about why their data is collected, the kinds of analyses and measurements their data will inform, and how this information is ultimately used for the overall improvement of patient care. When explained in the light of community benefit, patients may not only accept the reality of their information being used in large-scale data analyses, but actually seek out healthcare providers that engage in such activities because they are seen as continually improving care. Regardless, hospitals and healthcare organizations should have solid policies on this topic and be able to respond to patients' questions about secondary uses of their health information in minimal-risk research, quality improvement, and learning healthcare system activities.

REFERENCES

Anderson, E. E., Newman, S. B., & Matthews, K. A. (2017). Improving informed consent: Stakeholder views. *AJOB Empirical Bioethics*, 8(3), 178–188. DOI: 10.1080/23294515.2017.1362488 PMID: 28949896

Bankert, E. A. (2022). Assessing informed consent communication. In Bankert, E. A., Gordon, B. G., Hurley, E. A., & Shriver, S. P. (Eds.), *Institutional Review Board: Management and function* (3rd ed., pp. 317–322). Jones & Bartlett Learning.

Bhupathi, P. A., & Ravi, G. R. (2017). Comprehensive format of informed consent in research and practice: A tool to uphold the ethical and moral standards. *International Journal of Clinical Pediatric Dentistry*, 10(1), 73–81. DOI: 10.5005/jp-journals-10005-1411 PMID: 28377660

Bombard, Y., Baker, G. R., Orlando, E., Fancott, C., Bhatia, P., Casalino, S., Onate, K., Denis, J. L., & Pomey, M. P. (2018). Engaging patients to improve quality of care: A systematic review. *Implementation Science : IS*, 13(1), 98. Advance online publication. DOI: 10.1186/s13012-018-0784-z PMID: 30045735

Carrithers, J., & Morain, S. (2022). Pragmatic clinical trials and comparative effectiveness research. In Bankert, E. A., Gordon, B. G., Hurley, E. A., & Shriver, S. P. (Eds.), *Institutional Review Board: Management and function* (3rd ed., pp. 671–681). Jones & Bartlett Learning.

Denison, T. L., Sorensen, K. U., Blanton, M. P., Johnson, L., Byrd, T., Pass, S. E., Philips, L., Miller, J., McMahon, L. R., & Cherry, B. (2024, October). (2024). The Quality Improvement Review Board: An innovative approach to oversight of projects that do not meet criteria of human subject research. *Quality Management in Health Care*, 33(4), 269–277. Advance online publication. DOI: 10.1097/QMH.0000000000000446 PMID: 38414295

Eeckhout, D., Aelbrecht, K., & Ver Der Straeten, C. (2023). Informed consent: Research staff's perspectives and practical recommendations to improve research staff-participant communication. *Journal of Empirical Research on Human Research Ethics; JERHRE*, 18(1–2), 3–12. DOI: 10.1177/15562646221146043 PMID: 36562147

Eyal, N., & Magalhaes, M. (2022). The functions of informed consent. In Bankert, E. A., Gordon, B. G., Hurley, E. A., & Shriver, S. P. (Eds.), *Institutional Review Board: Management and function* (3rd ed., pp. 247–256). Jones & Bartlett Learning.

Faden, R. R., Beauchamp, T. L., & Kass, N. E. (2014). Informed consent, comparative effectiveness, and learning health care. *The New England Journal of Medicine*, 370(8), 766–768. DOI: 10.1056/NEJMhle1313674 PMID: 24552325

Faden, R. R., & Kass, N. E. (2022). Learning health care and the obligation to participate in research. *The Hastings Center Report*, 52(3), 29–31. DOI: 10.1002/hast.1393 PMID: 35763203

Faden, R. R., Kass, N. E., Goodman, S. N., Pronovost, S. T., & Beauchamp, T. L. (2013). An ethics framework for a leaning health care system: A departure from traditional research ethics and clinical ethics. *The Hastings Center Report*, 43(s1), S16–S27. DOI: 10.1002/hast.134 PMID: 23315888

Finkelstein, J. A., Brickman, A. L., Capron, A., Ford, D., Gombosev, A., Greene, S. M., Iafrate, R. P., Kolaczkowski, L., Pallin, S. C., Pletcher, M. J., Staman, K. L., Vazquez, M., & Sugarman, J. (2015). Oversight on the borderline: Quality improvement and pragmatic research. *Clinical Trials*, 12(5), 457–466. DOI: 10.1177/1740774515597682 PMID: 26374685

Fiscella, K., Tobin, J. N., Carroll, J. K., He, H., & Ogedegbe, G. (2015). Ethical oversight in quality improvement and quality improvement research: New approaches to promote a learning health care system. *BMC Medical Ethics*, 16(1), 63. Advance online publication. DOI: 10.1186/s12910-015-0056-2 PMID: 26383770

Godskesen, T., Bjork, J., & Juth, N. (2023). Challenges regarding informed consent in recruitment to clinical research: A qualitative study of clinical research nurses' experiences. *Trials*, 24(1), 801. Advance online publication. DOI: 10.1186/s13063-023-07844-6 PMID: 38082434

Grady, C. (2022). Improving informed consent. In Bankert, E. A., Gordon, B. G., Hurley, E. A., & Shriver, S. P. (Eds.), *Institutional Review Board: Management and function* (3rd ed., pp. 341–350). Jones & Bartlett Learning.

Grady, C., & Wendler, D. (2013). Making the transition to a learning health care system. *The Hastings Center Report*, 43(1), S32–S33. DOI: 10.1002/hast.137 PMID: 23315892

Hall, S., Lee, V., & Haase, K. (2020). Exploring the challenges of ethical conduct in quality improvement projects. *Canadian Oncology Nursing Journal*, 30(1), 64–68. PMID: 33118977

Hunt, D. F., Dunn, M., Harrison, G., & Bailey, J. (2021). Ethical considerations in quality improvement: Key questions and a practical guide. *BMJ Open Quality*, 10(3), e001497. Advance online publication. DOI: 10.1136/bmjoq-2021-001497 PMID: 34404685

Institute of Medicine. (2013). *Best care at lower cost: The path to continuously learning health care in America*. The National Academies Press., DOI: 10.17226/13444

Johnson, N., Vermeulen, L., & Smith, K. M. (2006). A survey of academic medical centers to distinguish between quality improvement and research activities. *Quality Management in Health Care*, 15(4), 215–220. DOI: 10.1097/00019514-200610000-00003 PMID: 17047495

Jones, R. D., Krenz, C., Gornick, M., Griffith, K. A., Spence, R., Bradbury, A. R., De Vries, R., Hawley, S. T., Hayward, R. A., Zon, R., Bolte, S., Sadeghi, N., Schilsky, R. L., & Jagsi, R. (2020). Patient preferences regarding informed consent models for participation in a learning health care system for oncology. *JCO Oncology Practice*, 16(9), e977–e990. DOI: 10.1200/JOP.19.00300 PMID: 32352881

Kass, N. E., Faden, R. R., Goodman, S. N., Pronovost, P., Tunis, S., & Beauchamp, T. L. (2013). The research-treatment distinction: A problematic approach for determining which activities should have ethical oversight. *The Hastings Center Report*, 43(1), S4–S15. DOI: 10.1002/hast.133 PMID: 23315895

Markman, K. M., Weicker, N. P., Klein, A. K., & Sege, R. (2023). Community-engaged training in informed consent. *Journal of Clinical and Translational Science*, 7(1), e108. Advance online publication. DOI: 10.1017/cts.2023.534 PMID: 37313379

Morain, S. R., Kraft, S. A., Wilfond, B. S., McGuire, A., Dickert, N. W., Garland, A., & Sugarman, J. (2022). Toward meeting the obligation of *respect for persons* in pragmatic clinical trials. *The Hastings Center Report*, 52(3), 9–17. DOI: 10.1002/hast.1391 PMID: 35763201

National Commission for the Protection of Human Subjects in Biomedical and Behavioral Research. (1979). *The Belmont report: Ethical principles and guidelines for the protection of human subjects in biomedical and behavioral research.* U.S. Department of Health and Human Services. https://www.hhs.gov/ohrp/regulations-and-policy/belmont-report

Nerenz, D. R., Stoltz, P. K., & Jordan, J. (2003). Quality improvement and the need for IRB review. *Quality Management in Health Care*, 12(3), 159–170. DOI: 10.1097/00019514-200307000-00006 PMID: 12891960

Nuremburg Code. (1949). *Trials of war criminals before the Nuremburg Military Tribunals under Control Council Law No. 10.* (Vol. 2); 181-182. U.S. Government Printing Office. https://history.nih.gov/display/history/Nuremburg+Code Office of Civil Rights. (1996). Health Insurance Portability and Accountability Act. https://www.hhs.gov/hipaa/for-professionals/privacy/laws-regulations/index.html

Office of Human Research Protection. (2024). *Participant-centered informed consent training video* [Video]. U.S. Department of Health and Human Services. https://www.hhs.gov/ohrp/education-and-outreach/human-research-protection- training/participant-centered-informed-consent-training/index.html

Perneger, T. V. (2004). Why we need ethical oversight of quality improvement projects. *International Journal for Quality in Health Care : Journal of the International Society for Quality in Health Care*, 16(5), 343–344. DOI: 10.1093/intqhc/mzh075 PMID: 15375092

Pietrzykowski, T., & Smilowska, K. (2021). The reality of informed consent: Empirical studies on patient comprehension – Systematic review. *Trials*, 22(1), 57. Advance online publication. DOI: 10.1186/s13063-020-04969-w PMID: 33446265

Pradeep, R., & Deora, K. [REMOVED HYPERLINK FIELD]Danca, E., Young, T., Mott, K., Chezik, A., Rocca, N., & Kudelka, A. P. (August 8-10, 2024). *Unveiling the impact: Exploring malnutrition rates in patients with cancer and its association with anticancer treatments* [poster]. ASCO Breakthrough Meeting, Yokohama, Japan. https://conferences.asco.org/breakthrough/welcome

Protection of Human Subjects. 45 C.F.R. § 46 (2009). http://www.hhs.gov/ohrp/regulations-and-policy/regulations/regulatory-text/index.html

Protection of Human Subjects. 45 C.F.R. § 46 (2018). http://www.hhs.gov/ohrp/regulations-and-policy/regulations/45-cfr-46/revised-common-rule-text/index.html

Stiegler, M. P., & Tung, A. (2017). Is it quality improvement or is it research? Ethical and regulatory considerations. *Anesthesia and Analgesia*, 125(1), 342–344. DOI: 10.1213/ANE.0000000000001815 PMID: 28207587

World Medical Association. (2018). *Declaration of Helsinki—Ethical principles for medical research involving human subjects.* www.wma.net/policies-post/wma-declaration-of-helsinki-ethical-principles-for-medical-research-involving-human-subjects/

Zondag, A. G. M., Groenhof, K. J., van der Graf, R., van Solinge, W. W., Bots, M. L., & Haitjema, S.UCC-CVRM study group. (2023). Asking informed consent may lead to significant participation bias and suboptimal cardiovascular risk management in learning healthcare systems. *BMC Medical Research Methodology*, 23(1), 98. Advance online publication. DOI: 10.1186/s12874-023-01924-6 PMID: 37087415

KEY TERMS AND DEFINITIONS

Informed Consent: Informed consent (IC) refers to the process of explaining procedures and protocols to potential research participants including responding to questions that prospective participants or their legal representatives may have. Often accompanied by a form that is reviewed and signed by the participant or legally authorized representative, informed consent can take various forms so that it is tailored to the needs of the target audience of potential participants. Although the Office of Human Research Protection outlines specific criteria for what IC documents should include, there is some leeway in minimal risk research that allow for alteration or waiver of IC, foregoing participant signature on IC forms, and alternative delivery such as verbal or electronic consent. Broad consent is a new option that allows participants to consent to future research without being re-contacted for consent.

Healthcare Research: Healthcare research involves systematic investigations into human health. It includes a wide span of research-like activities ranging from rigorously designed and regulated drug and device studies to comparative effectiveness studies to quality improvement. In the approval of drugs and devices, and other research considered to present more than minimal risk, randomized, multi-center, controlled clinical trials have been the gold standard of research. Other types of research, quality improvement initiatives, evidence-based implementations, and scholarly activity may also occur in hospital or ambulatory settings and could be considered under the umbrella of healthcare research.

Learning Healthcare Systems: Learning Healthcare Systems (LHSs) compile and analyze routinely-collected patient health data for secondary uses that are ultimately applied toward changes and improvements in healthcare delivery. Using data collected through the electronic medical record or other high-powered data storage platforms, LHSs continuously analyze patient data in order to identify problem areas, propose solutions, and monitor the effectiveness of changes made as a result of these analyses.

Minimal Risk Research: Minimal risk research is research that includes no more than the risk encountered in everyday life. The federal regulations (45 CFR 46.102) expand minimal risk as "the probability and magnitude of harm or discomfort anticipated in the research are not greater in and of themselves than those ordinarily encountered in daily life or during the performance of routine physical or psychological examinations or tests." Research that is considered to be minimal risk is often exempted from some of the rules and processes that guide research regulations, particularly the regulations regarding informed consent.

Quality Improvement: Quality Improvement refers to the practical application of process changes that are evidence-based with the goal of improving predetermined performance measures. QI activities often follow a Plan-Do-Study-Act cycle in order to systematically implement changes, measure their effectiveness, and monitor results. Many QI initiatives utilize tools and techniques that are also applied to research, however, QI is seen as being a distinct practice due so differences in the goals or purpose of the initiative, the roles of the investigators, the lack of a control group or randomized samples, and the continuous implementation of findings toward the improvement of clinical outcomes. With the changing landscape of QI and the technological advances in patient care, QI practices that were once seen as minimal risk should be re-evaluated for protection of human subjects.

Research Ethics: Research ethics is the study of how research is conducted in relation to the foundational principles derived from Greek and Roman philosophers and manifested in the Belmont Report of 1979. These principles, and those from the Nuremburg Code and Declaration of Helsinki, were then applied in the Human Subjects Research Act and the Protection of Human Subjects that was codified in the Code of Federal Regulations 45 Part 46. Whereas the federal regulations on the conduct of research were based on these principles, research ethics extends beyond the regulations to a higher level to encourage investigators to pursue moral ideals.

The Revised Common Rule: Revised Common Rule is the post-2018 regulations on the conduct of human subjects in research. Originating within the US Department of Health and Human Services, the rules regarding biomedical and social-behavioral research have been adopted by 16 federal agencies, which is why it is referred to as the "Common Rule." The original set of rules from 1991 has been updated and revised to take into account current research technologies and trends. This was the first major revision of the Common Rule since its adoption.

Chapter 4
Ethical Considerations in Graduate Information Technology Qualitative Research:
Human Research Protections and Considerations

Michalina Hendon
https://orcid.org/0009-0003-6088-2658
University of the Cumberlands, USA

Whitney Taylor
https://orcid.org/0009-0003-8050-7900
University of the Cumberlands, USA

Joy Waldrep
University of the Cumberlands, USA

ABSTRACT

This chapter discusses ethical challenges and barriers unique to graduate-level qualitative information technology (IT) research faced by computer science, systems, and management researchers. IT research presents several challenges stemming from a discrepancy between the researcher's approach to ethical research design and the characteristics of qualitative methodology, bias, and discrimination. The review of the ethical considerations considers current implementations and recommendations in safeguards and risk mitigation strategies to protect research participants' rights, privacy, and well-being. The chapter explores ethical reflections in research design,

DOI: 10.4018/979-8-3693-3848-3.ch004

the regulatory landscape governing IT research, and current studies addressing the general approach to applied ethics, focusing on education and knowledge of compliance requirements and the consequences of non-compliance in human subject research (HSR). Ultimately, this chapter underscores the need for researchers to prioritize ethical considerations throughout the IT research lifecycle, fostering ethical research practices.

INTRODUCTION

As technology advances at a dynamic pace, research plays a pivotal role in pushing the boundaries of what is possible. Conducting sound research is a moral and ethical responsibility that requires protecting the interests and well-being of study participants. Unique ethical challenges and barriers are presented when Graduate research surrounds IT systems, data management, and computer science from a social science perspective. Ethical constructs surrounding technology use and application are discussed in most research, yet ethical questions are raised regarding the methodological creation of research design (Levitt et al., 2021). In general, when speaking of the design and construction of ubiquitous technology use and acceptance (blockchain, data mining, user experience, and artificial intelligence), graduate student research is a point of entry to learning and honing their research skills to scientifically question or understand the moral and philosophical implications regarding what the research may unearth.

This chapter aims to provide ethical considerations that graduate researchers in technical fields face when choosing a qualitative research design during the beginning phases of their dissertation journey. This chapter is not intended to provide a complete overview of the ethical issues and regulations related to qualitative technical research. Instead, the chapter contains common issues drawn from important existing and current literature that offer opportunities for reflection on ethical considerations that can be incorporated into the conception of the technical graduate study. The highlighted considerations can be applied to qualitative inquiry to improve research preparation, experience, data collection, and outcomes; however, education in ethical reflection is key to implementation.

BACKGROUND

Current Ethical Landscape of Qualitative Technology Graduate Research

As the call for ethical regulations in the use of and application of technology surmounts in areas of ubiquitous technology as well as the creation and enactment of comprehensive and universal technical guidelines for sociality and government use, the question of specific education of ethical research practices should be a growing focus for a graduate researcher. When conducting quantitative studies using anonymous surveys, researchers often aim to maintain a certain level of ambiguity regarding the pre-established tools employed in data collection and issues with misleading statistical analysis intentionally or unintendedly. Sawka (2020) emphasizes the importance of acknowledging that "unintentional mistakes in statistical practices may stem from a lack of normalization" while also pointing out that "the absence of sufficient supervision, education, or training can frequently act as a catalyst for failures in the mechanisms and safeguards needed to resist the pressures highlighted earlier, ensuring that statistical techniques are employed with honesty, integrity, and impartiality" (pp. 97-105).

Within the same contextual understanding, a qualitative methodological approach relies on narrative synthesis and experiential understanding to discern a collective understanding while acknowledging the societal consequences. Research outside of academia is steadily gaining a more holistic understanding of the societal impact of research as a core principle tied to creating knowledge and research (Derrick, 2018). The nature of qualitative research is dynamic, and unexpected situations may arise that require decisions beyond basic ethical principles, necessitating the researcher to adapt ethical norms to the specific context while prioritizing participant needs (Shaw, 2023). The researcher is embedded in the study context, influencing and being influenced by it, confronting ethical issues throughout the research process, from conceptualization to dissemination, as ethics and scientific quality are inseparable. Despite not involving direct risks, qualitative studies can potentially harm participants physically, psychologically, morally, intellectually, socially, and culturally at any stage.

Current technical research increasingly involves diverse stakeholders like companies, researchers, policymakers, and end-users from varied cultural backgrounds. Even if there is a lack of reconciliation of the different value systems and ethical priorities of and within research and research groups, ethical frameworks for emerging technologies remain an ongoing challenge. However, the concern of ethical considerations is growing as disciplines such as technical graduate research intend to challenge the researcher in ubiquitous technology to use not only their technical skills but also their reflective skills. Understanding the application and considerations of ethical research can stem from the blending of professional technical associations (e.g., Association for Computing Machinery (ACM) and Institute of Electrical and Electronics Engineers (IEEE)) ethical guidelines and codes, universities' research ethic requirements, and personal codes of conduct. The connection of professional, academic, and personal ethical education should be emphasized and required throughout the technical graduate student's educational research experience. Ensuring that ethical understanding and considerations are detailed within research processes, noting ethical principles are addressed and challenged, is essential to implementing a holistic approach to research ethics that focuses on technology research and conceptually runs parallel to ethical review and assessment of ethical reflectiveness within the protocol application.

In qualitative research, establishing trust is a fundamental goal that is pursued through the scientific method. However, this trust can be compromised by the researcher and other participants involved in the study. Qualitative research relies heavily on personal experiences and insights, aiming to present the perspectives of individuals without manipulating the truth. This approach allows readers to gain a deeper understanding of the subject matter from various viewpoints. However, it is essential to recognize that research can have real-world consequences. A lack of trust in the findings or the research process can lead to unintended negative outcomes for participants, particularly in cases of disclosures or unethical research practices. The intent of the research should be to convey the lived experience; however, there is also an important element of safeguards implemented in research that focuses on technology as the lived experience is channeled through the researcher's reflection and the perspective of the tech solutionism discussion Lenhart and Owens (2020) define tech solutionism as "the presumption that technology can fix social, cultural, and structural problems" (p. 2). This concept raises critical questions about the intentions behind technological innovations. While inventors may have good intentions, the ethical implications of technology—whether it is ultimately beneficial or harmful—remain complex. Similarly, qualitative researchers must navigate these complexities, exploring the intricate relationship between technology and the human environment, which cannot be adequately captured through quantitative measurements alone. Trust in both the research process and the researcher must be carefully managed to avoid ethical pitfalls. While implementing appropriate safeguards, researchers can better convey the richness of lived experiences while maintaining the integrity of their work.

In industry, the use of ethical considerations that are not adhered to sufficiently can burden the organization both in reputation as well as possible legal ramifications. Biable et al. (2022) identified that a graduate student's self-assessment of their knowledge and ethical judgments in research is influenced by a combination of training, environmental factors, and individual characteristics. As technical graduate students advance to higher levels of research responsibility within both corporate and academic settings, the quality and depth of their training in the research process contribute to their personal understanding of ethical considerations of research practices. The personal resolve of an individual's ethical position in research is crucial for ensuring the protection of human subjects and managing the extensive applications and disclosures associated with research and development (Taquette & Borges Da Matta Souza, 2022). In academic settings, just as in organizational research and development, there is a pressing need for ethical training. Such training ensures that researchers and technologists are aware of and adhere to ethical standards related to data privacy, informed consent, and the responsible use of technology. This is crucial to prevent technological advancements' misuse and promote responsible practices that protect individuals' rights and privacy. Practical research ethical training helps bridge the gap between theoretical knowledge and practical application, fostering a culture of integrity and responsibility in both academic and organizational environments.

Importance of Education in Technical Research Ethics

As research in computer science and related fields expands multidisciplinary, research ethics should also be increasingly focused on the ethical and social impacts of computing with the integration of ethical classes or content focus at the undergraduate level. Fiesler et al. (2020) discovered that less than 20% of undergraduate and graduate courses covered research ethics (p. 5). As research ethics concepts are considered universal, teaching ethics in technology research aims to enhance understanding of issues closely linked to the field of computer science. Individuals with technical training often prioritize the functionality of or the use of technology, potentially overlooking the broader philosophical or social implications of their research, which can lead to overlooking their research's broader philosophical or social implications. In an article on Inside Higher Ed by D'Agostino (2023), Yoshua Bengio, the scientific director of the Mila-Quebec AI Institute and a professor of computer science at the University of Montreal, highlighted the lack of ethical integration in education. He stated, "In most computer science departments, there are zero mandatory classes in anything related to society — whether ethics or social impact" (para. 7). The statement points out that universities need to do more to encourage researchers to consider the ethical implications of their technical work. Ethics should be a fundamental part of education at every level, as technological advancements rely on various forms of research (D'Agostino, 2023). The identified shortcomings are the lack of a comprehensive approach to technical ethics in research, practice, and education.

Professional ethical codes in computer science serve to educate researchers and provide a moral foundation for the profession such ethical codes. Komić et al. (2015) provided context by determining the range of research ethics from research integrity to questionable research practices. Research integrity includes the ideal process and education to assess the risks and implications of research sufficiently. Questionable research practices include deliberate misconduct, whereas the researcher knowingly conducts violations, including "fabrication, falsification, and plagiarism" (p. 2). In the middle of this polarity, categories found include violating widely accepted norms through inaccuracies, misrepresentations, and biases in research are found. Previous literature shows that ethical questions that stem from the creation of the research design are often the beginning (Samuel et al., 2019). The need to assess the middle category is the point of ethical reflection by a researcher to ensure an understanding of the possibility of harm. Without identification and mitigation, the researcher could find themselves in an ethical pitfall that could have been previously assessed through forethought of the ethical implications of the research.

The Ethical Ecosystem defined by Samuel et al. (2019) applied to a dissertation process framework is the interconnectedness of ethical perspectives and guidance within the research process. In the dissertation process, there should be a primary mentor or Chair of the researcher, the committee members of the researcher, and an administration that should have their institutional ethics checked through their Institutional Review Board (IRB) as well as quality checks at milestones within their research. As Samuel et al. asserts, "Within this ecosystem and through a series of interconnections, actors all rely on each other to govern and encourage ethical research behavior, as well as dissuade unethical behavior through a variety of evaluation-governed gateways" (p. 322). The ethical judgment of research cannot be placed entirely in the association between the researcher's ethical education and ability to apply the guidelines comprehensively to the research direction, design, and application to the institution's ethics review board (Serpico, 2024). Within this academic ethical ecosystem, through the lens of research integrity, the question of understanding the research in enough depth to guide the perspective or discussion of the ethical ramifications of the research exists for both the researcher and the ethics reviewers. When the researcher is not required to submit their academic protocol for an ethical review, the study is passed through the ethical ecosystem and may not be addressed by an outside ethical perspective; the researcher could suffer not only in the experience of identification and mitigation of risks, which will be important when addressing organizational ethical practices but the researcher's work if published, is open to scrutiny.

ETHICAL CONSIDERATIONS IN RESEARCH

Conducting ethical research is a cornerstone of credible and responsible academic inquiry (Carpenter, 2018; Iphofen & Tolich, 2018). While Institutional Review Board (IRB) approval is important, more is needed to ensure the ethical integrity of a study (Hunter, 2018). As Shaw (2008) finds, "ethics of qualitative research design pose distinctive demands on principles of informed consent, confidentiality and privacy, social justice, and practitioner research" (p. 400). As Shaw outlined overarching categories of ethical considerations, the guidance offered by the administrative support found within the ethical ecosystem will question the framework of Carniel et al. (2023) in a practical ethics review approach, which includes the evaluation of methodological decisions and strategies proposed by the researcher(s) to understand the direction and methods that will be used to guide the researcher. The next consideration includes the recruitment process, specifically the recruitment process of participants in dissertation research, while upholding ethical standards and maximizing the quality of data collected.

The recruitment of participants and use of data is of particular interest to social and behavioral technology research as the use of interviews, data collected through external sources, and researchers' observations as the considerations of implications that stem from the widespread agreement regarding the desired outcomes of data-driven technologies: they should serve the greater good without causing harm, prioritize human autonomy over machine autonomy, promote justice instead of discrimination, and be transparent and accountable (Samuel et al. 2019). Through a researcher's development of a personal code of ethics that respects human rights and cultural complexities, striving for an "aspirational" ethical stance that goes beyond procedural requirements should be a goal of attainment (Lahman, 2018, pp. 2-6). This section explores the importance of culturally responsive, relational, and reflexive ethics in research, the concept of "Othering," and the considerations necessary when researching vulnerable populations.

The Need for a Personal Code of Ethics

According to Lahman (2018), researchers should aspire to ethical standards that transcend the basic procedural ethics required by IRBs. The aspirational approach to research described by Lahman involves anticipating ethical dilemmas that may take time to be apparent to the researcher. Developing a personal code of ethics helps researchers maintain a commitment to social justice, ensuring that their work respects the rights and dignity of all participants (Carpenter, 2018; Lahman, 2018). Building an individualized ethical stance is particularly important in culturally complex research settings, where the risk of inadvertently imposing one's values on others is high. An example of Personal Code of Ethics can be found in the IEEE Code of Ethics, which can be summarized in the following points:

- Uphold integrity, responsible behavior, and ethical conduct in professional activities, prioritizing public safety, health, welfare, ethical design, and sustainable development practices.
- Foster an understanding of technological capabilities and societal implications, including intelligent systems, and disclose conflicts of interest transparently.
- Refrain from unlawful conduct and bribery while seeking, accepting, and offering honest criticism, crediting others appropriately, and maintaining and improving technical competence.
- Treat all individuals fairly and respectfully, avoiding discrimination and refraining from harassment or bullying behavior.
- Encourage colleagues and co-workers to uphold the IEEE Code of Ethics, support its adherence, and refrain from retaliating against those reporting violations (IEEE, 2020).

Research practices and implications are demonstrated through specific prescriptions of how technologists should communicate within a community, which can be found in a personal code of ethics. An individual's personal code of ethics can inform the actions of others to have a view of doing the best with the technology with the intention of no harm and, at minimum, a minimization of harm. As Neukrug et al. (1996) found when discussing ethical codes in counselors, "Although an ethical code can provide a useful framework for the sometimes-arduous decision-making process, such a set of guidelines alone is often not fully adequate in helping the counselor with tough ethical dilemmas." (p. 99). The finding that ethical frameworks only allow for decision-making in a set of constraints can also be applied to most technical disciplines. The need to further the ethical scope of technology research to include the transparency of the procedures within the research is essential to understanding the context of the study as well as the research outcomes.

CONSIDERATIONS OF THE ROLE OF THE RESEARCHER AND SOCIAL JUSTICE

Built on a personal ethical code, the researcher's role and application of social justice should be evaluated for the credibility of the research. Technical researchers have a crucial role in promoting social justice through their work. Research should adopt a personal code of ethics that goes beyond procedural requirements, as Lahman

(2018) highlighted. Researchers must strive for an aspirational ethical stance that anticipates and addresses ethical dilemmas (Kara, 2018; Lahman, 2018). Defining a personal approach to ethics helps ensure that research complies with ethical standards by respecting participants and valuing all participants' diverse perspectives and individual experiences (Kara, 2018; Lahman, 2018; Tuhiwai-Smith, 2021). The role of researcher reflection is to allow the researcher to detail their ontology and epistemology as informed by the study. The understanding of the researcher's positionality provides the technology researcher the ability to describe the implications of their findings to a broad community and can also detail the limitations of the researcher as well as the scope of the research.

An example of researchers' epistemology is found in the social justice lens. Social justice is a fundamental principle emphasizing fair treatment of all individuals, ensuring everyone has the same opportunities and resources (Kara, 2018; Lahman, 2018). In the context of research, social justice involves conducting studies in a way that respects the rights, dignity, and cultural backgrounds of all participants, particularly those from marginalized or vulnerable groups (Bryant, 2019; Gordon, 2020). The concept of social justice extends beyond mere compliance with ethical guidelines; it requires a proactive stance toward recognizing and addressing inequalities and injustices that may arise in the research process (Lahman, 2018). Using a social justice lens can include ensuring that your interview pool is diverse and using the lens to describe the research participants. For example, a researcher should understand the optimal participant for the study context, which can be detailed to provide a participant persona. However, the participant persona may sometimes be identifiable to the reader. The possibility of deidentification should be acknowledged and protected throughout the ethical ecosystem. The social justice lens can also be found within the practice of research reflection.

Culturally Responsive and Reflexive Research Practices

Cultural responsiveness and reflective research practices built into the study by the researcher are vital to understanding the considerations made during the research process. Culturally responsive and reflective research in technology can be diverse as it relates to the research; however, for the purpose of this Chapter, the terms are pointed toward the identification of the researcher's understanding and bias. To promote social justice, researchers must be culturally responsive and reflexive in their practices (Birch & Miller, 2002). Culturally responsive research respects and honors the cultural backgrounds of participants, avoiding the imposition of the researcher's values and norms (Tuhiwai-Smith, 2021). Lahman (2018) emphasizes the importance of not "recolonizing others into values that are not their own" (p. 2), which means that researchers must be vigilant in recognizing and respecting cultural differences. Cultural responsiveness aligns with and can stem from the researcher's reflexivity.

A researcher who integrates culturally responsive ethics seeks to be involved and accept understanding while respecting cultural differences. Cultural differences may be generationally inherited values impacting the participants' worldviews (Tuhiwai-Smith, 2021). Researchers should foster a sense of trust and rapport with participants and be reflexive, constantly examining their biases and assumptions. Moreover, it implements reflexivity throughout the interview process, which may aid researchers in recognizing their cultural positioning and how it influences their interactions and interpretations (Lahman, 2018; Nii Laryeafio & Ogbewe, 2023). Lahman (2018) emphasizes that research should honor cultural diversity and avoid "recolonizing others into values that are not their own" (p. 2). A balance must be struck between respecting human rights and honoring cultural heritage. Research suggests developing an ongoing dialogue with participants to ensure the representation of their perspectives and values (Lahman, 2018). By honoring cultural diversity, the researcher recognizes and respects cultural differences without imposing the researcher's cultural norms and values.

Reflexivity is critical in promoting social justice and is a tool to understand how the researcher positions within the support of the study's ontology and epistemology. Reflexivity involves researchers continuously examining their biases, assumptions, and positionality (Lahman, 2018). By being reflexive, researchers can mitigate the risk of "*Othering*," a process described by Johnson et al. (2004) as identifying and categorizing individuals as different from oneself, which can reinforce positions of domination and subordination (p. 254). For researchers, Othering can manifest as a subtle separation between the researcher and the participants, leading to biased interpretations and potentially unethical practices. To avoid '*Othering*', researchers must self-reflect, acknowledge their positionality, and view participants as equals with valuable insights gained through personal experiences (Johnson et al., 2004). Reflexivity helps researchers view participants as equals with valuable insights, fostering an environment of mutual respect and understanding (Carpenter, 2018; Lahman, 2018). Pousti et al. (2021) found three broad categories of reflexivity should be applied within a study: "(1) reflexivity as a theoretical position; (2) reflexivity as a fabric of research design; and (3) reflexivity as a way of informing research practice and application" (p.11). When looking at technology research in qualitative research using interviews and focus groups, the researcher inquiry should understand both the use of the technology and the social implications of the use of the technology; thus, in the reflexivity of the theoretical position, the research should be open to understanding the use and divisiveness of the technology at hand.

The reflexivity in the research design should position what the researchers seek to inquire about regarding the development or use of the technology. Therefore, the researcher should be open to choosing the research structure based on the best design to fit the nature of the inquiry. The researcher should then question how the chosen design will communicate the findings in the most scientific, factual, and non-biased manner with considerations of the ethical protection of participants. The researcher's reflexivity could also encompass the use of computer-assisted qualitative data analysis (CAQDAS), as the researcher should choose and research a tool that can provide the most direction in deriving correct themes from the transcripts. While a researcher cannot fully predict what will occur in the future, it is equally important for the researcher to question both in a manner that strengthens, rather than undermines, the accurate depiction of reality and ensures the context of the information while also integrating ethical education in how to manage unintentional disclosure as well as ensuring context of the information gathered before analysis.

Addressing Vulnerabilities and Power Imbalances

Social justice in research also requires addressing the participant's vulnerabilities and power imbalances between the researcher and participants. For example, the researcher's role should avoid creating a sense of ruling the research agenda, making participants feel inferior, or showing a sense of superiority (Råheim et al., 2016). Vulnerable populations are at a higher risk of exploitation and harm. Vulnerable populations may include individuals with cognitive impairments, individuals experiencing homelessness, or those living in poverty (Tuhiwai-Smith, 2021). Researchers must carefully assess the specific vulnerabilities of their participants and take steps to protect their well-being. Taking steps to protect participants may include ensuring informed consent, avoiding coercion, and being sensitive to the potential impact of authority figures on participants' willingness to participate and express their views freely (Carpenter, 2018; Hammond, 2018; Lahman, 2018). By treating all populations as potentially vulnerable, researchers can adopt a mindset that prioritizes the protection and respect of participants. Taking steps to protect participants aligns with the principles of social justice, ensuring that the research process is inclusive, equitable, and respectful of all individuals, regardless of their background or circumstances (Gordon, 2020). A consideration in technology research would be the ability of the researcher to screen appropriately the sample or target population for the study. An additional example could involve using their role and education within a technology to steer the interview in a manner that may unconsciously filter the participants' responses through their own educational and experiential lens, leading to a skewed understanding of the usability issues and limitations of the technology for marginalized users, (meaning users that may not have the same skill or knowledge at the time of the interview) of the discussed technology.

ETHICAL CONSIDERATIONS FOR VULNERABLE POPULATIONS

The following will identify approaches to considering vulnerable populations in technology qualitative reflexive feedback, considering decisions that researchers must make at different stages of the research process and their ethical implications. Researching vulnerable populations requires heightened ethical sensitivity (Gordon, 2020; Kara, 2018; Lahman, 2018; McCarry, 2012). Vulnerable populations are at a higher risk of exploitation and harm (Liamputtong, 2007; Menih, 2013; Valado & Amster, 2012). Researchers must assess the specific vulnerabilities of potential participants to protect their well-being and autonomy (Gordon, 2020). For example, participants' cognitive function should be evaluated to ensure they can provide informed consent (Hammond, 2016). Additionally, special populations like armed forces personnel, prisoners, and students may feel coerced by authority figures, impacting their willingness to participate and freely express personal opinions (Lahman, 2018). Vulnerable populations in IT research can consider their population sample's ability to answer the questions within the interview process with a reference of thought for older adults, people with disabilities, racial/ethnic minorities, and low-income communities who face significant barriers to accessing and utilizing digital technologies. Additionally discussed are the processes of research content with medical patients; ethical considerations must include the potential impact of the research on their health and well-being (Hammond, 2016). Homeless or impoverished participants may face additional challenges that must be addressed, such as mental health concerns, to ensure their participation is voluntary and informed (Menih, 2013; Morton & Cunningham-Williams, 2009). Researchers should adopt a mindset that treats all populations as potentially vulnerable, continuously seeking to identify and mitigate any factors that could compromise ethical standards (Gordon, 2020).

Ethical Sensitivity in Researching Vulnerable Populations (Understanding Vulnerability in Research Contexts)

Vulnerable populations may include individuals who may be at greater risk of exploitation and harm due to various factors, including cognitive impairments, socio-economic status, or institutional environments. Research involving these groups necessitates a heightened level of ethical sensitivity to ensure their well-being, dignity, and autonomy throughout the study. Protecting participants involves adhering to ethical guidelines and adopting a compassionate approach to identifying and addressing specific vulnerabilities (Gordon, 2020; Lahman, 2018).

Assessing Cognitive Function for Informed Consent

Informed consent is a component of ethical research, requiring that participants fully understand the nature, purpose, and potential risks of the study. For individuals with cognitive impairments, this process is particularly critical. Researchers should carefully assess the cognitive function of potential participants to ensure they can comprehend the information provided and make an informed decision about their participation (Bracken-Roche et al., 2017; Gordon, 2020). Steps to aid participants in making an informed decision might involve simplifying consent forms, using visual aids, or involving caregivers or legal guardians in the consent process. The researcher should ensure that participants understand the informed consent, that participants will not be harmed or exploited, and that the research process protects individual rights (Bracken-Roche et al., 2017).

Addressing Coercion in Special Populations

Special populations such as armed forces personnel, prisoners, and students often operate within hierarchical structures where authority figures can significantly influence their decisions. These individuals might feel pressured to participate in research due to perceived or actual power dynamics, which can compromise the voluntariness of their consent. Researchers must be aware of these dynamics and take steps to mitigate any coercive influences. Mitigating coercive influences could involve ensuring confidentiality, emphasizing the voluntary nature of participation, and providing alternative ways for individuals to decline participation without fear of reprisal (Liamputtong, 2007; Menih, 2013; Valado & Amster, 2012).

Ethical Considerations for Medical Patients

Medical patients constitute a vulnerable group due to their health conditions and the potential impact of research interventions on their well-being. Researchers should prioritize minimizing harm and enhancing potential benefits when designing studies involving medical patients. This includes carefully assessing the risks and ensuring that the research does not interfere with or worsen the patients' medical treatment. Researchers should also provide clear and comprehensive information about the study, including potential risks and benefits. Allowing participants to make an informed decision about their participation in a study (Bracken-Roche et al., 2017).

Challenges Faced by Homeless and Impoverished Participants

Homeless or impoverished individuals face unique challenges that researchers must address to ensure ethical participation (Valado & Amster, 2012). These challenges include a need for more stable housing, limited access to healthcare, and potential mistrust of institutions. Researchers should approach these participants with sensitivity and respect, ensuring voluntary participation and not influenced by immediate needs or desperation (Menih, 2013). Steps might involve offering additional support, such as transportation or compensation, to facilitate their participation while respecting their dignity and autonomy (Liamputtong, 2007).

Developing a Personal Ethical Stance

Creating a personal ethical stance involves a commitment to ongoing ethical reflection and adaptation (Birch & Miller, 2002; Lahman, 2018). Researchers should strive to understand the cultural differences that may arise in their work and develop strategies to navigate these differences respectfully (Hilário et al., 2020). Developing a personal ethical stance includes being aware of one's values and how they might influence the research process (Hilário et al., 2020; Kara, 2018). By fostering a mindset built on cultural responsiveness, researchers may ensure a standard within the study that is ethical and respectful of their participants' diverse backgrounds and experiences.

To ensure ethical research practices in graduate research, the need to not only cross-check the protocol for IRB approval, it is also important to have a thorough and thoughtful approach that considers cultural awareness and builds respectful relationships. Researchers can navigate the complexities of ethical research by developing a personal code of ethics. By understanding how to avoid the process of '*Othering*', researchers can foster an inclusive and respectful research environment that values the perspectives and experiences of all participants. Ultimately, the role of the researcher in promoting social justice is to ensure that their work contributes positively to society, upholding the dignity and rights of all individuals.

QUALITATIVE RESEARCH DESIGN AND ETHICAL CONSIDERATIONS

Qualitative research provides rich, in-depth insights into complex human behaviors, experiences, and social phenomena (Flick, 2018). However, designing an ethical qualitative study requires considering various factors, including ethical standards, methodological rigor, and practical constraints (Flick, 2007). Qualitative research design criteria encompass a clear direction and understanding of their research in the academic framework. The study is also informed by the philosophy of the researcher. This is because the impact of science on society largely occurs through its application as technology. When conceptually developing the research, the researcher should begin with the philosophical questions of ontology and epistemology to derive the lens choice of the research. A qualitative researcher would describe the manner in which the situation will be framed, the researcher's reflexology, as well as the methods of the research to inform the reader of the "epistemology" or the foundation of the researcher's belief system in a factual nature (Lincoln & Guba, 1985). In contrast, the researcher would explain their ontology or the research's understanding of their epistemology as it fits within society and the environment. Understanding the lens of the research, which can provide the researcher with finding a theory or framework, will guide the researcher to use the chosen theory or framework to inform their scientific methods of deriving the questions that can be assessed with the support or to challenge the theory and or framework posed. This section explores the intricacies of crafting an ethical and feasible qualitative study, collecting reliable and valid data respectfully, and addressing potential challenges such as researcher bias, discrimination, subjectivity, time constraints, limited generalizability, replication difficulties, and ambiguous findings.

Designing a Qualitative Study and Ethical Considerations in Non-Compliance

The study's ethics and feasibility guide the study's development in a qualitative research design. Feasibility involves practical considerations such as time, resources, and participant access. Researchers must realistically assess their ability to conduct the study within these constraints. A well-designed study includes clear, achievable research questions and objectives, a thorough literature review, and a well-defined methodological approach (Maxwell, 2013). Ethical considerations revolve around ensuring participants' rights, dignity, and well-being. This includes obtaining informed consent, ensuring confidentiality, and minimizing harm. Researchers must obtain approval from institutional review boards (IRBs) and adhere to institutional or governmental oversight.

The IRB assesses the risks and benefices of research participation from both the lens of the benefits of the study and the risks of the study, considering factors like population vulnerability and topic sensitivity. There may be examples where the researcher's understanding of the protocol may not always align with the ethical considerations of the researchers and participants involved. However, the IRB review is aimed at safeguarding the well-being of human subjects and upholding ethical standards in research endeavors; however, in graduate research within the discipline of technology, the need of the reviewer also to be versed in the proposed studies' risks that may not be stated within the protocol descriptions can be the disconnect to the researcher not receiving a comprehensive evaluation. Additionally, it is important that both the researcher and the ethics reviewer acknowledge that vulnerability and sensitivity do not necessarily indicate risk and promote a more participant-centered approach to research ethics (Tara Opsal et al., 2016). Head (2020) avers a purely administrative review process is incapable of ensuring the ethical conduct of research, particularly qualitative research. In the interviews of 47 institutional officials across the United States, McIntosh et al. (2023) found the top reasons for noncompliance in research were lack of knowledge or training, failure to provide research team supervision, and researcher attitudes toward compliance. The top three courses of action in response to noncompliance were retraining in compliance or research integrity, follow-up and hands-on involvement with the researcher, and mandated oversight or mentoring. This is not the intention. However, using a researcher's ethical eco-system should provide a strategy where participant and research risk mitigation is the goal.

Although qualitative research faces several inherent challenges, including trustworthiness, subjectivity, time consumption, limited generalizability, replication difficulties, ambiguous findings, lack of thematic development, lack of homogeneity in data, and small sample sizes. All qualitative designs should consider factors that ensure the study proceeds with guiding principles, enabling an understanding of the researcher's ethics and reflexivity (Ahmed, 2024). The following topics ensure that the limitations of the methodology are addressed correctly.

Collecting Reliable, Valid, and Respectful Data

Data collection in qualitative research demands techniques that ensure reliability and validity while respecting participants. Methods such as interviews, focus groups, and participant observations should be carefully planned and executed. Researchers must build rapport with participants, fostering an environment of trust and openness. Furthermore, building rapport involves being culturally sensitive and aware of power dynamics that might influence interactions (Creswell & Poth, 2018). As Hennessy et al. (2022) noted, there can be issues not only with the capturing of the interview data but also with the process by which transcription should be assessed. Outsourcing transcription, as well as bias in transcription, can lead to issues within the research outcomes that directly affect the reliability and validity of the data.

Triangulation is a key strategy for enhancing reliability and validity and providing a method of communal understanding. Researchers can cross-verify findings using multiple data sources, methods, or investigators, reducing the risk of bias (Patton, 2015). Member checking, where participants review and validate the data or findings, further enhances the study's credibility (Lincoln & Guba, 1989). Finally, in Respectful Data, in studies with technical professionals, it is important to discuss the technology within the ethical boundaries of what can be disclosed without violating the participant's privacy and/or unintended disclosure of proprietary organizational information or procedures (Redmiles, 2022).

Potential Influence of Researcher Bias

Limiting bias in research should be a priority for the researcher no matter the methodology, however, bias creep can be more prevalent in qualitative studies. Researcher bias is a significant concern in qualitative research (Pannucci & Wilkins, 2010). Bias can stem from personal beliefs, experiences, and expectations that unconsciously influence data collection, analysis, or interpretation (Florczak, 2021). An important process in qualitative methodology is the researcher's reflection on the comprehensive nature of the study. Reflexivity, where researchers critically reflect on personal positionality and its impact on the research, aids in identifying personal bias (Galdas, 2017). Maintaining a reflexive journal throughout the study can help identify and mitigate biases. Additionally, employing a diverse research team and engaging in peer debriefing can provide alternative perspectives and challenge assumptions (Maxwell, 2013). Transparent documentation of the research process allows others to assess the influence of potential biases that can be found within the study.

Discrimination

Discrimination can subtly or overtly influence qualitative research, affecting participants and findings. Researchers should be trained in cultural competence, ensuring they know and can appropriately address discrimination-related issues (Creswell & Poth, 2018). Researchers should be vigilant against any form of discrimination, such as race, gender, socioeconomic status, or cultural characteristics of the participants. By implementing an inclusive research design that consciously includes diverse voices and perspectives, the research can reduce personal bias within the study (Patton, 2015). Ethical recruitment practices, ensuring voluntary participation, and avoiding leading questions that may perpetuate stereotypes are essential.

Subjectivity

While subjectivity can enrich qualitative research, providing deep insights into individual experiences, it can lead to biased interpretations. Balancing subjectivity with systematic analysis methods, such as coding and thematic analysis, can help manage this challenge (Handa, 2022). Acknowledging subjectivity is awareness of biases, beliefs, emotions, and opinions that influence how an individual interacts with the world as well as the lens described to identify a perspective.

Time Consumption

Qualitative research is often time-consuming due to the extensiveness of data collection and analysis. Effective project management and realistic timelines are essential. Prioritizing key research questions and employing efficient data analysis software can also mitigate time constraints (Creswell & Poth, 2018). However, as Hesse et al. (2019) identified, "Regrettably, the scholarship of amateur qualitative researchers is characterized often by a lack of engagement with core social science principles and lack of awareness or dismissal of theory and established bodies of social science scholarship" (p.566). The need for a connection between the technology knowledge base and the social science principles that are needed to guide the research may leave the IT researcher with the need to focus on more social science theoretical underpinnings to convey the research and inform the research practices. One can look at the generalizability of time consumption as either within the research protocol and process or the research that is needed to inform the

Limited Generalizability

Qualitative data's rich, context-specific nature often limits its generalizability. As noted by Maxwell (2013), a lack of generalizability stems from a lack of "being representative of a larger population, [instead] a qualitative study may provide an account of a setting or population that is illuminating as an extreme case or ideal type" (p. 246). However, transferability can be enhanced by providing detailed descriptions of the research context, enabling others to assess the applicability of findings to different settings (Lincoln & Guba, 1989). Replicating qualitative studies can be challenging due to the uniqueness of contexts and participants. Thorough documentation of the research process and transparent reporting of methods and findings can facilitate replication efforts (Maxwell, 2013).

Ambiguous Findings and Lack of Thematic Development

Ambiguity in findings can stem from insufficient thematic development. Rigorous data analysis methods, such as grounded theory or thematic analysis, can help identify clear patterns and themes. Iterative data collection and analysis cycles, where data is revisited and reanalyzed, can also enhance thematic clarity (Roberts et al., 2019).

Lack of Homogeneity in Data

Qualitative research often deals with heterogeneous data, reflecting the diversity of human experiences. While this can enrich the study, it can also complicate analysis. Researchers should embrace this diversity, using it to uncover deeper insights while clearly defining the scope and boundaries of their study (Patton, 2015).

Small Sample Size and Reaching Saturation. Small sample sizes are commonplace in qualitative research. While this limits statistical generalizability, qualitative research aims for depth rather than breadth (Flick, 2018). Researchers should focus on achieving saturation, where no new themes emerge, rather than expanding the sample size (Guest et al., 2006; Hennink et al., 2016). Researchers can determine if the number of participants is sufficient to reach code and meaning saturation. Guidelines for achieving both include a "richly textured understanding of issues" (Hennink et al., 2016, pp. 591-592). Hesse et al. (2019) have also warned against overlooking the importance of research with small sample sizes, which is common in qualitative data. Many believe that a key advantage of qualitative research is its focus on improving understanding of social interactions; thus, a smaller sample size allows for an understanding on a small scale.

Designing an ethical and feasible qualitative study requires meticulous planning and a deep commitment to ethical principles and methodological rigor. Collecting reliable, valid, and respectful data involves building participant trust and employing triangulation and member-checking strategies. Addressing challenges such as researcher bias, discrimination, subjectivity, time constraints, limited generalizability, replication difficulties, and ambiguous findings necessitates a reflective and transparent approach. Despite these challenges, qualitative research's rich, contextual insights make it an invaluable tool for understanding complex human phenomena.

PROMOTING ETHICAL LANDSCAPES GOVERNING IT RESEARCH

There is a growing concern that the regulation of research ethical oversight may impede the speed of research and innovation. However, society has not always reaped the benefits, just the opposite, as in the Ring example, the company amassed data to train proprietary algorithms, the data collected was used to advance the technology, and implications seemed to be short-sided in consideration of proprietary data of the user. As Surmiak and Męcfal, (2024) state, "Codes of ethics and ethics committees are intended, on the one hand, to help researchers make the best ethical research decisions and, on the other, to control their behavior during research." (p. 3) As the ethical landscape changes, the vulnerability of data and the epistemological view of personal privacy are common concerns in everyday life, and not just in participation in research; the importance of checks and balances in research practices is essential.

The consequences of noncompliance with standards in ethical research are presented from multiple viewpoints. When examining ethical perspectives in research, it is crucial to evaluate the enforcement mechanisms and oversight protocols in place. The degree of stringency or leniency in these measures can significantly influence the application of ethical frameworks. In tracing the rationale behind research governance, two foundational principles emerge as guiding forces: utilitarianism and deontology (Head, 2020). These principles shape the rules, codes, and ethical tenets that govern research oversight bodies.

Utilitarianism, rooted in the notion of maximizing overall well-being, often serves as a justification for actions that prioritize collective benefits. In contrast, deontological ethics emphasize adherence to moral duties and obligations, potentially constraining actions that may violate individual rights, even if they yield favorable outcomes for the majority. The interplay between these two philosophical underpinnings manifests in the varying degrees of enforcement and oversight witnessed across research domains. Utilitarian or consequentialist ethics are derived from the belief that the benefit from the research justifies the possibility of negative consequences (Vearrier & Henderson, 2021). In utilitarian ethics, focusing on maximizing overall happiness can lead to ethical concerns when the well-being of marginalized or vulnerable populations is compromised. The deontology lens focuses on duty rather than on possible outcomes. The Common Rule does not directly regulate information technology (IT) or data not part of a research study involving human subjects. However, if IT systems or data are used in the context of a human subject research study funded or conducted under the Common Rule, regulations would apply to handling and protecting that data. The issue in this scenario would be the forward thought of the ethical eco-system to allow the researcher to ponder the external reach of the social benefits, risks, and risk mitigation in the research design and outcomes.

There is a global concern regarding safeguards that prevent data scrapers, miners, and analysts from deriving an individual's identity associated with publicly accessible data (Throne & Hendon, 2024). The United States has not passed a singular law governing all data types (Jamison, 2019). Instead, there are agencies that regulate data by sector; for example, the Health Insurance Portability and Accounting Act (HIPAA) regulates the use and disclosure of medical records at a high level. The European Union has begun the process of implementing protections and litigation against the disclosure of personally identifiable information (PII) via the General Data Protection Regulation (GDPR) in research applied broadly and by sector. The GDPR establishes a lawful basis for processing data. GDPR does, however, make provision for scientific research with guards in place (Staunton et al., 2019). Head (2020) avers that advances in research deserve new ethical considerations for big data and the surrounding topics of privacy, sharing, linkage, and new technology. Artificial Intelligence (AI) is another fast-growing technology that is raising ethical concerns (Mehan, 2022). Though there are innumerable articles and books on the use of AI, the considerations of ethics, social, and security issues are secondary, if considerations at all. When considering the ethics and laws surrounding AI, Mehan (2022) asserts that only the humans designing the algorithms can be held accountable for the actions of AI. This author takes great pains to delineate the difference in ethics, laws, and morals. Mehan (2022) asserts that the three fundamental laws assigned to robotics in the 1940s could apply to AI today. Do not injure a human being or, through inaction, allow a human to be injured. Obey the orders given by human beings unless violating law number one. Protect their own existence as long as that does not conflict with the first or second law. The lawmakers' challenge with AI is to protect the privacy of individuals without limiting AI development.

An increased global interest in the regulation of AI has also piqued current interest in data privacy, which occurred in the wake of COVID-19. The National Security Commission on Artificial Intelligence (NSCAI) presented a report to Congress in 2021 to have the US AI-ready by 2025 (Meehan, 2022). This report contained steps the NCSAI considered necessary to protect privacy and promote safety in AI technology. Within all the ambiguity of laws and regulations, there must be a preponderance of data to uphold an accusation of noncompliance or unethical research. The focus of investigations into noncompliance or unethical research is on the participants' rights, not punitive actions toward the researcher (Bierer and Barnes, 2014). IRBs typically do not publicize findings of noncompliance unless required by agencies but rather focus on a corrective action plan. This might include retraining research team members, auditing ongoing or future research, updating ongoing protocols to reduce risk, and notifying subjects about new risks or potential harms resulting from noncompliance. The publication may be rescinded, in which case, there is a focus on alerting the journal of the tainted research. Again, this is not a punitive action for the author but a protection of the participants. The disclosure of the reason for the recension may matter to other nonimplicated authors in the publication. However, privacy is still the focus for the participants.

Head (2021) concludes that ethical research (in education) is more a question of morals rather than an administrative process. To produce credible qualitative research, Seymore and Meehan (2019) recommend comprehensive documentation on the processes employed at every phase of the research and artifacts to attest to the rigor and transparency throughout the entire study, creating an auditable trail. Klietzman (2011) states that IRBs have the ability to focus on research ethics, but there is no meaningful empirical research to explore this concept. The overseers of institutional review boards are variant in their protocols and processes (Mrema et al., 2023; Head, 2021; Klietzman, 2011). Punitive actions are not the focus of investigations; rather, the emphasis is on the participants' privacy (Bierer and Barnes, 2014). Laws and regulations surrounding IT in the US have gaps (Jamison, 2019). There are many lenses through which to admire the problem of noncompliance in qualitative research in the IT domain (Bozeman et al., 2021). The variance of consequences and the lack of consequences result from the multiple lenses by which to view the problem.

CONCLUSION

The chapter authors discussed unique ethical challenges and barriers faced by computer science, systems, and management researchers conducting qualitative information technology (IT) research. Navigating cultural differences and balancing ethical concerns with research objectives can vary for each researcher and research study. However, when a researcher possesses advanced knowledge, it raises questions about how this expertise should inform their approach to ethical research design. While qualitative research faces inherent challenges, the considerations can be addressed through rigorous study design and execution. By implementing strategies to enhance trustworthiness, mitigate bias, respect participants, and ensure ethical practices, researchers can conduct high-quality qualitative studies that yield rich, insightful findings. Ultimately, the value of qualitative research lies in its ability to provide deep, nuanced understandings of complex human experiences and phenomena. By embracing the diversity of perspectives and contexts, qualitative researchers can uncover insights that quantitative methods alone cannot capture.

As the field of qualitative research continues to evolve and include human and computer interaction, which is underpinned in social science, researchers must remain vigilant in addressing the ethical considerations outlined in this chapter. The importance of IT researchers to evaluate their practices ethically as well as realistically reflect on the communication, dissemination, and growth that are within the research life cycle and beyond. Through ongoing methodological refinement, critical self-reflection, and a commitment to ethical and inclusive practices, qualitative research can substantially contribute to understanding the human experience in all its complexity.

REFERENCES

Ahmed, S. K. (2024). The pillars of trustworthiness in qualitative research. *Journal of Medicine, Surgery, and Public Health*, 2, 100051. Advance online publication. DOI: 10.1016/j.glmedi.2024.100051

Biable, S. E., Garcia, N. M., Midekso, D., & Pombo, N. (2022). Ethical issues in software requirements engineering. *Software*, 1(1), 31–52. DOI: 10.3390/software1010003

Birch, M., & Miller, T. (2002). Encouraging participation: ethics and responsibilities. In *Ethics in Qualitative Research* (pp. 92–106). SAGE. DOI: 10.4135/9781849209090.n5

Bracken-Roche, D., Bell, E., Macdonald, M. E., & Racine, E. (2017). The concept of "vulnerability" in research ethics: An in-depth analysis of policies and guidelines. *Health Research Policy and Systems*, 15(1), 8–8. DOI: 10.1186/s12961-016-0164-6 PMID: 28173859

Bryant, A. (2019). Indigeneity. In *The varieties of grounded theory* (pp. 92–103). SAGE. DOI: 10.4135/9781529716542.n6

Burles, M. C., & Bally, J. M. G. (2018). Ethical, practical, and methodological considerations for unobtrusive qualitative research About personal narratives shared on the internet. *International Journal of Qualitative Methods*, 17(1), 1. Advance online publication. DOI: 10.1177/1609406918788203

Carpenter, D. (2018). Ethics, reflexivity and virtue. In *The SAGE Handbook of Qualitative Research Ethics* (pp. 35–50). SAGE. DOI: 10.4135/9781526435446.n3

Creswell, J. W., & Poth, C. N. (2016). *Qualitative inquiry and research design: Choosing among five approaches*. Sage publications.

D'Agostino, S. (2023). *In class, some colleges overlook technology's dark side*. Inside Higher Ed. https://www.insidehighered.com/news/tech-innovation/teaching-learning/2023/11/16/some-colleges-overlook-technologys-dark-side

Fiesler, C., Garrett, N., & Beard, N. (2020). What do we teach when we teach tech ethics?: A syllabi analysis. *Proceedings of the 51st ACM Technical Symposium on Computer Science Education*, 289–295. DOI: 10.1145/3328778.3366825

Gordon, B. G. (2020). Vulnerability in research: Basic ethical concepts and general approach to review. *The Ochsner Journal*, 20(1), 34–38. DOI: 10.31486/toj.19.0079 PMID: 32284680

Hammond, W. (Ed.). (2016). *Informed consent: procedures, ethics and best practices*. Nova Publishers.

Hennessy, M., Dennehy, R., Doherty, J., & O'Donoghue, K. (2022). Outsourcing transcription: Extending ethical considerations in qualitative research. *Qualitative Health Research*, 32(7), 1197–1204. DOI: 10.1177/10497323221101709 PMID: 35599615

Hesse, A., Glenna, L., Hinrichs, C., Chiles, R., & Sachs, C. (2019). Qualitative research ethics in the Big Data Era. *The American Behavioral Scientist*, 63(5), 560–583. DOI: 10.1177/0002764218805806

Hilário, A. P., Augusto, F. R., Hilário, A. P., & Augusto, F. R. (2020). *Practical and ethical dilemmas in researching sensitive topics with populations considered vulnerable*. MDPI.

Hine, C. (2021). Evaluating the prospects for university-based ethical governance in artificial intelligence and data-driven innovation. *Research Ethics*, 17(4), 464–479. DOI: 10.1177/17470161211022790

Hunter, D. (2018). Research ethics committees – what are they good for? In *The SAGE Handbook of Qualitative Research Ethics* (pp. 289–300). SAGE. DOI: 10.4135/9781526435446.n20

Hussain, M. I., Figueiredo, M. C., Tran, B. D., Su, Z., Molldrem, S., Eikey, E. V., & Chen, Y. (2020). A scoping review of qualitative research in JAMIA: Past contributions and opportunities for future work. *Journal of the American Medical Informatics Association: JAMIA*, 28(2), 402–413. DOI: 10.1093/jamia/ocaa179 PMID: 33225361

IEEE. (n.d.). *IEEE Code of Ethics*. Retrieved May 24, 2024, from https://www.ieee.org/about/corporate/governance/p7-8.html

Iphofen, R., & Tolich, M. (2018). Foundational issues in qualitative research ethics. In *The SAGE Handbook of Qualitative Research Ethics* (pp. 1–17). SAGE., DOI: 10.4135/9781526435446.n1

Johnson, J. L., Bottorff, J. L., Browne, A. J., Grewal, S., Hilton, B. A., & Clarke, H. (2004, April). Othering and being othered in the context of health care services. *Health Communication*, 16(2), 255–271. DOI: 10.1207/S15327027HC1602_7 PMID: 15090288

Kara, H. (2018). Democratizing research in practice. In *The SAGE Handbook of Qualitative Research Ethics* (pp. 103–112). SAGE. DOI: 10.4135/9781526435446.n7

Komić, D., Marušić, S. L., & Marušić, A. (2015). Research integrity and research ethics in professional codes of ethics: Survey of terminology used by professional organizations across research disciplines. *PLoS One*, 10(7), e0133662. Advance online publication. DOI: 10.1371/journal.pone.0133662 PMID: 26192805

Lahman, M. (2018). Becoming ethically responsive researchers: introduction. In *Ethics in Social Science Research: Becoming Culturally Responsive* (pp. 3-12). SAGE. DOI: 10.4135/9781071878750.n1

Langlais, P. J., & Bent, B. J. (2018). Effects of training and environment on graduate students' self-rated knowledge and judgments of responsible research behavior. *Ethics & Behavior*, 28(2), 133–153. DOI: 10.1080/10508422.2016.1260014

Levitt, H. M., Morrill, Z., Collins, K. M., & Rizo, J. L. (2021). The methodological integrity of critical qualitative research: Principles to support design and research review. *Journal of Counseling Psychology*, 68(3), 357–370. DOI: 10.1037/cou0000523 PMID: 34043379

Liamputtong, P. (2007). Researching the vulnerable. *SAGE*. DOI: 10.4135/9781849209861

Lincoln, Y. S., & Guba, E. G. (1985). Naturalistic inquiry. *SAGE*.

McCarry, M. (2012). Who benefits? A critical reflection of children and young people's participation in sensitive research. *International Journal of Social Research Methodology*, 15(1), 55–68. DOI: 10.1080/13645579.2011.568196

Menih, H. (2013). Applying ethical principles in researching a vulnerable population: Homeless women in Brisbane. *Current Issues in Criminal Justice*, 25(1), 527–539. DOI: 10.1080/10345329.2013.12035979

Morton, L. G.II, & Cunningham-Williams, R. M. (2009). The capacity to give informed consent in a homeless population with developmental disabilities. *Community Mental Health Journal*, 45(5), 341–348. DOI: 10.1007/s10597-009-9184-9 PMID: 19263221

Nii Laryeafio, M., & Ogbewe, O. C. (2023). Ethical consideration dilemma: Systematic review of ethics in qualitative data collection through interviews. *Journal of Ethics in Entrepreneurship and Technology*, 3(2), 94–110. DOI: 10.1108/JEET-09-2022-0014

Opsal, T., Wolgemuth, J., Cross, J., Kaanta, T., Dickmann, E., Colomer, S., & Erdil-Moody, Z. (2016). (2016). "There are no known benefits. . ." Considering the risk/benefit ratio of qualitative research. *Qualitative Health Research*, 26(8), 1137–1150. DOI: 10.1177/1049732315580109 PMID: 25857654

Pousti, H., Urquhart, C., & Linger, H. (2021). Researching the virtual: A framework for reflexivity in qualitative social media research. *Information Systems Journal*, 31(3), 356–383. DOI: 10.1111/isj.12314

Råheim, M., Magnussen, L. H., Sekse, R. J. T., Lunde, Å., Jacobsen, T., & Blystad, A. (2016). Researcher-researched relationship in qualitative research: Shifts in positions and researcher vulnerability. *International Journal of Qualitative Studies on Health and Well-being*, 11(1), 30996–12. DOI: 10.3402/qhw.v11.30996 PMID: 27307132

Redmiles, E. M. (2022). The need for respectful technologies: Going beyond privacy. In Werthner, H., Prem, E., Lee, E. A., & Ghezzi, C. (Eds.), *Perspectives on Digital Humanism* (pp. 309–313). Springer. DOI: 10.1007/978-3-030-86144-5_42

Samuel, G., Chubb, J., & Derrick, G. (2021). Boundaries between research ethics and ethical research use in artificial intelligence health research. *Journal of Empirical Research on Human Research Ethics; JERHRE*, 16(3), 325–337. DOI: 10.1177/15562646211002744 PMID: 33733915

Samuel, G., Derrick, G. E., & Van Leeuwen, T. (2019). The ethics ecosystem: Personal ethics, network governance and regulating actors governing the use of social media research data. *Minerva*, 57(3), 317–343. DOI: 10.1007/s11024-019-09368-3 PMID: 31501635

Sawka, K. (2020). The use and misuse of statistics. In Gruber, C. W. (Ed.), *The Theory of Statistics in Psychology* (pp. 95–110). Springer., DOI: 10.1007/978-3-030-48043-1_7

Serpico, K. (2024). Defining institutional review board application quality: Critical research gaps and future opportunities. *Research Ethics*, 20(1), 19–35. DOI: 10.1177/17470161231187099

Shaw, I. (2008). Ethics and the practice of qualitative research. *Qualitative Social Work: Research and Practice*, 7(4), 400–414. DOI: 10.1177/1473325008097137

Shaw, I. (2023). *Research and social work in time and place: Crossing boundaries* (1st ed.). Routledge., DOI: 10.4324/9781003306740

Surmiak, A., & Męcfal, S. (2024). Ethical challenges in contemporary social research [editorial]. *Diametros: a Journal of Philosophy*, 1–6. DOI: 10.33392/diam.1979

Taquette, S. R., & Borges da Matta Souza, L. M. (2022). Ethical dilemmas in qualitative research: A critical literature review. *International Journal of Qualitative Methods*, •••, 21.

The United States Federal Trade Commission. (2023, May 31). *FTC Says Ring employees illegally surveilled customers, failed to stop hackers from taking control of users' cameras*. Federal Trade Commission. https://www.ftc.gov/news-events/news/press-releases/2023/05/ftc-says-ring-employees-illegally-surveilled-customers-failed-stop-hackers-taking-control-users

Throne, R., & Hendon, M. (2024). Belmont 2.0: Ethical research expectations for PII in AI, ML, and data mining/scraping. In Demircioğlu, A. (Ed.), (pp. 1–19). Advances in information quality and management. IGI Global., DOI: 10.4018/979-8-3693-1726-6.ch001

Tuhiwai-Smith, L. (2021). *Decolonizing methodologies: Research and indigenous peoples* (3rd ed.). Bloomsbury., DOI: 10.5040/9781350225282

Valado, T., & Amster, R. (2012). *Professional lives, personal struggles: ethics and advocacy in research on homelessness* (1st ed.). Rowman & Littlefield.

Vearrier, L., & Henderson, C. M. (2021). Utilitarian Principlism as a framework for crisis healthcare ethics. *HEC Forum*, 33(1–2), 45–60. DOI: 10.1007/s10730-020-09431-7 PMID: 33449232

KEY TERMS AND DEFINITIONS

Codes of Ethics: A set of ethical behavior principles to serve as guidelines for individuals in the computing field, including those currently in the profession, as well as educators, learners, influencers, and any users of computing technologies.

Ethical Ecosystem: Samuel et al., (2019) coined as an evaluation network of interconnected actors existing at a number of levels within the academic system and at different points of the research production process

Othering: Identifying and categorizing individuals as different from oneself, which can reinforce positions of domination and subordination (Johnson et al., 2004, p. 254)

Institutional Review Board: An organization or institution with an established administrative board with the purpose of safeguarding the rights and welfare of individuals who are recruited as research subjects for studies carried out under the organization or institutional purview.

Chapter 5
Ethics for Responsible Data Research:
Integrating Cybersecurity Perspectives in a Digital Era

Sheetal Temara
https://orcid.org/0009-0006-2221-6605
University of the Cumberlands, USA

ABSTRACT

In today's digital age, ethical issues are a principal concern during responsible data research which requires thoughtful consideration and approaches for managing them. As technological advances materialize to increase the standard of living for people across the world, ethics must be prioritized by cybersecurity professionals to ensure these technologies are not being misused or inflicting harm to the general population, especially to underrepresented and underprivileged communities. Cybersecurity subject matter experts must develop awareness regarding ethical ramifications of their research endeavors to ensure security is balanced with moral standards. Observance of and adherence to ethical based policies, principles, and security best practices will delineate cybersecurity professionals from threat actors.

INTRODUCTION

The rapid evolution of technology has brought forth unprecedented opportunities and challenges in this digital era. Ethical issues in responsible data research have become a principal concern among these challenges necessitating thoughtful consideration and proactive management (Esmer & Arıbaş, 2022). As data becomes a pivotal asset for organizations and individuals alike, the methodologies and practices

DOI: 10.4018/979-8-3693-3848-3.ch005

encompassing its collection, analysis, and storage must be scrutinized to ensure ethical compliance (Edquist et al., 2022). The importance of ethics in responsible data research has been amplified within the realm of cybersecurity where researchers are tasked with securing sensitive information and protecting the privacy of individuals and organizations. Cybersecurity is not just about defending against malicious attacks but also about maintaining the trust of stakeholders through ethical conduct and moral lapses in cybersecurity can result in ramifications such as breaches of confidentiality, loss of data integrity, and violation of public trust (Macnish & Van der Ham, 2020). This implies providing a comprehensive understanding of the ethical deliberations cybersecurity researchers must navigate to balance security and moral standards in an effective manner. The examination of case studies, regulatory frameworks, and best practices can equip cybersecurity practitioners with the knowledge and tools required to uphold ethical standards in their work alongside underscoring the need for continuous ethical vigilance in a rapidly evolving digital landscape while ensuring that the data protection standards align with the broader societal values of fairness, privacy, and respect for individuals' rights (Miller & Bossomaier, 2024).

The objective of this chapter is to delve into the exploration of the ethical dimensions of responsible data research while emphasizing the integration of cybersecurity perspectives such as privacy and confidentiality, informed consent, data integrity & manipulation, and misuse of technology. The convergence of ethics in responsible data research and cybersecurity creates a complex landscape and broad overview of how the researchers must navigate legal, technical, and moral challenges (Pattison, 2020). This also highlights the importance of integrating ethical principles into cybersecurity practices to protect confidentiality and maintain trust in digital systems.

Figure 1. Word cloud of ethical principles in cybersecurity practices

Several information technology organizations including the Association for Computing Machinery (ACM) and the Institute of Electrical and Electronics Engineers (IEEE) have established a code of ethics and a list of behavioral requirements for researchers. There is a list of activities that organizations can perform to maintain ethical behaviors of personnel as they conduct responsible data research for meeting business needs. Ethical concerns are commonly experienced cybersecurity researchers during the performance of their job functions with each having potential negative implications and coinciding responsible actions which should be taken to safeguard morale behavior (Raul, 2021). Cybersecurity practitioners are routinely exposed to ethical issues due to their responsibility of securing and protecting the data and livelihood of individuals and organizations in a technological environment which is experiencing unprecedented growth, data proliferation to a variety of devices, significant design complexity, and an ever-expanding threat landscape (Dunn-Cavelty, 2018). In addition to considering risk and expense of controls while designing security solutions, cybersecurity researchers must also consider the ethical consequences of architectural decisions as well as decisions to accept risks. Cybersecurity researchers have responsibilities to protect the organizations including employees, investors, customers, and stakeholders to which they are aligned but also possess social accountability to protect the society itself (Richards et al., 2020).

BACKGROUND

For nearly three decades, the digital age is characterized by significant technological advancements that have transformed various aspects of human life. These advancements have improved the various facets of life including the standard of living, enhanced communication, and facilitated global connectivity ranging from cloud computing and artificial intelligence (Ademola, 2020). The technology has permeated every sector by creating massive amounts of data that drive both innovation and efficiency. Education, Entertainment, Finance, and Healthcare are just a few of these areas that have been revolutionized by digital technologies leading towards creating more personalized and efficient services. As digital devices started becoming more integrated into daily life, they generate and transmit large volumes of data and this data while valuable for improving services and fostering innovation also poses significant privacy and security risks (Aderibigbe, 2021). Some of the critical concerns arising from this extensive amounts of data proliferation can lead to unauthorized access to personal data, data breaches, and information misuse.

This speedy digital transformation has also introduced complex ethical dilemmas specifically in responsible data research in cybersecurity. The proliferation of data across multiple devices combined with the increasing sophistication of cyber threats demands a robust ethical framework to guide best cybersecurity practices (Allahrakha, 2023). Cybercriminals use sophisticated techniques to exploit vulnerabilities in systems resulting in potentially devastating consequences for individuals, businesses, and governments. These threats range from identity theft and financial fraud to large-scale attacks on critical infrastructure. Due to the evolving nature of these threats, cybersecurity measures must constantly adapt to safeguard sensitive information and maintain public trust (Artz, 2008).

A robust ethical framework is critical to provide for an orderly conduct of cybersecurity practices. Ethical guidelines will help ensure that the collection, storage, and data usage are conducted in a manner that respects individuals' privacy rights and conforms with the legal standards (Atapour-Abarghouei et al., 2020). They also provide a groundwork for developing policies and procedures that address the ethical challenges posed by new technologies and cyber threats. Establishing and maintaining such a robust ethical framework requires a multifaceted approach which involves creating policies that promote transparency and accountability in managing information, ensuring that individuals are informed about how their data is being utilized and protected (Bauer et al., 2020). Additionally, it calls for the implementation of security measures that prevent unauthorized access and mitigate the impact of cyber-attacks. Continuous education and training awareness for cybersecurity researchers are also critical as they must stay updated on the latest threats and ethical standards (Bynum, 2001). Ultimately, the goal is to balance the benefits of digital transformation with the need to protect individuals' rights and maintain the integrity of digital systems. Organizations can navigate the complexities of the digital age and build a safer and trustworthy digital environment for all by fostering a culture of ethical responsibility (Chen et al., 2023).

ETHICS FOR RESPONSIBLE DATA RESEARCH

Cybersecurity researchers are positioned at the confluence of technology and ethics. They experience ethical challenges that curtail from their dual responsibilities which includes protecting sensitive information and safeguarding the security of technological systems (Christen et al., 2020). These challenges are multifaceted and require a nuanced understanding of both technological capabilities and ethical principles including but not limited to issues related to consent, data manipulation and integrity, privacy, and the potential technology misuse. Ethical failures in cybersecurity can lead to severe consequences such as infringement of trust, financial damages, and individual harm (Edquist et al., 2022).

One of the primary ethical challenges in cybersecurity is consent. Data is collected and used without the explicit consent of the individuals concerned in many cases (Ademola, 2020). This lack of transparency can lead to mistrust and ethical breaches. Cybersecurity researchers must strive to implement security practices that ensure informed consent where individuals are fully aware of what data is being collected, how it will be used, and the measures in place to protect it (Dupuis & Renaud, 2021). Data manipulation and integrity can pose added ethical challenges. The accuracy and reliability of data is crucial as manipulated or corrupted data can lead to misinformation and potentially harmful choices. Cybersecurity researchers must uphold stringent standards to prevent unauthorized modifications to data and ensure its integrity throughout its lifecycle (Esmer & Arıbaş, 2022). Privacy is another critical ethical issue. Cybersecurity researchers frequently possess access to vast amounts of personal and sensitive information. Ensuring the confidentiality of this data is paramount and yet the appropriate measures taken to secure it can sometimes disregard on individual privacy (Herrmann & Pridöhl, 2020). For instance, intrusive monitoring and surveillance practices can be effective for security purposes but they can also violate privacy rights if not properly regulated and justified. The potential technology misuse can also present significant ethical problems (Familoni, 2024)). Innovative technologies such as artificial intelligence and machine learning can be powerful in defending against cyber threats. However, these same technologies can be misused for malicious purposes such as crafting sophisticated malware or conducting intrusive reconnaissance. Cybersecurity researchers must navigate the ethical implications of developing and deploying these technologies in order to ensure that these technologies are used in a responsible and ethical way (Formosa et al., 2021). Ethical failures in cybersecurity can also lead to severe consequences since security is everyone's responsibility. Breaches of trust can occur when organizations fail to protect sensitive information resulting in significant reputational damage and loss of customer confidence (Frohmann, 2000). Financial losses can be substantial for individuals whose data is compromised and for organizations experiencing the counteract and recovering from data breaches. Furthermore, the individual harm can be profound including identity theft, financial fraud, and other forms of cybercrime (Esmer & Arıbaş, 2022).

The ethical challenges in cybersecurity are complex and multifaceted often requiring a delicate balance between technological capabilities and ethical principles (Hawamleh et al., 2020; Joanna, 2021). In order to address these ethical challenges, cybersecurity researchers must adhere to established ethical frameworks and continuously update their knowledge and skills to keep pace with evolving threats and technologies. This includes fostering a culture of ethical awareness and responsibility within their organizations and ensuring that the ethical considerations are integral to all cybersecurity practices and decisions (Bauer et al., 2020).

Cybersecurity researchers can protect sensitive data and technological systems while maintaining public trust and safeguarding individual rights by upholding high ethical standards and continuously adapting to new challenges

CODE OF ETHICS

Code of Ethics provide a foundation for addressing these issues by offering guidelines for ethical decision-making and help delineate acceptable behavior from unethical practices. By adhering to these guidelines, cybersecurity researchers can navigate complex ethical dilemmas, make informed decisions that balance technical requirements with ethical considerations, and align their actions with moral standards (Lee, 2020).

Two prominent organizations including the Association for Computing Machinery (ACM) and the Institute of Electrical and Electronics Engineers (IEEE) have established codes of ethics and behavioral requirements for researchers in various fields. These frameworks provide comprehensive guidelines for ethical conduct related to fairness, honesty, and respect for privacy. Despite these guidelines, the dynamic nature of the digital landscape requires continuous evaluation and adaptation of ethical standards to address the emerging challenges around responsible data research (Loi & Christen, 2020). The ACM Code of Ethics and Professional Conduct guidelines document was designed to inspire and guide the ethical conduct of all computing researchers. It highlights the importance of contributing to society and human well-being, avoiding harm to others, being honest and trustworthy, and respecting privacy and confidentiality (Anderson, 2018). These principles guide researchers in making decisions that prioritize and uphold ethical standards in their work. For instance, the ACM code emphasizes the need for transparency in data practices and the protection of users' personal information. Similarly, the IEEE Code of Ethics (2020) outlines the commitment of engineers to ethical practices. It includes tenets such as improving the understanding of technology and its potential consequences, maintaining and improving technical competence, and avoiding real or perceived conflicts of interest (IEEE, 2020). For instance, when confronted with a decision about whether to implement a particular security measure or not, researchers can refer to these codes to evaluate the potential impact on privacy and fairness. The IEEE (2020) code also stresses the importance of treating all persons fairly and with respect, which includes safeguarding the privacy and dignity of individuals in the digital realm. Also, the IEEE Code of Ethics stresses the importance of transparency and accountability in professional activities encouraging engineers to act in ways that are honest, impartial, and fair.

Despite the robust guidelines provided by these leading organizations, the dynamic nature of the digital landscape requires continuous evaluation and adaptation of ethical standards. The rapid development of technological advancements mean that the new ethical dilemmas are constantly emerging (Lehtonen, 2021). For instance, the advent of artificial intelligence and machine learning introduces questions about culpability, bias, and the ethical use of autonomous systems. Ethical frameworks must be flexible and responsive and require ongoing dialogue among researchers, ethicians, policymakers, and the general public to ensure that the ethical guidelines remain pertinent and current in order to address these emerging challenges (Macnish & Van der Ham, 2020). Continuous professional development and education are crucial for cybersecurity researchers to stay abreast of the latest ethical issues and best practices (Lonsdale, 2020). Organizations must foster a culture of ethical awareness and accountability. This involves not only adhering to established codes of ethics but also encouraging open discussions about ethical dilemmas and supporting ethical decision-making processes. Internal policies and procedures should reflect the organization's commitment to ethical standards and support procedures should be implemented to address and resolve ethical concerns (Lucas, 2017).

Code of Ethics provide a structured approach to addressing ethical dilemmas by ensuring that the decisions are well-reasoned, fair, and aligned with societal values. This approach not only protects individuals and organizations but also fosters a culture of ethical responsibility within the cybersecurity profession (Mahfood et al., 2005). By adhering to these ethical principles, cybersecurity researchers can make informed decisions that not only meet technical requirements but also uphold the highest standards of ethical conduct.

ETHICAL ISSUES IN CYBERSECURITY RESEARCH

Ethical issues highlight the need for fairness and non-discrimination. The researcher should analyze whether the technology disproportionately affects certain groups or individuals and take steps to mitigate any biases (Manjikian, 2017). They must ensure the technology is applied rightfully and does not unjustly target or exclude specific population. They also stress the importance of transparency and accountability in research activities. The researcher would document their decision-making process, provide clear rationales for their choices, and be open to scrutiny (Mbinjama-Gamatham & Olivier, 2020). This transparency helps build trust with stakeholders and demonstrates a commitment to ethical standards. A key principle in ethical guidelines is to avoid harm (Onyancha, 2015). For instance, a researcher would need to weigh the potential security benefits of the surveillance technology against the possible harm to individuals' privacy and civil liberties. They should strive to find a solution that enhances security without causing undue harm or infringing on rights (Miller & Bossomaier, 2024).

Intersection of Cybersecurity and Research with Human Subjects

The intersection of cybersecurity and research with human subjects is a critical area where ethical considerations overlap while presenting unique challenges and opportunities. Research involving human subjects often requires collecting, storing, and analyzing sensitive personal data making it susceptible to cyber threats such as data breaches, unauthorized access, and misuse. Cybersecurity principles and codes of ethics were developed to protect data and digital infrastructure and this can be extended to enhance the ethical framework governing research with human subjects (Fiesler et al., 2024). Institutions can better protect participants' privacy, uphold their rights, and ensure the responsible use of data by integrating cybersecurity standards into research practices. The integration of cybersecurity practices into human subjects research is not just a technical requirement but a fundamental ethical obligation that helps ensure the responsible use of data in an increasingly digital world.

Applying Ethical Codes in Real-World Scenarios

Research with human subjects frequently involves the collection of sensitive data including personal identifiers, health records, and behavioral information which are valuable targets for cyber attackers (Fiesler et al., 2024). This intersection necessitates a dual focus on safeguarding the data and protecting the individuals behind it. Cybersecurity codes such as the Association for Computing Machinery (ACM) and the Institute of Electrical and Electronics Engineers (IEEE) focus on principles like fairness, transparency, accountability, and avoiding harm which align closely with the ethical guidelines for human subjects research.

The IEEE Code of Ethics outlines the commitment to "protect privacy and respect confidentiality" and to "avoid real or perceived conflicts of interest" which are principles that directly apply to managing human subjects data ethically in the context of cybersecurity (IEEE, 2020). These principles can be operationalized in research settings by implementing robust data protection measures and maintaining transparency about how data is stored and secured.

Application of Cybersecurity Codes in Research with Human Subjects

Data Protection and Privacy for Safeguarding Personal Data

One of the most significant intersections between cybersecurity and human subjects research is the shared commitment to data protection and privacy (Andrews et al., 2024). Cybersecurity codes advocate for the implementation of strong encryption methods, secure data storage solutions, and access controls to prevent unauthorized access (Maalem Lahcen et al., 2020). Institutions conducting research must adopt these cybersecurity measures to prevent data breaches and unauthorized disclosures (Dalkıran, 2024). This can be applied in real-world scenarios by using secure platforms for data collection that employ end-to-end encryption to ensure that data is protected from the point of collection to storage and analysis. One scenario would be researchers making use of secure web forms with built-in encryption features when conducting surveys or interviews that involve collecting sensitive information. In another instance cybersecurity principles suggest performing regular audits of data security practices to ensure they are up-to-date and resilient against emerging cyber threats to reinforce the need for continual improvement of security measures in research settings.

Informed Consent for Transparency About Cybersecurity Risks

Cybersecurity codes stress the importance of transparency which overlaps with ethical requirements for informed consent in human subjects research. Participants must be fully informed not only about the nature of the research but also about the cybersecurity measures in place to protect their data. Participants should be made aware of potential cybersecurity risks, including the possibility of data breaches or unauthorized access as part of the informed consent process (De Roche & Silver,

2024; Pirani, 2024). Researchers can enhance informed consent documents in practice to include detailed information about data security protocols such as encryption methods, data anonymization, and how personal data will be stored and accessed. This approach guarantees that participants have a clear understanding of how their data will be protected and what measures are in place to mitigate risks by aligning with both cybersecurity standards and human subjects research ethics.

Accountability and Ethical Data Use for Preventing Misuse of Data

Cybersecurity codes emphasize accountability in data handling and require that data is used ethically to maintain researcher accountability for any actions that compromise data security (Macnamara, 2024). Institutions should implement strict access controls and ensure that data is used solely for the purposes outlined in the research (Parthasarathy et al., 2023). This directly applies to research with human subjects where the misuse of data can lead to significant ethical breaches. Accountability can be reinforced in real-world applications by implementing role-based access controls that limit data access to authorized personnel only ensuring that sensitive data cannot be accessed or used by individuals not directly involved in the research (Andrews et al., 2024). Employing logging and monitoring systems that track who accesses data and when the access occurred helps maintain accountability and provides a record that can be reviewed if any ethical or security concerns arise.

Avoiding Harm and Mitigating Cybersecurity Risks to Protect Participants

Avoiding harm is a shared principle in both cybersecurity and human subjects research. This means implementing measures that prevent data loss, unauthorized access, and other cyber threats that could harm research participants. The critical need for robust cybersecurity protocols to protect participant data from being compromised prevents potential psychological, financial, or social harm (McKee, 2024; Stegenga et al., 2023). Institutions can operate this by conducting regular risk assessments to identify potential vulnerabilities in their data handling and storage practices. For instance, researchers should select platforms with strong security credentials and configure settings to maximize data protection when using cloud storage solutions (Gunathilake, 2024). Also, employing data minimization techniques to collect only the data necessary for the study can reduce the risk of harm in case of a data breach.

SOLUTIONS AND RECOMMENDATIONS

The following principles are fundamental to ethics for responsible data research in cybersecurity. These principles ensure that researchers conduct their work with integrity, fairness, and a strong commitment to protecting individuals' rights (Pattison, 2020). By embedding these values into their daily practices, cybersecurity researchers can navigate ethical challenges effectively and uphold the highest standards of ethical conduct in their field.

- **Addressing Current Ethical Challenges:** This provides researchers with the tools to effectively address current ethical challenges (Rajasekharaiah et al., 2020). In the face of issues like data breaches, privacy concerns, and the ethical use of emerging technologies, having a robust ethical foundation helps researchers make informed and principled decisions (Morgan & Gordijn, 2020). For instance, when faced with a decision about using a new technology that could enhance security but also raise privacy concerns, ethical guidelines provide a basis for weighing the benefits and risks in a balanced manner.
- **Alignment with Societal Values:** Cybersecurity practices must be aligned with broader societal values such as justice, fairness, and respect for individual rights. By grounding their actions in these values, researchers can navigate ethical challenges in a way that respects societal norms and expectations (Raul, 2021). For instance, prioritizing the protection of user privacy and data security reflects the societal value placed on personal privacy and trust in digital systems.
- **Accountability:** Cybersecurity and research are closely tied to transparency and involve holding researchers responsible for their actions and decisions throughout the research lifecycle. Accountability can be operationalized by establishing clear roles and responsibilities for each team member to implement regular ethical audits and maintain adherence to established ethical standards (Dalkıran, 2024). It is essential to have mechanisms in place such as peer review and ethical oversight committees that actively monitor research activities and intervene when ethical breaches are identified. These structures help ensure that researchers are not only aware of ethical norms but also consistently apply them in their work.

Accountability in research means that researchers are answerable for their actions and decisions. Operationalizing accountability involves setting up structures and processes that monitor and evaluate research activities. Ethical review boards, regular audits, and clear reporting lines within research teams are critical mechanisms that enforce accountability. Regular training on ethical standards and peer review processes helps researchers remain vigilant about their responsibilities and remain compliant with best practices (Stegenga et al., 2023).

Accountability requires researchers to uphold the principles of responsible authorship and data stewardship. This includes acknowledging the contributions of all team members accurately and ensuring that data is stored securely and handled according to ethical guidelines. Accountability mechanisms should facilitate prompt corrective actions including retractions, public disclosures, and disciplinary measures when ethical breaches occur.

- **Consistency:** This offers a set of standardized guidelines that researchers can follow in order to ensure consistency in decision-making across various scenarios. This consistency helps in maintaining a uniform approach to ethical issues and reducing the risk of arbitrary or biased decisions (Richards et al., 2020). For instance, when dealing with data breaches, researchers can rely on established protocols that emphasize prompt disclosure, mitigation efforts, and transparency, while making sure that the responses are predictable and fair.
- **Fairness:** Cybersecurity practices should not disproportionately harm or disadvantage any group. Fairness guides researchers to consider the broader societal implications of their actions and strive for justifiable outcomes (Sakka & Spyrou, 2015). This requires implementing bias checks at various stages including data collection, model development, and data interpretation. Fairness can be enhanced through standardized protocols that minimize the potential for bias by ensuring that all groups are equitably represented in the data (Pirani, 2024). Research teams must regularly audit their methodologies and algorithms to identify any systemic biases and adjust their processes accordingly to avoid unjust outcomes.

Fairness in research entails ensuring that all participants and groups are treated equitably and without bias (Miteu, 2024). This requires deliberate efforts to include diverse populations in studies to avoid the exclusion or overrepresentation of any group which can lead to skewed results and unethical outcomes. Fairness can be operationalized by establishing clear criteria for participant selection and consistently reviewing these criteria to prevent discrimination (Pirani, 2024). Researchers should employ randomized sampling techniques and carefully monitor their methodologies to avoid biases that might affect vulnerable or underrepresented groups. Fairness also extends to the equitable distribution of research benefits. This involves ensuring that the outcomes of research, such as medical treatments or technological innovations, are accessible to all groups rather than only benefiting certain demographics or geographic locations (Karunarathna et al., 2024; Richards et al., 2020). Implementing community engagement strategies and feedback loops with study populations can help researchers understand the needs and expectations of different groups by tailoring their approaches to be more inclusive and fairer. Researchers help prevent biases and ensure that their actions promote justice and equality, contributing to a more inclusive and equitable digital environment by striving for the principle of fairness (Schlehahn, 2020; Miteu, 2024).

- **Honesty:** Ethics for responsible data research must emphasize the need for honesty in all research activities (Schultz, 2005). This principle requires cybersecurity researchers to be truthful about the capabilities and limitations of cybersecurity measures. This includes transparent communication of potential risks and vulnerabilities to stakeholders which would ensure them to have an accurate understanding of the security landscape (Sharan & Boruah, 2016). Honesty also involves accurately reporting findings and incidents without exaggeration and/or omission. By adhering to the principle of honesty, researchers build trust with clients, colleagues, and the public thereby fostering a culture of integrity and reliability.
- **Preparation for Future Developments:**Shou (2012) states that as technology and cyber threats evolve, so do the ethical challenges associated with them. Ethical frameworks prepare researchers to adapt to these changes by promoting continuous learning and ethical vigilance (Shou, 2012). This proactive approach ensures that researchers are not only reactive to ethical issues but also prepared to anticipate and address new challenges as they arise. For instance, staying informed about the ethical implications of artificial intelligence and machine learning in cybersecurity allows researchers to integrate these technologies responsibly and ethically.
- **Respect for Privacy:** Respecting privacy is a core tenet of ethical conduct in cybersecurity. Ethics of conduct guide researchers to implement measures that protect individuals' personal information and uphold confidentiality (Wolf et al., 2024). This involves adopting practices that minimize data collection in order to make sure that only necessary information is gathered and stored in a secure fashion (Shukla et al., 2022). Researchers must also be vigilant in protecting data from unauthorized access, breaches, or misuse. Respect for privacy requires ongoing diligence to uphold privacy rights and balancing the need for security with the imperative to protect individuals' personal information (Van den Hoven, 2010). By prioritizing the principle of privacy, cybersecurity researchers help maintain public trust and confidence in digital systems.

Institutions conducting research with human subjects that leverage data on humans face significant challenges in preserving privacy and protection in the digital age. The proliferation of digital technologies and vast data collection capabilities has intensified the need for robust ethical standards and protective measures. Institutions must navigate complex legal, technical, and ethical landscapes to safeguard participants' rights, privacy, and data security. Implementing comprehensive frameworks that address these challenges can help institutions maintain trust, compliance, and ethical integrity.

Informed consent is a cornerstone of ethical research involving human subjects. Institutions must ensure that participants are fully aware of how their data will be used, stored, and shared. This process must extend beyond traditional consent forms to include detailed information on data security, potential risks, and how data privacy will be maintained. It is important to emphasize clear and accessible consent forms that outline the scope of data use, potential risks, and the measures taken to protect participant privacy (De Roche & Silver, 2024). Digital consent platforms can enhance transparency by allowing participants to view and control their consent status dynamically and to adapt their permissions as needed. Institutions should also consider incorporating ongoing consent mechanisms to periodically inform participants about how their data is being used and provide the opportunity to withdraw consent if desired. This approach allows the participants to maintain control over their data throughout the research lifecycle by aligning with ethical standards that prioritize individual autonomy and privacy.

Institutions must employ advanced data anonymization techniques that prevent the identification of individuals from datasets to protect the privacy of human subjects. Data anonymization involves removing or encrypting personal identifiers and implementing techniques such as data aggregation and differential privacy to further obscure individual data points (Parthasarathy et al., 2023). Differential privacy adds statistical noise to datasets to help prevent re-identification of individuals while preserving the overall utility of the data for analysis. Institutions must also regularly audit anonymization practices for effectiveness against evolving re-identification risks as computational techniques for data mining become more sophisticated. Researchers should avoid collecting more personal information than necessary and adopt a "privacy by design" approach to integrate privacy considerations into the research process from the outset.

Data security is paramount when handling sensitive information about human subjects. Institutions must implement robust cybersecurity measures including encryption, secure data storage, and stringent access controls to protect data from unauthorized access, breaches, and misuse. Institutions should employ multi-layered security protocols such as two-factor authentication, regular security audits, and intrusion detection systems to safeguard data (Dalkıran, 2024). Access to sensitive data should be restricted to authorized personnel who are adequately trained in data security and ethical handling of human subjects' information. Role-based access controls (RBACs) enforce that individuals only have access to the data necessary for their specific research tasks. Institutions should implement logging and monitoring systems that track data access and usage allowing for prompt detection and response to unauthorized activities.

Institutions must establish comprehensive data governance policies that outline standards for data collection, use, sharing, and retention. These policies should be aligned with legal requirements such as the General Data Protection Regulation (GDPR) and other relevant data protection laws. The importance of data governance frameworks that prioritize ethical considerations, such as the minimization of data collection and the protection of participant confidentiality (Schneble Butz, 2024). Data governance policies should also specify the protocols for data sharing with third parties to ascertain that any shared data remains secure and that recipients adhere to the same ethical and legal standards. Institutions can implement data use agreements that clearly define the permissible uses of shared data and the responsibilities of all parties involved in maintaining data security and privacy.

Preserving privacy and protections for human subjects also requires ongoing training for researchers and staff on data protection best practices and ethical standards. Institutions should provide regular training sessions that cover the latest developments in data privacy laws, cybersecurity measures, and ethical research practices. Continuous education helps researchers stay informed about the evolving ethical landscape and reinforces the importance of protecting participant data (Pirani, 2024). Institutional Review Boards (IRBs) play a critical role in monitoring research involving human subjects. These boards should conduct regular reviews of research protocols to ensure compliance with privacy standards and assess whether the measures taken to protect data are adequate. Enhanced oversight mechanisms including post-approval monitoring and audits can help identify potential ethical violations early and ensure that corrective actions are implemented quickly.

Privacy-Enhancing Technologies (PETs) such as encryption, secure multi-party computation, and homomorphic encryption provide additional layers of protection for data used in research. These technologies enable researchers to perform data analysis without directly accessing sensitive information to reduce the risk of data breaches and privacy violations (McKee, 2024). Institutions can enhance data security and ensure that privacy protections are maintained even when data is shared or analyzed by adding PETs into their data handling processes.

Avoiding Harm: This requires proactive measures to protect research participants from physical, psychological, and data-related risks. The importance of informed consent and data anonymization is highlighted as a means to prevent harm (De Roche & Silver, 2024). Researchers must assess the potential impacts of their studies and take steps to mitigate risks such as securing data against misuse and providing participants with clear information about how their data will be used. The principle of avoiding harm assures that findings are not misused in ways that could disadvantage specific groups or lead to harmful policy decisions. This principle can be operationalized through rigorous risk assessments and the implementation of safety protocols that minimize potential harms. Acquiring informed consent is important and involves clearly communicating the risks, benefits, and purpose of the research to participants allowing them to make fully informed decisions about their involvement (Pirani, 2024).

Researchers must maintain that data privacy and confidentiality are upheld throughout the research process (Wolf et al., 2024). This includes using secure data storage solutions, anonymizing datasets, and restricting access to sensitive information. Researchers are also advised to consider the broader societal implications of their findings because misuse of data or misinterpretation of results can lead to harmful consequences beyond the immediate study context. Developing clear guidelines for data use and regularly reviewing these policies helps mitigate the risk of harm, particularly in areas involving sensitive or high-stakes data (Mbinjama-Gamatham & Oliver, 2020).

Transparency: Transparency is crucial for maintaining trust between researchers and participants. Institutions should provide participants with clear information about how their data will be used, the measures in place to protect their privacy, and their rights regarding data access and withdrawal. Open communication and the provision of easy-to-understand privacy notices can significantly enhance participants' confidence in the research process (Mozersky et al., 2020). Institutions can also consider implementing participant portals that allow individuals to view their consent status, access information about ongoing studies, and control the use of their data. Such platforms empower participants and promote a collaborative relationship between researchers and the communities that are studied. Key operational steps include openly sharing information about research goals, methodologies, data handling procedures, and potential conflicts of interest. Transparency can be enhanced through comprehensive documentation and the public sharing of research protocols and data analysis plans, allowing others to scrutinize and replicate the research (Parthasarathy et al., 2024). This level of openness helps to demystify research processes and fosters accountability by making researchers' actions observable and subject to evaluation.

Transparency should also extend to the results and limitations of the study. Researchers are encouraged to report negative results and disclose uncertainties in their findings to maintain transparency in reporting to avoid misinterpretation and misuse of research data while protecting public interests (Dalkıran, 2024). Creating clear, accessible summaries of research outcomes and sharing data responsibly through open-access platforms also contribute to greater transparency.

Organizations can adopt several strategies to maintain ethical conduct and integrate ethical principles into everyday operations and decision-making processes. These strategies include developing clear policies, conducting regular audits and assessments, and providing whistleblower protections (Wylde et al., 2022). Developing clear policies involves establishing comprehensive and actionable guidelines that outline ethical standards and expectations for all employees, and covering key areas such as conflict of interest, data privacy, informed consent, and responsible use of technology (Van de Poel, 2020). Some essential steps include the creation of procedures where policies are easily accessible to all employees, implementing training sessions to educate employees about these policies, and regularly reviewing and updating policies in conjunction with evolving ethical standards and emerging challenges (Yaokumah et al., 2020). Regular audits and assessments including compliance audits to assess adherence to ethical guidelines and risk assessments to identify potential ethical risks help organizations address issues in a proactive manner. Utilizing audit and assessment results to identify areas for improvement and sharing findings with relevant stakeholders to maintain transparency and accountability further demonstrate the organization's commitment to ethical conduct. Providing whistleblower protections such as secure and confidential reporting mechanisms, strong anti-retaliation policies, support for whistleblowers, and clear follow-up procedures for handling reports of unethical behavior encourages reporting and supports ethical behavior (Zhang & Ghorbani, 2021).

Enhancing Ethical Awareness

Cultivating ethical behavior among cybersecurity researchers is crucial to raise awareness about the ethical implications of their work. Comprehensive training programs that cover beyond technical skills to encompass ethical principles and their real-world applications are essential alongside addressing topics such as data privacy, informed consent, and the ethical use of emerging technologies (Wiafe et al., 2020). Continuous learning through workshops, seminars, and online courses or webinars keeps researchers well-versed on evolving ethical standards and new challenges while case studies and simulations illustrate ethical dilemmas and appropriate responses underlining the consequences of decisions and the importance of ethical behavior (Loi & Christen, 2020). Ethical leadership within organizations is vital with senior management modeling integrity and ethical decision-making, developing and enforcing clear policies, and fostering open communication about ethical concerns without fear of retribution. Community engagement involving diverse stakeholders, underrepresented and underprivileged communities is critical in relation to ethical cybersecurity practices, conducting public awareness campaigns, and creating collaborative platforms for dialogue between researchers, policymakers, and community members (Joanna, 2021; Lee, 2020). Organizations can significantly enhance ethical awareness among cybersecurity researchers by implementing these strategies in order to ensure that cybersecurity practices not only protect data and systems but also uphold the highest standards of integrity and respect for individual rights.

Ethical Decision Making

Adopting ethical decision-making models helps researchers evaluate the ethical dimensions of their actions and well-considered decisions aligned with ethical principles. These models provide a structured approach to navigating complex ethical dilemmas involving three key steps of identifying ethical issues, evaluating alternatives, and making informed decisions (Lucas, 2017). The first step involves recognizing and clearly defining the ethical issues involved, understanding the context in which they arise, and formulating the ethical dilemma by considering the interests of all relevant stakeholders (Formosa et al., 2021). The second step entails generating a range of possible actions, analyzing the potential outcomes and impacts on stakeholders, and weighing these alternatives against established ethical principles, such as integrity, transparency, and respect for individual rights (Anderson, 2018; IEEE, 2020). The final step requires choosing the action that best aligns with ethical principles and promotes consistency, developing a clear implementation plan, transparent communication of the decision and its rationale, and reviewing and reflecting on the outcomes to improve future ethical decision-making (Hawamleh et al., 2020).

Organizations can effectively implement ethical policies and practices, create a robust ethical framework that promotes integrity, transparency, and accountability within the organization and significantly enhance ethical awareness among cybersecurity researchers while upholding the highest standards of integrity and trust (Hermann & Pridöhl, 2020) by adopting these strategies.

REGULATORY AND LEGAL FRAMEWORKS GOVERNING CYBERSECURITY AND HUMAN RESEARCH PROTECTIONS

Several key regulatory and legal frameworks play a critical role in shaping how data is managed, protected, and ethically used. These frameworks set standards that help institutions and researchers safeguard participant data, maintain compliance with ethical norms, and navigate complex legal landscapes. Understanding the impact of these regulations on ethical decision-making and the measures necessary for compliance is essential for researchers conducting studies that involve human data.

The ethical and legal frameworks governing human research protections are essential for safeguarding the rights, dignity, and welfare of research participants. These frameworks provide guidelines that researchers must follow to ensure that human subjects are treated ethically, their data is handled responsibly, and their privacy is protected. Understanding these frameworks is crucial for researchers who handle sensitive personal data. This section provides a detailed discussion of key ethical and legal frameworks such as the Common Rule, HIPAA, and GDPR, and explores how these regulations apply to cybersecurity and research involving human subjects.

General Data Protection Regulation (GDPR)

The GDPR is one of the most comprehensive data protection laws enacted by the European Union (EU) to regulate the processing of personal data and protect the privacy rights of individuals. It applies to any organization that handles data of EU citizens regardless of where the organization is located. The GDPR emphasizes principles such as data minimization, consent, transparency, lawfulness, fairness, and accountability which align closely with ethical standards in both cybersecurity and human research protections (Bakare et al., 2024). The GDPR impacts ethical decision-making by requiring researchers to justify their data collection practices and maintain neutrality to the intended research outcomes. Researchers must obtain explicit consent from participants while detailing how their data will be used and stored (De Roche & Silver, 2024). This regulation also enforces the right to data access, adjustment, and removal compelling researchers to establish clear processes for responding to participants' data requests.

GDPR places stringent requirements on obtaining informed consent, ensuring data subject rights, and implementing strong data protection measures for research involving human subjects. Researchers must provide clear information about how personal data will be used, stored, and protected. The emphasis of GDPR on data minimization means that researchers should only collect data that is necessary for the specific research purpose and avoid excessive data collection. GDPR compliance in research settings requires the implementation of technical measures such as data anonymization, encryption, and pseudonymization to protect personal data from unauthorized access (Dupuis & Renaud, 2021). Researchers must also conduct Data Protection Impact Assessments (DPIAs) when data processing is likely to result in high risks to data subjects, particularly when using new technologies or handling sensitive data.

Researchers must establish transparent data handling practices including clear data processing agreements and participant rights management processes to comply with GDPR. Institutions should designate Data Protection Officers (DPOs) responsible for overseeing compliance and conducting regular audits of data security practices. GDPR also mandates that researchers report data breaches to authorities within 72 hours which requires the development of robust incident response plans (Bakare et al., 2024). GDPR compliance is not just a legal requirement but an ethical imperative to protect participants' data for IRBs and researchers. GDPR mandates that data collection must be lawful, transparent, and limited to what is necessary for the research purpose. Researchers are also required to obtain explicit consent from participants detailing the specific purposes for which their data will be used.

IRBs can ensure GDPR compliance by implementing robust consent processes that are clear and specific about data usage. Consent forms should outline data protection measures detailing how data will be stored, who will have access, and how long the data will be retained. IRBs should regularly review consent procedures and data protection practices to ensure they meet GDPR standards (Bakare et al., 2024). IRBs can provide researchers with templates and guidelines that include GDPR compliant language to reduce the burden on researchers navigating these complex requirements independently.

Health Insurance Portability and Accountability Act (HIPAA)

HIPAA is a U.S. law that sets national standards for the protection of health information. It governs how healthcare providers, insurers, and researchers handle sensitive patient data. The Privacy Rule of HIPAA addresses how personally identifiable health information must be protected and mandates safeguards to prevent unauthorized access (Tangudu et al., 2024). The law's Privacy Rule and Security Rule are particularly relevant to research as they establish guidelines for how personal health information (PHI) must be protected during and after data collection. The stringent requirements of HIPAA for data security and privacy have direct implications for ethical decision-making for researchers. The Security Rule requires that researchers must implement administrative, physical, and technical safeguards to ensure that health data is protected from breaches (Bakare et al., 2024). HIPAA requires that researchers obtain waivers of authorization if full consent cannot be obtained which will involve rigorous review by IRBs.

Researchers and institutions must develop comprehensive data management and security plans that outline how PHI will be protected throughout the research process (Dalkıran, 2024). These plans must include protocols for data anonymization, secure data storage, and restricted access to sensitive information in digital environments where data breaches are a significant concern. Researchers can adhere to HIPAA by conducting regular risk assessments and implementing risk management strategies that address potential vulnerabilities in data handling practices. Training staff on data protection policies and maintaining audit trails of data access are crucial steps for meeting the security requirements. The establishment of data use agreements that define the permissible use of PHI in research further strengthens compliance efforts and aligns with HIPAA's focus on minimizing unauthorized data use.

Common Rule (45 CFR 46)

The Common Rule governs research involving human subjects in the U.S. It establishes the requirements for IRB review, informed consent, and the protection of vulnerable populations. The revised version introduced new provisions for informed consent that emphasizes greater transparency and includes concise explanations of risks and benefits. Its primary purpose is to ensure that research is conducted ethically, with respect for persons, beneficence, and justice as its core principles.

The Common Rule directly impacts how ethical decisions are made in research involving human subjects concerning the adequacy of consent and the protection of participants' privacy. It mandates that IRBs must review and approve the adequacy of privacy protections and data security measures in research proposals to align closely with cybersecurity principles. This integration underscores the need for researchers to be vigilant about both ethical and legal obligations to guarantee that all aspects of data management are compliant and protective of participants' rights (Bakare et al., 2024).

The Common Rule applies to all federally funded research involving human subjects requiring IRB review to assess the ethical implications of research proposals. This regulation is relevant in research settings involving cybersecurity where sensitive data about human subjects is often collected, stored, and analyzed. The Common Rule mandates that IRBs evaluate the adequacy of privacy protections and data security measures to minimize risks to participants. Researchers are required to implement robust data protection protocols such as encryption and access controls to comply with these ethical standards. Compliance with the Common Rule involves preserving data privacy measures align with the principles of respect for persons and beneficence, protecting participants from harm while preserving their autonomy (Pirani, 2024). This includes employing additional safeguards such as anonymizing data and securing informed consent specifically for data collection and security risks in cybersecurity research.

Federal Information Security Management Act (FISMA)

FISMA establishes a framework for protecting government information, operations, and assets against natural or human-made threats. It requires federal agencies that are involved in research to develop, document, and implement information security programs (Dhablia, 2024). Compliance with FISMA means implementing stringent cybersecurity protocols to safeguard data integrity and confidentiality for researchers conducting federally funded studies. FISMA enforces a culture of proactive risk management as the impact on ethical decision-making is significant. Researchers must continuously assess potential vulnerabilities in their data handling processes and update their security measures accordingly. This approach fosters an ethical commitment to maintaining the highest standards of data security to prevent harm to participants resulting from data breaches (Dhablia, 2024).

IMPACT OF REGULATIONS ON ETHICAL DECISION-MAKING

The regulations previously discussed impose rigorous requirements that shape the ethical landscape of research involving human subjects as they enforce accountability requiring researchers to actively consider the potential impacts of their work on participants' rights and welfare.

One of the central ethical dilemmas influenced by the regulations is balancing the need for data utility with the obligation to protect participant privacy. The emphasis of GDPR on data minimization requires researchers to collect only the data necessary for their studies which challenges traditional research practices that often prioritize comprehensive data collection (Capili & Anastasi, 2024). This regulation requires ethical decision-making that prioritizes participants' privacy while still achieving meaningful research outcomes. Regulations such as the GDPR and the Common Rule mandate transparent and comprehensive informed consent processes. This has significant ethical implications as researchers must attest that participants are fully aware of how their data will be used and what risks are involved. The requirement for clear communication not only fulfills legal obligations but also aligns with ethical standards that emphasize respect for participant autonomy (Chen et al., 2023). Regulations foster accountability by mandating rigorous oversight through mechanisms such as IRB reviews, data protection impact assessments (DPIAs), and security audits. Ethical oversight ensures that researchers are held responsible for the ethical implications of their work which promotes a culture of transparency and integrity. The need for regular assessments encourages researchers to reflect continuously on their practices and adjust them to meet evolving standards (Christen et al., 2020).

Ensuring Compliance and Addressing Ethical and Legal Challenges

Researchers must develop detailed data management plans (DMPs) that outline how data will be collected, stored, analyzed, and shared to adhere to compliance with regulatory requirements. These plans should specify security measures such as encryption, access controls, and data anonymization techniques. DMPs are essential for operationalizing both ethical and legal obligations as they serve as living documents that guide researchers in their daily practices (Parthasarathy et al., 2024). Researchers and institutional staff must receive ongoing training on current regulatory requirements and ethical standards. Regular workshops and seminars can help keep researchers updated on changes to laws like GDPR and HIPAA, as well as best practices for data security and privacy. Education is critical for fostering a culture of compliance and ethical awareness among research teams (Resnik et al., 2024)

Proactive engagement with IRBs and data protection officers (DPOs) is essential for navigating the ethical and legal complexities of research involving human subjects. These oversight bodies provide valuable guidance on compliance issues and can help researchers identify potential ethical challenges early in the research process. Researchers can ensure that their projects align with both regulatory standards and ethical best practices by fostering open communication and collaboration. Researchers must have clear response protocols in place in the event of a data breach or ethical violation. Immediate actions should include notifying affected participants, conducting thorough investigations, and implementing corrective measures to prevent future incidents. Ethical decision-making in such scenarios requires transparency, accountability, and a commitment to rectifying harm (Edquist, 2022).

The Role of Institutional Review Boards (IRBs) in Human Research Protections

IRBs are critical oversight bodies that evaluate research proposals to ensure that ethical and legal standards are met. They actively assess the adequacy of data privacy and security measures to ensure that the participant data is handled responsibly and ethically.

IRBs review research proposals to manage data privacy concerns which includes examining how informed consent will be obtained, the types of data collected, the security measures in place, and how data will be stored and shared. IRBs often require researchers to implement specific privacy-enhancing technologies such as secure data storage and anonymization techniques to minimize risks to participants (Friesen et al., 2023). IRBs also evaluate cybersecurity risks in research involving human subjects when digital data collection or online platforms are used. They assess whether the proposed data security measures align with current cybersecurity standards and whether the research team is adequately prepared to handle potential data breaches (Dalal et al., 2022). IRBs may request additional safeguards or modifications to research protocols if the proposed cybersecurity measures are deemed insufficient (Peled-Raz et al., 2021).

INTERSECTION OF CYBERSECURITY ETHICS AND HUMAN RESEARCH PROTECTIONS

The integration of cybersecurity ethics into human research protections is becoming increasingly vital as research methodologies evolve to incorporate advanced digital technologies. IRBs serve as critical gatekeepers to make sure that research involving human subjects is conducted ethically with robust measures in place to protect participants' data from cyber threats. The unique challenges posed by cybersecurity in research settings need a clearer understanding of how IRBs specifically address these concerns. A better understanding of how IRBs operate within this evolving landscape can be developed by recognizing the safety and rights of participants. IRBs are traditionally responsible for evaluating the ethical dimensions of research involving human subjects including aspects such as informed consent, risk minimization, and the protection of vulnerable populations. As digital data collection has become more prevalent, IRBs have had to expand their oversight to include cybersecurity-related concerns in recent years. This shift reflects the growing recognition that data security is not just a technical issue but a fundamental ethical consideration in human research.

One of the primary ways IRBs address cybersecurity concerns is by rigorously evaluating the data security measures proposed in research protocols. IRBs require researchers to outline comprehensive data management plans that include technical safeguards such as encryption, secure data storage, and access controls to protect sensitive information from unauthorized access (White, 2020). IRBs assess whether these measures are sufficient to prevent data breaches and other cyber threats by handling participant data responsibly throughout the research lifecycle. IRBs often mandate the use of encryption protocols to secure data both in transit and at rest to reduce the risk of data interception during online data collection or storage. Access to sensitive data is restricted to authorized personnel and IRBs may require role-based access controls to limit exposure to the minimum necessary individuals involved in the study.

Cybersecurity concerns extend to how data privacy and security risks are communicated to participants during the informed consent process. IRBs require informed consent documents to include clear information about data security measures and potential cybersecurity risks to allow participants to make fully informed decisions about their involvement in the research (De Roche & Silver, 2024). This transparency is vital in building trust between researchers and participants in studies involving sensitive data such as health information or behavioral patterns. IRBs guarantee that consent forms are not only comprehensive but also accessible often recommending layered consent models where detailed information about cybersecurity measures is provided in a clear and non-technical language.

Connecting Cybersecurity Ethics to Human Research Protections

The ethical principles of cybersecurity such as confidentiality, integrity, and availability intersect with the core tenets of human research protections. IRBs play a pivotal role in operationalizing these principles within research settings to align cybersecurity ethics with broader ethical standards in human research. One of the moral obligations in cybersecurity is the prevention of unauthorized data use and misuse which aligns with the mandate of Common Rule to minimize risks to participants by affirming that data is only used for the purposes outlined in the research protocol (Kavak et al., 2021). IRBs actively review the data handling procedures proposed by researchers to reinforce compliance with this principle. The importance of IRBs scrutinizing data sharing agreements, ensuring that data is not shared with third parties without explicit consent and appropriate data security measures (Barnes et al., 2020). IRBs also require that researchers implement audit trails and monitoring systems that track data access to enable prompt detection of any unauthorized activities. This approach not only protects participants but also upholds the ethical integrity of the research by preventing data from being exploited in ways that could cause harm.

AI and ML technologies pose unique ethical challenges in research related to bias and fairness. IRBs have started to address these issues by incorporating ethical reviews of AI and ML models into their standard evaluation processes. IRBs now require researchers to demonstrate how they will assess and mitigate biases in AI-driven analyses to assure that outcomes do not unfairly disadvantage specific groups (McKee, 2024). This involves scrutinizing the datasets used to train AI models and not perpetuate existing biases. IRBs also evaluate whether researchers are employing explainable AI techniques which make the decision-making processes of AI systems more transparent and understandable to both researchers and participants (Davison et al., 2024).

Implementing Failsafes for Enhancing Protections Against Cybersecurity Threats

IRBs can implement a range of failsafes that enhance protections against cybersecurity threats. These measures help establish that both institutions and researchers adhere to ethical standards even as digital technologies continue to evolve (Stoudt et al., 2024). IRBs can encourage the use of PETs which allow data to be analyzed without exposing sensitive information. These technologies align with ethical standards that prioritize participant privacy and reduce the risk of data breaches. PETs are valuable in research settings that involve large datasets and complex data sharing arrangements (Knight et al., 2024) Dynamic consent models offer a more adaptive approach to managing participant consent that allows individuals to update their consent preferences as the research evolves (Stoudt et al., 2024). IRBs can recommend these models to enhance participant autonomy and certify that the consent remains relevant as data use changes over time (Yaokumah et al., 2020). This is important in studies involving AI and ML where new data applications may emerge after the initial consent is obtained.

IRBs should implement continuous monitoring of research projects by conducting post-approval audits to verify compliance with data security standards and consent agreements. This ongoing oversight helps identify potential cybersecurity vulnerabilities early and stipulates that corrective actions can be taken (Zhang & Ghorbani, 2021). The value of continuous monitoring in maintaining the integrity of research involving human subjects in data-centric and technologically complex studies (Wylde et al., 2022).

Human Research Protections, Data Security, Privacy, and AI Challenges

AI and ML technologies offer powerful tools for analyzing large datasets and generating insights that were previously unattainable (Ajwang & Ikoha, 2024). These technologies pose significant ethical challenges when they are used in research involving human subjects. The use of AI-driven models can raise issues of data misuse, privacy violations, and biases that may result in harmful or discriminatory outcomes. There is a need for IRBs to closely examine the development and application of AI and ML models within research protocols to ensure they do not inadvertently harm participants or reinforce existing biases (Ajwang & Ikoha; McKee, 2024).

The rapid integration of AI & ML into research practices has transformed how data is collected, analyzed, and applied that created both opportunities and new ethical challenges (Buchanan, 2020). This technological shift has amplified concerns about data security, privacy, and the protection of human research participants. IRBs play a crucial role in overseeing these emerging complexities directing that the research involving human subjects adheres to ethical and legal standards. There is a critical need for human research protections alongside robust data security measures while growing challenges posed by AI and ML need to be addressed (Sontan & Samuel, 2024). IRBs can effectively manage these aspects and outline potential failsafes to safeguard human subjects codifying that institutions and researchers maintain the highest standards of ethical conduct. IRBs must expand their focus to include the ethical implications of data security, privacy, and algorithmic decision-making. IRBs are now required to evaluate not only the traditional aspects of human subject protections but also the complex data flows and potential risks associated with modern research methodologies (Friesen et al., 2023).

Data Security and Privacy Oversight

As research increasingly relies on digital tools that collect and store sensitive data, data security and privacy are foundational to ethical research involving human subjects. IRBs must enforce that researchers have implemented comprehensive security measures including encryption, secure access controls, and regular audits to prevent unauthorized data access and protect participant confidentiality. IRBs should rigorously assess data management plans to confirm that they meet regulatory standards like GDPR and HIPAA which set stringent requirements for data security and privacy protections (Dalkıran, 2024). IRBs must scrutinize how researchers obtain informed consent by establishing that participants are fully aware of data privacy risks and the measures taken to mitigate them. Transparent communication about data handling, storage, and potential cybersecurity threats is essential to uphold participants' trust and autonomy (Dalal et al., 2022).

Navigating the Ethical Complexities of AI and ML in Human Research

One of the major ethical concerns with AI is algorithmic bias where models trained on biased datasets produce skewed results that can disproportionately affect certain groups (Sontan & Samuel, 2024). IRBs should mandate that researchers conduct bias assessments and validation checks to validate that AI models are fair and representative of diverse populations. This aligns with the ethical principle of justice which requires equitable treatment of all research participants.

The opaque nature of many AI and ML models complicates ethical oversight because it can be challenging to understand how decisions are made (Traianou & Hammersley, 2024). IRBs should encourage researchers to adopt explainable AI techniques that offer insights into the decision-making processes of their models

(Lapid et al., 2023). This approach promotes transparency and allows participants and stakeholders to better understand the potential impacts of AI-driven research. IRBs help maintain that the research is accountable and ethically sound by requiring clear documentation of AI methodologies.

Data minimization refers to collecting only the necessary data for the research purpose which is a key strategy to reduce privacy risks. Research proposals must prioritize data minimization and employ de-identification techniques such as anonymization and pseudonymization to protect participants' identities as outlined by IRB (Miller & Bossmaier, 2024). These approaches are essential for maintaining privacy in large-scale data analysis and for safeguarding sensitive information, particularly when data sharing is involved (Barnes et al., 2020).

Traditional consent models may not fully capture the complexities of AI and data-driven research as the uses of data can evolve over time (Familoni, 2024). IRBs should recommend dynamic consent models to allow participants to update their consent preferences as the study progresses and new uses of data emerge. Layered consent approaches where information is presented in stages based on participants' needs and interests can enhance understanding and engagement (Davison et al., 2024). This adaptive approach aligns with ethical standards that prioritize respect for participants' autonomy and informed decision-making.

IRBs should implement continuous monitoring and post-approval audits of research projects to maintain ongoing compliance with ethical standards. This proactive approach allows IRBs to review data security practices, assess compliance with consent agreements, and verify that researchers adhere to privacy protocols throughout the study. Ongoing oversight is critical for identifying potential ethical violations early and maintaining the integrity of the research (Pirani, 2024).

IRBs can benefit from establishing AI ethics committees that include experts in data science, cybersecurity, ethics, and law. These committees can provide specialized guidance on the ethical challenges associated with AI and ML to help IRBs make informed decisions about complex research proposals (Knight et al., 2024). Engaging with experts affirms that IRBs remain up-to-date with technological advancements and are equipped to address the unique risks posed by AI in human research (Koali et al., 2024).

Institutional and Researcher Accountability

IRBs must enforce strict accountability measures to direct institutions and researchers to comply with ethical standards and legal regulations. This involves verification of research protocols that align with the requirements of key regulations such as the Common Rule, HIPAA, and GDPR (Bakare et al., 2024). Researchers must provide detailed data management plans that outline the security measures in place to protect participant data and preserve privacy. Institutions should support IRBs by providing access to training resources that enhance IRB members' understanding of AI, data security, and ethical research practices. Continuous education for IRB members is critical and enables them to evaluate the ethical implications of cutting-edge research methodologies (Stegenga et al., 2023).

Enhancing Practical Guidance for IRBs for Implementing Ethical Frameworks in Data-Driven Research

IRBs require practical tools and actionable strategies to effectively manage ethical issues in research involving human subjects in the realm of cybersecurity (Lapid et al., 2023). IRBs play a pivotal role in protecting participants by maintaining that research adheres to ethical standards and regulatory requirements. The rapidly evolving nature of data security, privacy, and AI technologies requires practical guidance on how to apply these frameworks in real-world scenarios. There are practical guidelines, tools, and examples that IRBs can use to address cybersecurity and ethical concerns in research to enhance their ability to safeguard human subjects.

IRBs can benefit from standardized checklists that help assess the adequacy of cybersecurity measures and ethical considerations in research protocols. These checklists serve as practical tools to assure that all critical aspects of data protection and participant safety are addressed.

Data Security Measures:

- Is data encrypted both in transit and at rest? (Yes/No)
- Are access controls implemented to restrict data access to authorized personnel only? (Yes/No)
- Are data storage solutions compliant with regulatory standards such as GDPR and HIPAA? (Yes/No)

Informed Consent:

- Does the consent form clearly outline data privacy risks and security measures? (Yes/No)
- Are participants informed of how their data will be used, stored, and shared? (Yes/No)

Considerations of AI and ML:

- Are potential biases in AI and ML models assessed and mitigated? (Yes/No)
- Is the AI model explainable, allowing participants to understand how their data contributes to outcomes? (Yes/No)

Incident Response and Data Breach Protocols:

- Are there clear procedures in place for responding to data breaches? (Yes/No)
- Is there a plan for notifying participants in the event of a data breach? (Yes/No)

This checklist aligns with recommendations which emphasize the need for systematic evaluations to identify and mitigate cybersecurity risks in research involving human subjects.

IRBs can employ decision-making frameworks that guide ethical evaluations when dealing with complex cybersecurity and AI-related concerns. These frameworks provide a structured approach for assessing the ethical implications of research protocols to help IRBs navigate challenging decisions.

Identify Ethical and Cybersecurity Concerns:

- Define the ethical issues related to data security, privacy, and AI.
- Assess the potential risks to participants and data integrity.

Evaluate Risks and Benefits:

- Weigh the benefits of the research against the potential risks.
- Determine whether the data security measures in place adequately mitigate identified risks.

Ethical Guidelines and Regulations:

- Reference relevant ethical frameworks, such as the Common Rule, GDPR, and HIPAA, to ensure compliance.
- Use guidance which provides standards for assessing data privacy and security measures.

Engage Stakeholders:

- Involve researchers, data security experts, and legal advisors in discussions to explore ethical concerns.
- Consult with participants when feasible to understand their perspectives on data security and privacy.

Implement Safeguards and Monitor Compliance:

- Recommend additional safeguards if necessary.
- Establish monitoring protocols to ensure ongoing compliance and address emerging ethical concerns.

IRBs can learn from successful policies and practices implemented at other institutions. Incorporating dynamic consent models allows participants to adjust their consent preferences as the study evolves which will enhance participant autonomy and engagement. Dynamic consent models are adaptable and responsive to participants' evolving needs while enhancing ethical oversight in data-centric research (Davison et al., 2024). Such models provide an additional layer of protection for participants to remain informed and in control of their data throughout the research lifecycle.

IRBs must be equipped with the knowledge and skills necessary to evaluate cybersecurity and ethical concerns effectively (Lapid et al., 2023). Continuous training and access to specialized resources can enhance IRB members' ability to address complex issues in research involving digital technologies. IRBs should participate in targeted training programs that cover key areas such as data security protocols, AI ethics, and regulatory compliance. Training modules should include cybersecurity best practices, AI and ML ethics, and regulatory compliance. Cybersecurity best practices should cover encryption, secure data storage, and access control measures (Formosa et al., 2021). AI and ML ethics should include assessing algorithmic bias, transparency, and the implications of automated decision-making (Familoni, 2024). Regulatory compliance should be comprised of navigating GDPR, HIPAA, and other relevant legal frameworks (Bakare et al., 2024). IRBs can enhance their decision-making capabilities by collaborating with cybersecurity experts, data scientists, and legal advisors who can provide specialized insights into the ethical implications of research technologies (Frohmann, 2000). Establishing AI ethics committees within IRBs can also provide valuable support to prepare IRBs to evaluate the complexities of data-driven research (Haneef & Agarwal, 2024).

CONCLUSION

The ethical considerations in cybersecurity are paramount especially in the context of responsible data research (Pirani, 2024). The rapid evolution of technology requires that the cybersecurity researchers prioritize ethical principles in order to ensure that their work benefits the society without causing harm. This involves adhering to ethical frameworks, raising ethical awareness, and implementing robust policies to navigate the complex landscape of cybersecurity (Raul, 2021; Shukla et al., 2022). Emerging technologies like Artificial Intelligence (AI), Machine Learning (ML), and Internet-of-Things (IoT) present new ethical challenges that require innovative solutions and interdisciplinary approaches (Rajasekharaiah, 2020). Integrating insights from philosophy, sociology, and law can provide a comprehensive understanding of these dilemmas while global perspectives are essential to address cultural differences in ethical standards. By adopting ethical decision-making models and fostering a culture of ethical responsibility, cybersecurity researchers can make informed decisions that uphold the highest standards of integrity, fairness, and respect for individual rights (Liu & Murphy, 2020; Ramirez, 2024). Future research should continue to explore these areas alongside developing adaptive regulations and ethical guidelines that stay in tune with technological advancements. Ultimately, this holistic approach ensures that cybersecurity practices not only protect data and systems but also align with societal values fostering a secure and trustworthy digital environment.

Future Research Directions

Interdisciplinary research offers a holistic perspective on ethical dilemmas facilitating the development of robust and nuanced ethical frameworks. Integrating insights from various disciplines including philosophy, sociology, and law can significantly enhance the understanding of ethical issues in cybersecurity. Philosophical insights such as ethical theories like utilitarianism, deontology, and virtue ethics, provide a foundation for understanding ethical principles and enhancing moral reasoning skills among cybersecurity researchers (Lehtonen, 2021). Sociological perspectives examine the broader social implications of cybersecurity practices, addressing issues like the digital divide, access to technology, and social equity while considering cultural contexts to ensure culturally sensitive approaches (Lucas, 2017). Legal frameworks clarify regulatory requirements and the rights and responsibilities of stakeholders ensuring that cybersecurity practices comply with laws while upholding ethical standards (Christen et al., 2020). Encouraging collaborative projects that bring together experts from various fields and conducting interdisciplinary case studies on real-world incidents can provide comprehensive insights and effective solutions. Interdisciplinary research can also inform the development of policies that are ethically sound, socially responsible, and legally compliant while enhancing ethical guidelines for cybersecurity practices (Artz, 2008). Designing educational programs that incorporate interdisciplinary perspectives prepares cybersecurity researchers for ethical challenges, and ongoing professional development programs keep them updated on the latest insights and best practices. Adoption of interdisciplinary approaches in future research can significantly enhance the understanding and management of ethical issues in cybersecurity which can foster a more responsible and inclusive digital environment.

REFERENCES

Ademola, O. E. (2020). An ethical approach to understanding cyber security. *Cyber Security Practitioner's Guide*, 475-502.

Aderibigbe, N. A. (2021). Synopsis on cyberethics behaviour: A literature review. *Inkanyiso*, 13(2), 273–290. DOI: 10.4102/ink.v13i2.8

Ajwang, S. O., & Ikoha, A. P. (2024). Publish or perish in the era of artificial intelligence: Which way for the Kenyan research community? *Library Hi Tech News*. Advance online publication. DOI: 10.1108/LHTN-04-2024-0065

Allahrakha, N. (2023). Balancing cyber-security and privacy: Legal and ethical considerations in the digital age. *Legal Issues in the Digital Age*, 4(2), 78–121. DOI: 10.17323/10.17323/2713-2749.2023.2.78.121

Anderson, R. E. (2018). ACM Code of Ethics and Professional Conduct. *Communications of the ACM*, 35(5), 94–99. DOI: 10.1145/129875.129885

Andrews, J., Zhao, D., Thong, W., Modas, A., Papakyriakopoulos, O., & Xiang, A. (2024). Ethical considerations for responsible data curation. *InProceedings of the 37th International Conference on Neural Information Processing Systems, 36*, 55320-55360.

Artz, J. M. (2008). Addressing the central problem in cyber ethics through stories. In *Information Security and Ethics: Concepts, Methodologies, Tools, and Applications* (pp. 3824–3828). IGI Global. DOI: 10.4018/978-1-59904-937-3.ch260

Atapour-Abarghouei, A., McGough, A. S., & Wall, D. S. (2020). Resolving the cybersecurity data sharing paradox to scale up cybersecurity via a co-production approach towards data sharing. In *2020 IEEE International Conference on Big Data (Big Data)*, 3867-3876. IEEE. DOI: 10.1109/BigData50022.2020.9378014

Bakare, S. S., Adeniyi, A. O., Akpuokwe, C. U., & Eneh, N. E.Seun Solomon BakareAdekunle Oyeyemi AdeniyiChidiogo Uzoamaka AkpuokweNkechi Emmanuella Eneh. (2024). Data privacy laws and compliance: A comparative review of the EU GDPR and USA regulations. *Computer Science & IT Research Journal*, 5(3), 528–543. DOI: 10.51594/csitrj.v5i3.859

Barnes, M., Carrithers, J., & Sugarman, J. (2020). Ethical and practical concerns about IRB restrictions on the use of research data. *Ethics & Human Research*, 42(6), 29–34. DOI: 10.1002/eahr.500072 PMID: 33136330

Bauer, T. N., Truxillo, D. M., Jones, M. P., & Brady, G. (2020). Privacy and cybersecurity challenges, opportunities, and recommendations: Personnel selection in an era of online application systems and big data. *Big Data in Psychological Research*, 393–409. DOI: 10.1037/0000193-018

Buchanan, B. (2020). A national security research agenda for cybersecurity and artificial intelligence. *Center for Security and Emerging Technology Issue Brief, 7*.

Bynum, T. W. (2001). Computer ethics: Its birth and its future. *Ethics and Information Technology*, 3(2), 109–112. DOI: 10.1023/A:1011893925319

Capili, B., & Anastasi, J. K. (2024). Ethical research and the institutional review board: An introduction. *The American Journal of Nursing*, 124(3), 50–54. DOI: 10.1097/01.NAJ.0001008420.28033.e8 PMID: 38386835

Chen, J., Henry, E., & Jiang, X. (2023). Is cybersecurity risk factor disclosure informative? Evidence from disclosures following a data breach. *Journal of Business Ethics*, 187(1), 199–224. DOI: 10.1007/s10551-022-05107-z

Christen, M., Gordijn, B., & Loi, M. (2020). *The ethics of cybersecurity*. Springer Nature., DOI: 10.1007/978-3-030-29053-5

Dalal, R. S., Howard, D. J., Bennett, R. J., Posey, C., Zaccaro, S. J., & Brummel, B. J. (2022). Organizational science and cybersecurity: Abundant opportunities for research at the interface. *Journal of Business and Psychology*, 37(1), 1–29. DOI: 10.1007/s10869-021-09732-9 PMID: 33564206

Dalkiran, A (2024). Guidelines for scientific research and publication ethics. *International Journal of Aviation Science and Technology (IJAST)*, 1-14.

Davison, R. M., Chughtai, H., Nielsen, P., Marabelli, M., Iannacci, F., van Offenbeek, M., & Panteli, N. (2024). *The ethics of using generative AI for qualitative data analysis*. Wiley., DOI: 10.1111/isj.12504

De Roche, M., & Silver, R. C. (2024). Ethical issues in research with human subjects. *How Science Engages with Ethics and Why It Should. An Interdisciplinary Approach*, 61, 61–78. Advance online publication. DOI: 10.1515/9783111142463-005

Dhablia, A. (2024). The role of cybersecurity policies in ensuring compliance with US federal and state network security laws. *Network Security*, 2024(7), 14–21.

Dunn Cavelty, M. (2018). Cybersecurity research meets science and technology studies. *Politics and Governance*, 6(2), 22–30. DOI: 10.17645/pag.v6i2.1385

Dupuis, M., & Renaud, K. (2021). Scoping the ethical principles of cybersecurity fear appeals. *Ethics and Information Technology*, 23(3), 265–284. DOI: 10.1007/s10676-020-09560-0

Edquist, A., Grennan, L., Griffiths, S., & Rowshankish, K. (2022). Data ethics: What it means and what it takes. *McKinsey Report, 23.*

Esmer, Y., & Arıbaş, A. N. (2022). Information ethics in the context of current developments. In *Handbook of Research on Digital Violence and Discrimination Studies* (pp. 685–707). IGI Global. DOI: 10.4018/978-1-7998-9187-1.ch030

Familoni, B. T.Babajide Tolulope Familoni. (2024). Cybersecurity challenges in the age of AI: Theoretical approaches and practical solutions. *Computer Science & IT Research Journal*, 5(3), 703–724. DOI: 10.51594/csitrj.v5i3.930

Fiesler, C., Zimmer, M., Proferes, N., Gilbert, S., & Jones, N. (2024). Remember the human: A systematic review of ethical considerations in Reddit research. *Proceedings of the ACM on Human-Computer Interaction, 8*(GROUP), 1-33. DOI: 10.1145/3633070

Formosa, P., Wilson, M., & Richards, D. (2021). A principlist framework for cybersecurity ethics. *Computers & Security*, 109, 102382. DOI: 10.1016/j.cose.2021.102382

Friesen, P., Gelinas, L., Kirby, A., Strauss, D. H., & Bierer, B. E. (2023). IRBs and the protection-inclusion dilemma: Finding a balance. *The American Journal of Bioethics*, 23(6), 75–88. DOI: 10.1080/15265161.2022.2063434 PMID: 35482887

Frohmann, B. (2000). Cyber ethics: Bodies or bytes? *The International Information & Library Review*, 32(3-4), 423–435. DOI: 10.1080/10572317.2000.10762528

Gunathilake, S. (2024). *The role of ethics in shaping modern scientific research.* UVA Clinical Research.

Haneef, I., & Agrawal, M. (2024). Ethical issues in educational research. *Asian Research Journal of Arts & Social Sciences*, 22(5), 29–38. DOI: 10.9734/arjass/2024/v22i5535

Hawamleh, A. M. A., Alorfi, A. S. M., Al-Gasawneh, J. A., & Al-Rawashdeh, G. (2020). Cyber security and ethical hacking: The importance of protecting user data. *Solid State Technology*, 63(5), 7894–7899.

Herrmann, D., & Pridöhl, H. (2020). Basic concepts and models of cybersecurity. *The Ethics of Cybersecurity*, 11-44. https://doi.org/DOI: 10.7196/SAJBL.2024.v17i2.1671

IEEE. (2020). *IEEE Code of Ethics*. www.ieee.org

Joanna, K. (2021). Cyber ethics: Historical outline and basic problems. *Редакційна колегія*, 21.

Kara, H. (2024). Ethics for independent researchers. *NCIS Guide for Independent Scholars*, 98.

Karunarathna, I., Hapuarachchi, T., Ekanayake, U., Rajapaksha, S., Gunawardana, K., Aluthge, P., Bandara, S., Jayawardana, A., De Alvis, K., Gunasena, P., & Gunathilake, S. (2024). *The role of ethics in shaping modern scientific research*. UVA Clinical Research.

Kavak, H., Padilla, J. J., Vernon-Bido, D., Diallo, S. Y., Gore, R., & Shetty, S. (2021). Simulation for cybersecurity: State of the art and future directions. *Journal of Cybersecurity*, 7(1), 1–13. DOI: 10.1093/cybsec/tyab005

Knight, S., Shibani, A., & Vincent, N. (2024). Ethical AI governance: Mapping a research ecosystem. *AI and Ethics*, 1–22. DOI: 10.1007/s43681-023-00416-z

Koali, S., Khasoane, N., & Mongezi, M. (2024). Promoting research integrity through the lens of virtue ethics and deontological approach. *South African Journal of Bioethics and Law*, 17(2), 78–82.

Lapid, M. I., Ouellette, Y., Drake, M. T., & Clarke, B. L. (2023). Institutional Review Board (IRB): US perspectives. In *Handbook of Bioethical Decisions. Volume II: Scientific Integrity and Institutional Ethics*, 219-240. Springer International. DOI: 10.1007/978-3-031-29455-6_15

Lee, W. W. (2020). Ethical computing for data protection. [IJT]. *International Journal of Technoethics*, 11(1), 43–58. DOI: 10.4018/IJT.2020010104

Lehtonen, T. (2021). Ethics of security: From personal safety to cyber security. In *Multidisciplinary Approaches to Ethics in the Digital Era* (pp. 44–59). IGI Global. DOI: 10.4018/978-1-7998-4117-3.ch004

Loi, M., & Christen, M. (2020). Ethical frameworks for cybersecurity. *The Ethics of Cybersecurity*, 73-95.

Lonsdale, D. J. (2020). The ethics of cyber attack: Pursuing legitimate security and the common good in contemporary conflict scenarios. *Journal of Military Ethics*, 19(1), 20–39. DOI: 10.1080/15027570.2020.1764694

Lucas, G. R. (2017). *Ethics and cyber warfare: the quest for responsible security in the age of digital warfare*. Oxford University Press. DOI: 10.1093/acprof:oso/9780190276522.001.0001

Maalem Lahcen, R. A., Caulkins, B., Mohapatra, R., & Kumar, M. (2020). Review and insight on the behavioral aspects of cybersecurity. *Cybersecurity*, 3(1), 1–18. DOI: 10.1186/s42400-020-00050-w

Macnamara, J. (2024). Human research ethics review challenges in the social sciences: A case for review. *Journal of Academic Ethics*, 1-17. https://doi/org/DOI: 10.1007/s10805-024-09532-9

Macnish, K., & Van der Ham, J. (2020). Ethics in cybersecurity research and practice. *Technology in Society*, 63, 101382. DOI: 10.1016/j.techsoc.2020.101382

Mahfood, S., Astuto, A., Olliges, R., & Suits, B. (2005). Cyberethics: Social ethics teaching in educational technology programs. *Communication Research Trends*, 24(4), 3–23.

Manjikian, M. (2017). *Cybersecurity ethics: an introduction*. Routledge. DOI: 10.4324/9781315196275

Mathis, D. (2024). Preparation of an ethics application. *Arthroskopie*, 1-4(4), 254–257. Advance online publication. DOI: 10.1007/s00142-024-00684-9

Mbinjama-Gamatham, A., & Olivier, B. (2020). 'Dark technology', aggressiveness and the question of cyber-ethics. *Acta Academica*, 52(1), 99–120. DOI: 10.18820/24150479/aa52i1/1

McKee, K. R. (2024). Human participants in AI research: Ethics and transparency in practice. *IEEE Transactions on Technology and Society*, 5(3), 279–288. DOI: 10.1109/TTS.2024.3446183

Miller, S., & Bossomaier, T. (2024). *Cybersecurity, ethics, and collective responsibility*. Oxford University Press. DOI: 10.1093/oso/9780190058135.001.0001

Miteu, G. D. (2024). Ethics in scientific research: A lens into its importance, history, and future. *Annals of Medicine and Surgery (London)*, 86(5), 2395–2398. DOI: 10.1097/MS9.0000000000001959 PMID: 38694399

Morgan, G., & Gordijn, B. (2020). A care-based stakeholder approach to ethics of cybersecurity in business. *The Ethics of Cybersecurity*, 119-138.

Mozersky, J., Walsh, H., Parsons, M., McIntosh, T., Baldwin, K., & DuBois, J. M. (2020). Are we ready to share qualitative research data? Knowledge and preparedness among qualitative researchers, IRB members, and data repository curators. *IASSIST Quarterly*, 43(4), 1–23. Advance online publication. DOI: 10.29173/iq952 PMID: 32205903

Onyancha, O. B. (2015). An informetrics view of the relationship between internet ethics, computer ethics and cyberethics. *Library Hi Tech*, 33(3), 387–408. DOI: 10.1108/LHT-04-2015-0033

Parthasarathy, S., Panigrahi, P. K., & Subramanian, G. H. (2024). A framework for managing ethics in data science projects. *Engineering Reports*, 6(3), e12722. DOI: 10.1002/eng2.12722

Pattison, J. (2020). From defence to offence: The ethics of private cybersecurity. *European Journal of International Security*, 5(2), 233–254. DOI: 10.1017/eis.2020.6

Peled-Raz, M., Tzafrir, S. S., Enosh, G., Efron, Y., & Doron, I. (2021). Ethics review boards for research with human participants: Past, present, and future. *Qualitative Health Research*, 31(3), 590–599. DOI: 10.1177/1049732320972333 PMID: 33213280

Pirani, S. (2024). Navigating research ethics: Strategies for preventing and addressing research misconduct. *International Journal of Multidisciplinary Research & Reviews*, 3(02), 96–104. DOI: 10.56815/IJMRR.V3I2.2024/96-104

Rajasekharaiah, K. M., Dule, C. S., & Sudarshan, E. (2020). Cyber security challenges and its emerging trends on latest technologies. []. IOP Publishing.]. *IOP Conference Series. Materials Science and Engineering*, 981(2), 022062. DOI: 10.1088/1757-899X/981/2/022062

Ramírez, K. (2024). Ethical principles in dental research of behavioral and social factors. *Odovtos-International Journal of Dental Sciences*, 20-25. DOI: 10.15517/ijds.2024.59408

Raul, A. C. (Ed.). (2021). *The privacy, data protection and cybersecurity law review*. Law Business Research Limited.

Resnik, D. B., Antes, A., & Mozersky, J. (2024). Should researchers destroy audio or video recordings? *Ethics & Human Research*, 46(2), 30–35. DOI: 10.1002/eahr.500205 PMID: 38446098

Richards, D., Formosa, P., Ryan, M., Hitchens, M., & McEwan, M. (2020). A proposed AI-enhanced serious game for cybersecurity ethics training. In *Conference of the Australasian Institute of Computer Ethics (9th: 2020)* (pp. 1-9). Australasian Institute of Computer Ethics (AiCE).

Sakka, G., & Spyrou, I. (2015). CyberEthics Case Study. In *Human Rights and Ethics: Concepts, Methodologies, Tools, and Applications* (pp. 646-658). IGI Global. DOI: 10.4018/978-1-4666-6433-3.ch036

Schlehahn, E. (2020). Cybersecurity and the State. *The Ethics of Cybersecurity*, 205-225. Springer Open.

Schneble Butz, C. O. (2024). *Ethical and legal perspectives on big data in research* (Doctoral dissertation, University of Basel).

Schultz, R. A. (Ed.). (2005). *Contemporary issues in ethics and information technology*. IGI Global.

Sharan, R., & Boruah, B. (2016). Ethical Concerns of Human-Being, Cyber-Being and Cybertariat: An Educational Perspective. *International Journal of Information Ethics*, 25, 25. DOI: 10.29173/irie107

Shou, D. (2012). Ethical considerations of sharing data for cybersecurity research. In *Financial Cryptography and Data Security: FC 2011 Workshops* (pp. 169–177). Springer Berlin Heidelberg. DOI: 10.1007/978-3-642-29889-9_15

Shukla, S., George, J. P., Tiwari, K., & Kureethara, J. V. (2022). Data security. In *Data Ethics and Challenges* (pp. 41–59). Springer Singapore. DOI: 10.1007/978-981-19-0752-4_3

Sontan, A. D., & Samuel, S. V.Adewale Daniel SontanSegun Victor Samuel. (2024). The Intersection of Artificial Intelligence and Cybersecurity: Challenges and Opportunities. *World Journal of Advanced Research and Reviews*, 21(2), 1720–1736. DOI: 10.30574/wjarr.2024.21.2.0607

Spinello, R. (2010). *Cyberethics: Morality and law in cyberspace*. Jones & Bartlett Publishers.

Stegenga, S. M., Steltenpohl, C. N., Lustick, H., Meyer, M. S., Renbarger, R., Standiford Reyes, L., & Lee, L. E. (2024). Qualitative research at the crossroads of open science and big data: Ethical considerations. *Social and Personality Psychology Compass*, 18(1), e12912. DOI: 10.1111/spc3.12912

Stoudt, S., Jernite, Y., Marshall, B., Marwick, B., Sharan, M., Whitaker, K., & Danchev, V. (2024). Ten simple rules for building and maintaining a responsible data science workflow. *PLoS Computational Biology*, 20(7), 1–19. DOI: 10.1371/journal.pcbi.1012232 PMID: 39024267

Tangudu, A., Jain, S., & Aggarwal, A. (2024). Best Practices for Ensuring Salesforce Application Security and Compliance. *Journal of Quantum Science and Technology*, 1(2), 88–101. DOI: 10.36676/jqst.v1.i2.18

Traianou, A., & Hammersley, M. (2024). Interrogating the concept of vulnerability in social research ethics. *Diametros : a Journal of Philosophy*, 1–16. Advance online publication. DOI: 10.33392/diam.1891

Van de Poel, I. (2020). Core values and value conflicts in cybersecurity: beyond privacy versus security. *The Ethics of Cybersecurity*, 45-71. Springer Open

Van den Hoven, J. (2010). The use of normative theories in computer ethics. *The Cambridge handbook of information and computer ethics*, 59-76. Cambridge University Press.

White, M. G. (2020). Why human subjects research protection is important. *The Ochsner Journal*, 20(1), 16–33. DOI: 10.31486/toj.20.5012 PMID: 32284679

Wiafe, I., Yaokumah, W., & Kissi, F. A. (2020). Students' intentions on cyber ethics issues. In *Modern Theories and Practices for Cyber Ethics and Security Compliance* (pp. 105-121). IGI Global.

Wolf, L. E., Ram, N., Contreras, J., & Beskow, L. M. (2024). Certificates of confidentiality: Privileging research data. *Journal of Law and the Biosciences*, 11(1), lsae003. Advance online publication. DOI: 10.1093/jlb/lsae003 PMID: 38405099

Wylde, V., Rawindaran, N., Lawrence, J., Balasubramanian, R., Prakash, E., Jayal, A., & Platts, J. (2022). Cybersecurity, data privacy and blockchain: A review. *SN Computer Science*, 3(2), 127. DOI: 10.1007/s42979-022-01020-4 PMID: 35036930

Yaokumah, W., Rajarajan, M., Abdulai, J. D., Wiafe, I., & Katsriku, F. A. (Eds.). (2020). *Modern Theories and Practices for Cyber Ethics and Security Compliance*. IGI Global. DOI: 10.4018/978-1-7998-3149-5

Zhang, X., & Ghorbani, A. A. (2021). Human factors in cybersecurity: Issues and challenges in big data. *Research Anthology on Privatizing and Securing Data*, 1695-1725. IGI Global.

ADDITIONAL READING

Golyan, A., Panchal, S., Vaghasiya, D., & Parekh, H. (2024). Data ethics and privacy. In *Recent Trends and Future Direction for Data Analytics* (pp. 259–268). IGI Global. DOI: 10.4018/979-8-3693-3609-0.ch011

Kim, S. J. (2024). Research ethics and issues regarding the use of ChatGPT-like artificial intelligence platforms by authors and reviewers: A narrative review. *Science Editing*, 11(2), 96–106. DOI: 10.6087/kcse.343

Kumar, R., Joshi, A., Sharan, H. O., Peng, S. L., & Dudhagara, C. R. (Eds.). (2024). *The ethical frontier of AI and data analysis*. IGI Global. DOI: 10.4018/979-8-3693-2964-1

O'Keefe, K., & Brien, D. O. (2023). *Data ethics: practical strategies for implementing ethical information management and governance*. Kogan Page.

Verma, B., Singla, B., & Mittal, A. (Eds.). (2024). *Digital technologies, ethics, and decentralization in the digital era*. IGI Global. DOI: 10.4018/979-8-3693-1762-4

KEY TERMS AND DEFINITIONS

Artificial Intelligence (AI) and Machine Learning (ML) Ethics: Ethical considerations regarding the development and use of AI and ML focus on issues like bias, transparency, and the fair treatment of all individuals affected by these technologies.

Code of Ethics: A set of guidelines established by organizations like ACM and IEEE to define acceptable behavior for researchers in computing and engineering focusing on fairness, transparency, and respect for privacy.

Cybersecurity Ethics: The principles that govern the conduct of cybersecurity professionals which focus on the protection of data and privacy while maintaining fairness, transparency, and accountability in their practices.

Data Anonymization: Techniques used to protect individual privacy by removing or encrypting personal identifiers, making it difficult to trace data back to individuals, thereby safeguarding their privacy.

Data Integrity: The accuracy and reliability of data throughout its lifecycle.

Data Minimization: A principle emphasized by regulations like GDPR that requires collecting only the data necessary for the specific purpose of the research, thus reducing risks related to privacy violations.

Ethical Hacking: The practice of legally probing systems to find vulnerabilities before malicious actors can exploit them which protects the security and integrity of digital systems.

General Data Protection Regulation (GDPR): A European Union regulation that sets strict guidelines on data collection, use, and protection which emphasizes principles such as consent, transparency, data minimization, and accountability.

Informed Consent: A fundamental ethical principle in research where participants are fully aware of how their data will be used, stored, and protected, including the potential risks of unauthorized access or misuse.

Privacy by Design: A proactive approach that integrates privacy considerations into developing and deploying technologies from the outset, minimizing data collection and providing robust data protection.

Role-Based Access Control (RBAC): A cybersecurity measure that limits access to data based on the user's role in an organization to guarantee that only authorized personnel can access sensitive information.

Transparency: A key ethical principle requiring organizations to openly communicate their data collection, usage, and security measures to stakeholders so that individuals are informed and can trust in the responsible handling of their data.

Chapter 6

Current Trends in AI Ethics for Software as a Medical Device (SaMD)

Thanh Vu
Stanley Manne Children's Research Institute, USA

Robin Throne
https://orcid.org/0000-0002-3015-9587
University of the Cumberlands, USA

ABSTRACT

Past scholarship has noted the growing concerns and complex challenges surrounding the emergent use of ubiquitous digital technologies in medicine and healthcare, including artificial intelligence (AI) and machine learning (ML). The scholarship has also noted the challenges that human research protection programs (HRPP) and institutional review boards (IRB) have faced in an evolving regulatory climate to review human subjects research involving AI and ML for medical devices. This chapter presents the results of a trends analysis of the current scholarship surrounding compliance trends for AI ethics specific to the use of protected health information (PHI) and compliance trends specifically surrounding AI ethics in software as a medical device (SaMD) for new investigators seeking IRB review for AI/ML research activities. The trends offered in the current scholarship also present many opportunities for policy implications needed to assure ethical oversight for the protection of human subjects in research involving AI/ML for IRBs and HRPPs to review this research.

DOI: 10.4018/979-8-3693-3848-3.ch006

INTRODUCTION

Three-quarters of Americans say their greater concern is that health care providers will move too fast implementing AI in health and medicine before fully understanding the risks for patients.

~Pew Research Center[1]

The past scholarship has noted the growing concerns and complex challenges surrounding the emergent use of ubiquitous digital technologies in medicine and healthcare, including artificial intelligence (AI) and machine learning (ML) (Shearer et al., 2021) and Bourgeois and Espinoza (2023) noted that these challenges are often even more unique for medical device development for pediatric populations. Further, Ngiam and Khor (2019), among others, highlighted the advances AI and ML have contributed to complex healthcare datasets amidst significant ethical challenges for patient data privacy and security, especially for protected health information (PHI) and personally identifiable information (PII).

The 21st-century scholarship has well-documented the societal and sociocultural impact of the emergence of ubiquitous digital technologies in healthcare. As these advances continued into the third decade of the century, healthcare disruption has continued via an infusion of personal digital health technologies (Awad et al., 2021). Human research protection programs (HRPP) and institutional review boards (IRB) are often the responsible entities within research organizations for the ethical and regulatory review of research involving AI and/or ML in the design, development, and testing of medical devices for the diagnosing, curing, mitigating, or treatment of disease (U.S. Food and Drug Administration [FDA], 2018). Since the U.S. FDA (2021) joined with Health Canada and the United Kingdom's Medicines and Healthcare products Regulatory Agency (MHRA) to jointly identify 10 guiding principles to inform the development of Good Machine Learning Practice (GMLP) to promote safe, effective, and high-quality medical devices that use AI/ML, new guidance has been offered, and much continues to change.

Due to these rapid policy changes, new investigators using AI or ML in the design or development of medical or healthcare devices should continually consult their organization's HRPP and IRB requirements and policies for the review of human subjects research before the commencement of these research activities to ensure mitigation of patient risks and privacy/confidentiality assurances for PHI and PII.[2] The specific objectives of this chapter are to present the current compliance trends surrounding AI ethics in the use of protected health information (PHI) and compliance trends surrounding AI ethics in SaMD for investigators.

BACKGROUND

Navigating the process of receiving IRB or other ethics committee reviews by new investigators for medical devices involving AI or ML can be challenging. Blackman (2021), among many others, noted that the growing attention on AI fairness and inclusion, as well as AI algorithmic and model bias, has raised attention to the need for updated risk assessment by IRBs and other ethical review processes to ensure any PHI or other personal identifiable information (PII) assurances are provided to reduce risks for human subjects, or their data, involved in the research activities. Blackman (2021) called for organizations developing AI medical devices to use an "AI IRB" even when the organization may not have previously established one. Thus, it remains essential for new investigators to understand the IRB or other ethical review committee requirements and procedures within the organization prior to embarking on the design of research activities involving AI and/or ML, as not all AI research activities require IRB review. Nonetheless, AI ethics must remain at the forefront of consideration in medical device development involving AI and/or ML.

Likewise, Lysaght et al. (2019) stressed the potential for group harms arising from algorithmic bias in using AI and the varied levels of medical professionals responsible for AI ethics. Docrr and Meeder (2022) offered similar concerns, noting specifically the effect on public policy and the need to reconsider regulations in this climate. These ongoing discussions surrounding the potential for group harms have called for increased "accountability and transparency" in the ethics evaluation of AI-assisted or AI use in healthcare devices (Lysaght et al., 2019, p. 299). Still, others, such as Di Nucci (2019), highlighted the concern that medical AI introduces a threat to patient autonomy and raises more significant ethical considerations that affect healthcare foundationally: "Concluding, it is indeed a crucial ethical and political question—possibly one of the most important questions of this century—how much we should delegate to technology and in particular to complex technologies such as machine learning algorithms" (Di Nucci, 2019, p. 557). Yet, Varlamov et al. (2019) stressed that AI will not replace physician-patient empathy but require new communication and analytical skills amidst an AI-driven healthcare environment.

Despite the many concerns raised in the current decade regarding AI ethics and medical devices (Eto, 2022; Throne & Hendon, 2024), researchers continued to be excited over the past decade to introduce direct-to-consumer healthcare devices and mobile apps, pervasive sensor technologies, and advancements to enhance access to social network data that offers unprecedented personal health monitoring and disease treatment intervention (Nebeker et al., 2019). Over this time, Johnson (2019) echoed the calls for medical leadership across nations to address these advancing concerns and establish standardized guidelines for AI ethics in healthcare:

We must engage with all stakeholders locally and internationally to work on ethical guidelines for the implementation of AI and ML in medicine and health care. We will need to work closely with professionals who have expertise in data science, computer program design, and machine learning. Ultimately we will need regulation. (p. 440)

As Ngiam and Khor (2019) stated, whatever challenges are faced in the ethical and regulatory review of research activities involving AI/ML in medicine and healthcare must be met, as the advantages and benefits of healthcare advances far outweigh these barriers.

Specifically, FDA requirements and guidance for AI ethics and AI and SaMD have evolved significantly over recent years and require continued monitoring to ensure researchers and IRB professionals remain in step with current guidance. In the following sections, the trends analysis results of the current scholarship since 2020 are presented as guidance for new investigators seeking IRB or other ethics reviews of medical/healthcare device research activities, specifically for SaMD IRB review involving AI and/or ML.

COMPLIANCE TRENDS IN AI ETHICS AND PROTECTED HEALTH INFORMATION (PHI)

According to the FDA (2024, May), it has authorized 882 AI/ML-enabled medical devices, with 151 of the 882 final decisions rendered between August 2023 and March 2024[3]. Concurrently, authors and researchers have continued to report compliance concerns for PHI used for research that involves ethical AI/ML. For example, Anom (2020) acknowledged the benefits to society that research and development of medical devices offer yet noted the need for improved ethical oversight of these technologies to protect patient information. "To confront some of the challenges and perceive devils presented by the technologies, the bioethics and healthcare communities must establish sound data security, data privacy, and data governance strategies" (Anom, 2020, p. 10). Researchers have continued to recommend including measures for *transparency* as a critical construct in building ethical frameworks needed to govern AI/ML use, especially involving PHI (Anom, 2020; Eto, 2022; Reddy et al., 2020; Throne & Hendon, 2023). Dalton-Brown (2020) echoed this need for clear, substantive policies to oversee AI/ML in healthcare to ensure that ethics of care are not violated. Further, Anom (2020) recommended that "policies must be sufficiently established to ensure that all patients have total control over their own data and trust that the data will not be used to harm or discriminate against them" (p. 10).

In addition, Minssen et al. (2020) reported that the ongoing sustainable regulatory systems in the U.S. are encouraging. However, more attention is needed within these systems to ensure "respect for the rule of law, human rights, diversity, and fairness, as well as other societal and democratic values in the design of AI systems and devices" (Minssen et al., 2020, p. 18). Similarly, Morley et al. (2020) reported the results of a systematic literature review that showed multiple ethical concerns surrounding the continued incorporation of data-driven AI/ML healthcare that affects a range of stakeholders ranging from the individual to the societal domain. Findings included an overarching concern for distinguishing data privacy between individuals and groups, without which "ethical AI" discussions may be too simplistic (Morley et al., 2020, n.p.). Reddy et al. (2020) noted succinctly that the critical issue of AI systems in healthcare must "be protected from privacy breaches to prevent psychological and reputational harm to patients, and patients must provide explicit consent for their data to be used for any specific use" (p. 492). They recommended,

> *Where patient data may be shared with AI developers, there must be a process to seek fully informed consent from patients and if it is unrealistic to seek approval, data must be anonymized to that extent individual patient details cannot be recognized by the developers. The permissions to provide data should be rescindable. Also, differential privacy, a technological solution, which minimizes the risks of analyzing confidential and sensitive data should be considered. Through this approach, a high standard of data anonymization is achieved by shrinking the risks associated with reidentification, thus upholding the privacy of patients. (Reddy et al., 2020, p. 494)*

More recently, authors and researchers have called for improved ethics in regulatory considerations for AI/ML despite the significant gains in healthcare treatment, diagnostics, and patient care. Tiwari (2023) stressed "the importance of balancing innovation with the ethical and safe utilization of AI technologies" (p. 23), and Bouderhem (2024) stressed the continued need to consider transparency in AI/ML utilization within healthcare as well as the need to "protect health data, mitigate risks and regulate more efficiently the use of AI in healthcare" (p. 1). Specifically, from a global perspective, the author recommended,

> *Ethical and regulatory challenges posed by the novelty of AI systems in healthcare such as bias, data protection or explainability have to be addressed by states. European regulations – GDPR, Data Act, and AI Act – can provide reliable legal frameworks and established standards to be implemented by all stakeholders for ethical and responsible AI systems. WHO Members need to actively cooperate and elaborate new guidelines and legally binding rules under the IHR. (Bouderhem, 2024, p. 8)*

McGraw and Mandl (2021) reported the need for strong privacy protections that include accountability for de-identified or anonymized data definitions and the potential for harm to individuals and populations. These measures may also aid in ensuring the use of health data for societal benefit and to continue the advancements and benefits AI/ML has brought to healthcare while concurrently protecting human subjects. The following sections analyzed the current scholarship for trends specifically related to the need for IRB and HRPP oversight of AI/ML in medical device development, deployment, and ongoing use. These trends remain essential for continued examination and review by investigators involved in research using AI/ML in medical device design and development and seeking IRB review of the research.

Current Views on IRB Oversight of AI/ML and Data-Centric Human Subjects Research

While the research related to AI/ML-driven human subjects research remains limited, many have called out specifically the attention needed for the individual and group protections that must be addressed in regulatory reviews of this research (see Figure 1). Concerning oversight of AI/ML ethics in research that is subject to IRB or other ethics committee review oversight, Chapman et al. (2023) noted that U.S. governmental regulations must continue to address the potential for group harms and recommended three areas of the Common Rule be addressed: 1) Context: IRB assessment of risks and benefits; 2) disclosure requirements in the informed consent process; and 3) criteria for waivers of informed consent. The authors stressed that data-centric research of all forms should require group harms to be evaluated by IRBs or other ethical review committees (Chapman et al., 2023). As such, "research participants should be told about any potential group harm in the informed consent process. The preclusion statement and criteria for waivers of informed consent should also be reevaluated" (Chapman et al., 2023, p. 9). Similarly, Doerr and Meeder (2022) noted the need for regulatory reforms to ensure group harms are considered in research involving data-centric research:

> *An important consideration about the IRB's role in reviewing protocols for big health data research is an apparent conflict between the Common Rule's preclusion statement (at 45 C.F.R. 46.111) and the ethical principles outlined in The Belmont Report. The Common Rule states that "[t]he IRB should not consider possible longrange effects of applying knowledge gained in the research (e.g., the possible effects of the research on public policy) as among those research risks that fall within the purview of its responsibility." We contend that the preclusion statement unnecessarily limits IRBs from considering longer-term risks, such as group harms, that may result from big health data research.... The ambiguity around this aspect of the Common Rule leaves some research*

protections professionals debating whether to counsel their IRB members to go against regulations and consider long-range effects when it is clear there might be negative ramifications from a research study or to counsel them to ignore possible longrange effects, including potential societal harm. (p. 36)

Likewise, Ferretti et al. (2022) found in interviews with research ethics professionals that experience is lacking in data-centric research and insufficient expertise in data science. Amid these uncertainties, the ethical and regulatory review of data-centric research in many forms means more guidance, training, and education may be needed for IRBs and HRPPs to understand better how to mitigate these research risks. Makridis et al. (2023) summarized the health benefits AI/ML has brought to healthcare and the concurrent growth of clinical trials involving AI/ML in medical device development. However, the authors also raised concerns about the oversight of AI/ML in research and development. They said, "However, there are also many new risks that AI poses even before it is ever deployed—that is, in the R&D process. Chen and Esmaeilzadeh (2024) concluded that while medical research involving AI/ML is transformative, ethical oversight and collaborative consideration across the healthcare sector, including regulatory oversight with developers, are needed to ensure these technologies are employed responsibly.

Therefore, an IRB process to address these new, distinct risks is urgently needed" (Makridis et al., 2023, p. 6). Novel inquiry questions must be developed for IRB and HRPP professionals to consider the risks of these digital technologies in medical software development better. Like Eto (2022), the authors presented a new tool to assess better and evaluate risk determinations for healthcare technologies involving AI/ML. Finally, Winter and Davidson (2022) offered more specific concerns related to research involving patient-generated health data in the development of mobile devices. The authors called for improved regulatory harmony among various protections for health information, including the Health Insurance Portability and Accountability Act of 1996 (HIPAA)[4] and the EU's General Data Protection Regulation (GDPR). The authors recommended that U.S. regulatory regimes be harmonized to continue to consider best practices for the review of data-centric research and research involving AI/ML. "To help bring about these outcomes, data protection regulations are critical public policy tools to help balance stakeholders' interests in PGHD and thus to guide developments and innovations in socially desirable ways for all affected parties," they said (Winter & Davidson, 2022, p. 17).

Figure 1. Scholarship trends in ethical compliance for in AI/ML and data-centric research

Transparency/ Algorithim Transparency	Regulatory Harmonization	Reliability/Trustworthiness
Winter & Davidson (2022)	Bouderhem (2024)	Tiwari (2023)
Tiwari (2023)	Doerr & Meeder (2022)	Makridis et al. (2023)
Makridis et al. (2023)	Eto (2022)	Ferretti et al. (2022)
Ferretti et al. (2022)	FDA (2024)	Doerr & Meeder (2022)
Eto (2022)	Tiwari (2023)	Chen & Esmaeilzadeh (2024)
Chapman et al. (2023)	Winter & Davidson (2022)	Chapman et al. (2023)
		Bouderhem (2024)

COMPLIANCE TRENDS IN AI ETHICS AND SAMD FOR INVESTIGATORS

The FDA has adopted the term Software as a Medical Device (SaMD) to be defined by the International Medical Device Regulators Forum (IMDRF) as "software intended to be used for one or more medical purposes that perform these purposes without being part of a hardware medical device" (FDA, 2018). As of August 2024, the FDA has authorized 950 AI and/or ML-enabled medical devices.[5] Regarding AI ethics for SaMD, Bommu (2022) noted that identifying ethical challenges in the development and deployment of AI-powered medical device software is essential for promoting responsible innovation and safeguarding patient welfare. Key ethical challenges such as data privacy, transparency, accountability, bias, and equity arise in the context of AI-driven healthcare technologies. For example, Bommu (2022) emphasized that clear guidelines and frameworks are needed to allocate responsibility among developers, healthcare providers, regulatory bodies, and other stakeholders involved in designing, deploying, and using AI-powered medical device software. Further, establishing mechanisms for reporting adverse events, investigating incidents, and remedying harms is essential to ensure accountability and uphold patient safety. Promoting diversity and inclusivity in AI development teams and datasets can help mitigate biases and enhance fairness in AI-powered medical device software (Bommu, 2022).

Similarly, Gerke et al. (2020) questioned how the use of AI to assist with the care of patients interfaces with the principles of informed consent and stated the need to examine under what circumstances (if at all) the principles of informed consent should be deployed in the clinical AI space. Gerke questioned how much transparency is needed and what if the patient is reluctant to allow the use of specific categories of data (e.g., genetic data and family history). Furthermore, investigators must closely consider whether to include data from vulnerable populations (e.g., children and persons with impaired decision-making). If these individuals have diminished autonomy to provide informed consent without a parent, guardian, or legal representative, the IRB may not include their data. Thus, some authors have argued that only PHI should be used by adults empowered to decide to participate for themselves in the development and deployment of AI-powered medical device software. Data gathered via AI-SaMD from vulnerable groups must be gathered only after robust testing and revision. This was supported by Bommu (2022), who found that patients expressed concerns about the potential loss of control and autonomy in the context of AI-driven healthcare technologies, emphasizing the importance of informed consent as a means of empowering patients and preserving their rights (Katsaliaki & Mustafee, 2011, as cited in Bommu, 2022).

In addition, Gerke et al. (2020) emphasized that it remains vital for AI makers to be aware of these risks and to minimize potential biases at every stage in the process of SaMD product development. In the health sector, where phenotype- and sometimes genotype-related information are involved, biased AI could, for instance, lead to false diagnoses and render treatments ineffective for some subpopulations and thus jeopardize their safety (Gerke et al., 2020). More importantly, as Bommu (2022) noted, research in this area has shown that disparities in access to AI-driven healthcare technologies could exacerbate existing healthcare disparities, particularly among underserved or marginalized populations. Konda (2022) reported study results where participants expressed significant concerns with algorithmic bias as the top concern (45.6%), data privacy as the second concern (38.9%), and transparency as the third concern (29.8%) (see Table 1). Bommu (2022) stressed that promoting diversity and inclusivity in AI development teams and datasets can help mitigate biases and enhance fairness in AI-powered medical device software. Additionally, implementing fairness-aware algorithms and conducting bias audits can help identify and address biases that may arise during software and medical device development and deployment (Faheem & Dutta, 2023).

Table 1. Frequency distribution of ethical concerns in AI development

Ethical Concern	Frequency (%)
Algorithmic bias	45.6
Data privacy	38.9
Transparency	29.8
Accountability	25.4
Fairness	21.7

Note. Table 1 from Konda, S. R. (2022). Ethical considerations in the development and deployment of AI-driven software systems. *International Journal of Computer Science and Technology, 6*(3), 86-101.

Bommu (2022) stressed that protecting patient privacy and ensuring compliance with privacy regulations are paramount to maintaining trust and confidence in AI-driven healthcare systems. Implementing robust data protection measures, such as encryption, anonymization, access controls, and data minimization, can help mitigate privacy risks and safeguard patient information from unauthorized access or misuse. For data protection, researchers need to consider "information security," such as how The National Institute of Standards and Technology (n.d.) defines information security as the protection of information and information systems from unauthorized access, use, disclosure, disruption, modification, or destruction in order to provide confidentiality, integrity, and availability.

Zhou and Yang (2022) noted that although the application of artificial intelligence in medical device research usually exists in a closed network environment, hackers can still technically invade and penetrate it through firewalls. The result of executing these attackers' corrupted medical devices will seriously affect their security. Due to the unique nature of the medical device industry and the lagging legal system, AI in the medical device field lacks a comprehensive information security regulatory system, and this status quo will inevitably bring great uncertainty and instability to the long-term development of AI in the medical device field. Likewise, Bernstein et al. (2021) explained that AI research often falls outside the purview of existing feedback mechanisms such as the IRB or other ethical review committees designed to evaluate harms to human subjects rather than the overall harms to human society. Because of this, Stanford University has developed the Ethics and Society Review (ESR) board, a feedback panel that works with researchers to mitigate negative ethical and societal aspects of AI research (Bernstein et al., 2021).

Healthcare AI algorithms have reported several adverse outcomes. For example, a widely published IBM Watson for Oncology (2023) used AI algorithms to assess information from patient medical records and help physicians explore cancer treatment options for their patients. The AI algorithm reportedly gave "unsafe and incorrect" recommendations for cancer treatments as the AI algorithm was trained on synthetic cancer cases rather than actual patient data (Faheem & Dutta, 2023).

In March 2024, the FDA reported that its "traditional paradigm of medical device regulation was not designed for adaptive artificial intelligence and machine learning technologies. Many changes to artificial intelligence and machine learning-driven devices may need a premarket review" (FDA, 2024). This paralleled Gerke et al. (2020), who emphasized that a complete system approval and monitoring approach is out of reach for today's regulators, especially because of expertise, resources, political obstacles, and the difficulties in controlling human behavior. Similarly, Bommu (2022) reiterated that by embracing ethical principles and values in the development and deployment of AI-powered medical device software, stakeholders can navigate the complexities of AI in healthcare responsibly and ethically. Overall, researchers continue to press for guidance from stakeholders across the healthcare ecosystem, including regulatory agencies, during the AI development and deployment for this collaborative effect (Bommu, 2022; Gerke et al., 2020; Konda, 2022; Zhao & Yang, 2022).

GRAVITY OF CONCERNS FOR THE USE OF AI IN HEALTHCARE

Four recent systematic reviews were conducted to assess the gravity of concerns surrounding the use of AI in healthcare, specifically for concerns related to SaMD PHI privacy protections. The gravity of concerns remains diverse and complex, from the regulatory and ethical to the physical and psychosocial, as well as specific to the privacy implications. The authors of these reviews consistently described the multidimensional matrix of concerns for AI and its impact on healthcare while concurrently describing the immense promise it brings to contemporary medicine. Williamson and Prybutok (2024) reported that while AI stands to transform healthcare, the ethical and moral implications cannot be overlooked. They said, "It is imperative to consider the efficiency and effectiveness of AI and its applications' moral implications, particularly in sensitive areas like patient data handling and decision-making processes" (p. 44). Likewise, Udegbe et al. (2024) noted the myriads of concerns surrounding "AI in healthcare [that] requires navigating a complex landscape of challenges, including data privacy and security, ethical and legal considerations, interoperability and integration issues, scalability and accessibility concerns, and the dynamics of human-AI interaction" (p. 506).

Further, Younis et al. (2024) described these challenges as difficult and complex, and range from clinical settings to medical research and the technologies all employ. They said, "In the context of healthcare and the medical field, challenges can encompass a wide range of issues, including but not limited to medical advancements, patient care, resource allocation, healthcare access, disease prevention and control, chronic disease management, healthcare costs, medical ethics, data privacy and security, protecting patient data range of issues, including but not limited to medical advancements, patient care, resource allocation, healthcare access, disease prevention and control, chronic disease management, healthcare costs, medical ethics, data privacy and security, protecting patient data and ensuring compliance with privacy regulations, interdisciplinary collaboration, patient education, and public health initiatives" (p. 31). They additionally noted that specific AI tools for PHI, regulatory compliance, expansion of telemedicine capabilities, and advancement of medical education tools all must be closely examined and monitored (Younis et al., 2024). The chapter authors concur that these concerns parallel the needed inspection of AI in SaMD, as reported earlier. Similarly, Ara and Mifa (2024) offered a summary of the AI concerns for replicated bias and the ethical and moral issues highlighted by others but also stressed the importance of preventing AI from dehumanizing healthcare. They said, "While AI can provide valuable insights and efficiencies, it cannot replicate the empathy, judgment, and holistic understanding that human healthcare providers bring to patient care" (p. 9). Finally, Eto et al. (2024) called for AI developers to "ensure that each phase of development adheres to rigorous standards and that the data used to develop these products is ethically obtained, with consent and authorization as appropriate" and "each step in AI development must be carefully and thoughtfully built to ensure the final product is safe, effective, and trustworthy. (p. 9).

FUTURE RESEARCH DIRECTIONS AND CONCLUSION

The chapter's authors concur with the specific calls for future research as reported in the current scholarship for the protection of human subjects in all research involving AI/ML, especially for the documented risks of group harms that IRBs and other ethical review committees must consider. The calls for regulatory harmonization would also provide further guidance and frameworks for the ethical considerations needed for SaMD involving AI/ML. Chen and Esmaeilzadeh (2024) stressed that future research and development involving AI must "emphasize security and privacy to ensure the responsible and trustworthy use of these AI models in health care" (p. 14), and Chapman et al. (2023) called for more empirical research and normative work to critically assess and ensure groups and vulnerable communities are appropriately and adequately protected by the federal regulations on research involving human subjects. From a regulatory perspective, Doerr and Meeder (2022) noted that the path forward in the review of data-centric research must include regulatory changes for IRBs and HRPPs to offer guidance on the interpretation of methodologies for this research.

Similarly, Morley et al. (2020) pressed for "the policies, standards, and regulations in development in different healthcare systems across the globe to assess the extent to which these are likely to be effective at mitigating these ethical concerns" (p. 10). Furthermore, Tiwari (2023) called for more research to consider transparency in AI systems with parent and algorithmic comprehensibility so regulators and others overseeing ethical oversight can better comprehend regulatory reviews of this research. The trends identified in this chapter deserve ongoing examination and consideration for IRBs and HRPPs to consider better medical research and SaMD involving AI/ML. The trends offered in the current scholarship also present many opportunities for policy implications needed to assure ethical oversight for the protection of human subjects in research involving AI/ML and to inform the current limitations of the regulatory policies used by IRBs and HRPPs to review this research.

REFERENCES

Anom, B. Y. (2020). Ethics of Big Data and artificial intelligence in medicine. *Ethics, Medicine, and Public Health*, 15, 2352–5525. DOI: 10.1016/j.jemep.2020.100568

Ara, A., & Mifa, A. F. (2024). Integrating artificial intelligence and big data in mobile health: A systematic review of innovations and challenges in healthcare systems. *Global Mainstream Journal of Business, Economics, Development & Project Management, 3*(01), 01-16.

Awad, A., Trenfield, S. J., Pollard, T. D., Ong, J. J., Elbadawi, M., McCoubrey, L. E., Goyanes, A., Gaisford, S., & Basit, A. W. (2021). Connected healthcare: Improving patient care using digital health technologies. *Advanced Drug Delivery Reviews*, 178, 113958. DOI: 10.1016/j.addr.2021.113958 PMID: 34478781

Bernstein, M. S., Levi, M., Magnus, D., Rajala, B., Satz, D., & Waeiss, C. (2021). ESR: Ethics and society review of artificial intelligence research. https://arxiv.org/pdf/2106.11521

Blackman, R. (2021). If your company uses AI, it needs an institutional review board. *Harvard Business Review*.https://hbr.org/2021/04/if-your-company-uses-ai-it-needs-an-institutional-review-board

Bommu, R. (2022). Ethical considerations in the development and deployment of AI-powered medical device software: Balancing innovation with patient welfare. *Journal of Innovative Technologies*, 5(1), 1–7.

Bouderhem, R. (2024). Shaping the future of AI in healthcare through ethics and governance. *Humanities & Social Sciences Communications*, 11(1), 1–12. DOI: 10.1057/s41599-024-02894-w

Bourgeois, F. T., & Espinoza, J. C. (2023). Advancing equity in medical device development for children. *JAMA Pediatrics*, 177(6), 561–562. DOI: 10.1001/jamapediatrics.2023.0790 PMID: 37093608

Chapman, C. R., Quinn, G. P., Natri, H. M., Berrios, C., Dwyer, P., Owens, K., Heraty, S., & Caplan, A. L. (2023). Consideration and disclosure of group risks in genomics and other data-centric research: Does the Common Rule need revision? *The American Journal of Bioethics*, 1–14. DOI: 10.1080/15265161.2023.2276161 PMID: 38010648

Chen, Y., & Esmaeilzadeh, P. (2024). Generative AI in medical practice: In-depth exploration of privacy and security challenges. *Journal of Medical Internet Research*, 26, e53008. DOI: 10.2196/53008 PMID: 38457208

D'Antonoli, T. A. (2020). Ethical considerations for artificial intelligence: An overview of the current radiology landscape. *Diagnostic and Interventional Radiology (Ankara, Turkey)*, 26(5), 504–511. DOI: 10.5152/dir.2020.19279 PMID: 32755879

Dalton-Brown, S. (2020). The ethics of medical AI and the physician-patient relationship. *Cambridge Quarterly of Healthcare Ethics*, 29(1), 115–121. DOI: 10.1017/S0963180119000847 PMID: 31858938

Di Nucci, E. (2019). Should we be afraid of medical AI? *Journal of Medical Ethics*, 45(8), 556–558. DOI: 10.1136/medethics-2018-105281 PMID: 31227547

Doerr, M., & Meeder, S. (2022). Big health data research and group harm: The scope of IRB review. *Ethics & Human Research*, 44(4), 34–38. DOI: 10.1002/eahr.500130 PMID: 35802789

Eto, T. (2022). *Conducting an effective IRB review of artificial intelligence human subjects research (AI HSR)*. Tech In HSR. https://techinhsr.com/wp-content/uploads/2022/08/AI-HSR-WHITE-PAPER-TechInHSR-08.2022-1.pdf

Eto, T., Lifson, M., & Vidal, D. (2024). A novel, streamlined approach to the IRB review of artificial intelligence human subjects research (AI HSR). Preprint. https://purl.stanford.edu/zj025zw1714

Faheem, H., & Dutta, S. (2023). Artificial intelligence failure at IBM 'Watson for Oncology'. *IUP Journal of Knowledge Management*, 21(3), 47–75.

Ferretti, A., Ienca, M., Hurst, S., & Vayena, E. (2020). Big data, biomedical research, and ethics review: New challenges for IRBs. *Ethics & Human Research*, 42(5), 17–28. DOI: 10.1002/eahr.500065 PMID: 32937036

Ferretti, A., Ienca, M., Velarde, M. R., Hurst, S., & Vayena, E. (2022). The challenges of big data for research ethics committees: A qualitative Swiss study. *Journal of Empirical Research on Human Research Ethics; JERHRE*, 17(1-2), 129–143. DOI: 10.1177/15562646211053538 PMID: 34779661

Gerke, S., Minssen, T., & Cohen, G. (2020). Ethical and legal challenges of artificial intelligence-driven healthcare. In *Artificial intelligence in healthcare* (pp. 295–336). Academic Press. DOI: 10.1016/B978-0-12-818438-7.00012-5

IBM. (2023). 5725-W51 IBM Watson for oncology. https://www.ibm.com/docs/en/announcements/watson-oncology?region=US

Johnson, S. L. (2019). AI, machine learning, and ethics in health care. *Journal of Legal Medicine*, 39(4), 427–441. DOI: 10.1080/01947648.2019.1690604 PMID: 31940250

Konda, S. R. (2022). Ethical considerations in the development and deployment of AI-driven software systems. *International Journal of Computer Science and Technology*, 6(3), 86–101.

Makridis, C. A., Boese, A., Fricks, R., Workman, D., Klote, M., Mueller, J., Hildebrandt, I. J., Kim, M., & Alterovitz, G. (2023). Informing the ethical review of human subjects research utilizing artificial intelligence. *Frontiers of Computer Science*, 5, 1235226. DOI: 10.3389/fcomp.2023.1235226

McGraw, D., & Mandl, K. D. (2021). Privacy protections to encourage use of health-relevant digital data in a learning health system. *NPJ Digital Medicine*, 4(1), 2. DOI: 10.1038/s41746-020-00362-8 PMID: 33398052

Minssen, T., Gerke, S., Aboy, M., Price, N., & Cohen, G. (2020). Regulatory responses to medical machine learning. *Journal of Law and the Biosciences*, 7(1), 1–18. DOI: 10.1093/jlb/lsaa002 PMID: 34221415

Morley, J., Machado, C. C., Burr, C., Cowls, J., Joshi, I., Taddeo, M., & Floridi, L. (2020). The ethics of AI in health care: A mapping review. *Social Science & Medicine*, 260, 113172. DOI: 10.1016/j.socscimed.2020.113172 PMID: 32702587

National Institute of Standards and Technology. (n.d.). Information security. https://csrc.nist.gov/glossary/term/information_security

Nebeker, C., Torous, J., & Bartlett Ellis, R. J. (2019). Building the case for actionable ethics in digital health research supported by artificial intelligence. *BMC Medicine*, 17(1), 1–7. DOI: 10.1186/s12916-019-1377-7 PMID: 31311535

Ngiam, K. Y., & Khor, W. (2019). Big data and machine learning algorithms for health-care delivery. *The Lancet Oncology*, 20(5), e262–e273. DOI: 10.1016/S1470-2045(19)30149-4 PMID: 31044724

Reddy, S., Allan, S., Coghlan, S., & Cooper, P. (2020). A governance model for the application of AI in health care. *Journal of the American Medical Informatics Association : JAMIA*, 27(3), 491–497. DOI: 10.1093/jamia/ocz192 PMID: 31682262

Shearer, E., Cho, M., & Magnus, D. (2021). Regulatory, social, ethical, and legal issues of artificial intelligence in medicine. In *Artificial intelligence in medicine* (pp. 457–477). Academic Press. DOI: 10.1016/B978-0-12-821259-2.00023-5

Throne, R., & Hendon, M. (2024). Belmont 2.0: Ethical research expectations for PII in AI, ML, and data mining/scraping. In *Methodologies and ethics for social sciences research* (pp. 1-19). IGI Global.

Tiwari, A. (2023). Regulatory consideration in AI-driven healthcare. *FOCUS WTO*, 25(4), 23–31.

Udegbe, F. C., Ebulue, O. R., Ebulue, C. C., & Ekesiobi, C. S.Francisca Chibugo UdegbeOgochukwu Roseline EbulueCharles Chukwudalu EbulueChukwunonso Sylvester Ekesiobi. (2024). The role of artificial intelligence in healthcare: A systematic review of applications and challenges. *International Medical Science Research Journal*, 4(4), 500–508. DOI: 10.51594/imsrj.v4i4.1052

U.S. Department of Health and Human Services [HHS]. (2022, October 25). Guidance regarding methods for de-identification of protected health information in accordance with the Health Insurance Portability and Accountability Act (HIPAA) Privacy Rule. Health Information Privacy. https://www.hhs.gov/hipaa/for-professionals/privacy/special-topics/de-identification/index.html

U.S. Food and Drug Administration. [FDA]. (2018, December 4). Software as a medical device (SaMD). https://www.fda.gov/medical-devices/digital-health-center-excellence/software-medical-device-samd

U.S. Food and Drug Administration. [FDA]. (2021, October 27). Good machine learning practice for medical device development: Guiding principles https://www.fda.gov/medical-devices/software-medical-device-samd/good-machine-learning-practice-medical-device-development-guiding-principles

U.S. Food and Drug Administration. [FDA]. (2024, March 15). Artificial intelligence and medical products: How CBER, CDER, CDRH, and OCP are working together. https://www.fda.gov/media/177030/download?attachment

U.S. Food and Drug Administration. [FDA]. (2024, May 13). Artificial intelligence and machine learning (AI/ML)-enabled medical devices. https://www.fda.gov/medical-devices/software-medical-device-samd/artificial-intelligence-and-machine-learning-aiml-enabled-medical-devices

Varlamov, O. O., Chuvikov, D. A., Adamova, L. E., Petrov, M. A., Zabolotskaya, I. K., & Zhilina, T. N. (2019). Logical, philosophical and ethical aspects of AI in medicine. *International Journal of Machine Learning and Computing*, 9(6), 868–873. DOI: 10.18178/ijmlc.2019.9.6.885

Williamson, S. M., & Prybutok, V. (2024). Balancing privacy and progress: A review of privacy challenges, systemic oversight, and patient perceptions in AI-driven healthcare. *Applied Sciences (Basel, Switzerland)*, 14(2), 675. DOI: 10.3390/app14020675

Winter, J. S., & Davidson, E. (2022). Harmonizing regulatory regimes for the governance of patient-generated health data. *Telecommunications Policy*, 46(5), 102285. DOI: 10.1016/j.telpol.2021.102285

Younis, H. A., Eisa, T. A. E., Nasser, M., Sahib, T. M., Noor, A. A., Alyasiri, O. M., Salisu, S., Hayder, I. M., & Younis, H. A. (2024). A systematic review and meta-analysis of artificial intelligence tools in medicine and healthcare: Applications, considerations, limitations, motivation and challenges. *Diagnostics*, 14(1), 109. DOI: 10.3390/diagnostics14010109 PMID: 38201418

Yu, K. H., Healey, E., Leong, T. Y., Kohane, I. S., & Manrai, A. K. (2024). Medical artificial intelligence and human values. *The New England Journal of Medicine*, 390(20), 1895–1904. DOI: 10.1056/NEJMra2214183 PMID: 38810186

Zhao, H., & Yang, G. (2022). Information security and legal ethics of artificial intelligence medical devices. *Forest Chemicals Review*, 236-245.

ADDITIONAL READING

Bhambri, P., Soni, R., & Tran, T. A. (2024). *Smart healthcare systems: AI and IoT perspectives*. CRC Press. DOI: 10.1201/9781032698519

Bohr, A., & Memarzadeh, K. (Eds.). (2020). *Artificial intelligence in healthcare*. Elsevier Science.

Eto, T. (2024). Effective IRB review of AI human subjects research. Care-Q. https://careq.ucsd.edu/resources/index.html#Engagement-Webinars

Gerke, S., Minssen, T., & Cohen, G. (2020). Ethical and legal challenges of artificial intelligence-driven healthcare. In *Artificial intelligence in healthcare* (pp. 295–336). Academic Press. DOI: 10.1016/B978-0-12-818438-7.00012-5

Reddy, S. (Ed.). (2020). *Artificial intelligence: Applications in healthcare delivery*. CRC Press. DOI: 10.4324/9780429317415

Rizk, S. H. (2022). Ethical and regulatory challenges of emerging health technologies. In *Applied ethics in a digital world* (pp. 84–100). IGI Global.

World Health Organization. (2021). *Generating evidence for artificial intelligence-based medical devices: A framework for training, validation and evaluation*. https://www.who.int/publications/i/item/9789240038462

KEY TERMS AND DEFINITIONS

Artificial Intelligence: "Artificial Intelligence is a machine-based system that can, for a given set of human-defined objectives, make predictions, recommendations, or decisions influencing real or virtual environments. Artificial intelligence systems use machine- and human-based inputs to perceive real and virtual environments; abstract such perceptions into models through analysis in an automated manner; and use model inference to formulate options for information or action" (FDA, 2024)

Machine Learning: "Machine Learning is a set of techniques that can be used to train AI algorithms to improve performance at a task based on data" (FDA, 2024).

Protected Health Information (PHI): Protected health information (PHI) means individually identifiable health information maintained in electronic and non-electronic formats, including healthcare payment and insurance systems. The past, present, or future payment for the provision of health care to the individual, and that identifies the individual or for which there is a reasonable basis to believe can be used to identify the individual. Protected health information includes many common identifiers (e.g., name, address, birth date, Social Security Number) when they can be associated with the health information (HHS, 2022).

Software as a Medical Device (SaMD): "Software intended to be used for one or more medical purposes that perform these purposes without being part of a hardware medical device" (FDA, 2018).

ENDNOTES

1 Tyson, A., Pasquini, G., Spencer, A., & Funk, C. (2023). 60% of Americans would be uncomfortable with provider relying on AI in their own health care. Pew Research Center. https://www.pewresearch.org/

2 Tamiko Eto's White Paper from August 2022 offered an AI decision tree for human subjects research (HSR) as adapted from the Office for Human Research Protection (OHRP) post-2018 IRB review decision trees and intended to further the discussion of human subjects protections within AI for IRB capacity building. See https://techinhsr.com/wp-content/uploads/2022/08/AI-HSR-WHITE-PAPER-TechInHSR-08.2022-1.pdf

3 The FDA maintains publicly available, downloadable information on AI/ML-enabled devices by the date of the final decision at https://www.fda.gov/medical-devices/software-medical-device-samd/artificial-intelligence-and-machine-learning-aiml-enabled-medical-devices#resources

4 Dr. Barbara Evans provides an extensive analysis of HIPAA and AI/ML in clinical decision tools in the preprint available at https://ssrn.com/abstract=4316211

5 U.S. Food and Drug Administration [FDA]. (2024, August 7). Artificial intelligence and machine learning (AI/ML)-enabled medical devices. https://www.fda.gov/medical-devices/software-medical-device-samd/artificial-intelligence-and-machine-learning-aiml-enabled-medical-devices

Chapter 7
Institutional Animal Care and Use Committee:
The Keystone to Animal-assisted Interventions Research

Hemangi Narayan Narvekar
https://orcid.org/0000-0002-8958-8343
Christ University, India

ABSTRACT

Recently, there has been increased awareness of the health benefits of animal-assisted interventions. When it comes to using animals in research, this field is unique since it views animals as therapeutic agents. In this process, the animal's welfare is the most important consideration. Researchers do not understand the ethical procedures that govern research involving such therapeutic human-animal interactions well. This chapter discusses the Institutional Animal Care and Use Committee, which oversees the ethical use of animals in research like how the Institutional Review Board (IRB) oversees the ethical treatment of human participants. The chapter covers the committee's responsibilities, including reviewing protocols, monitoring animal programs, assessing professional standards, and overseeing other facets of animal care. The chapter is important for advising academics, researchers, and practitioners on how to carry out studies with the maximum advantages and the least amount of harm to both humans and animals.

DOI: 10.4018/979-8-3693-3848-3.ch007

INTRODUCTION

Research is at the core of building and advancing the society. Research does not only have implications for the scientific community but also has relevance in solving the practical aspects of life. Research is seen as the key driver of human progress. But research, like human endeavours, is prone to bias and partisanship. Therefore, ethics in research are of utmost importance as they help protect the rights of participants, reduce the risk of harm, and maintain the integrity of the study. Research ethics provides explicit guidelines for the responsible conduct of research studies. The researcher needs to follow the major ethical principles of honesty, objectivity, integrity, openness, respect for intellectual property, confidentiality, social responsibility, competence, and obeying laws.

Subjects play an important role in any kind of research. Especially human subjects in social science studies bring with it a host of ethical concerns. Over the years, the ethical considerations surrounding such research have changed with practitioners focusing more on the humane treatment of subjects. There are guidelines laid by different scientific communities and organizations on ethical ways of conducting research. And it starts from the first stage of research itself, which is drafting a research proposal and getting it sanctioned from the concerned authorities. The research proposal is sent to Institutional Review Boards (IRBs) which are entities that scrutinize the proposal idea and understand whether the risks involved in conducting the research are less than the larger goal of the research. It ensures that the participants in the study are protected, and their well-being is of utmost importance. This step is crucial for developing a sound and ethical research plan.

INSTITUTIONAL REVIEW BOARD AND ANIMAL-ASSISTED INTERVENTIONS

The Institutional Review Board plays an important role in protecting the research participants' welfare. It ensures that the research conducted is ethical and no participants are harmed before, during or after the research. It complies with national and international laws, regulations and codes to safeguard the human subjects in the study. However, there are recent fields that are emerging that involve animals as part of the intervention programs and an IRB body need not be an expert in such cases. For instance, the field of animal-assisted therapy (AAT) has gained recognition as Complementary and Alternative Medicine (CAM) and research has been well-established in this area providing evidence about the health benefits of AAT in different sets of populations. Animal-assisted therapy is defined as a structured, goal-defined intervention to improve the well-being of humans with the help of a trained animal-therapist team (IAHAIO, 2000). Similarly, there are also other forms of interventions such as animal-assisted activities (AAA), animal-assisted education (AAE), or animal-assisted coaching/counselling (AAC) which form a part of the larger field of animal-assisted interventions (AAIs) that help from animals to benefit humans. The International Association of Human-Animal Interaction Organizations (IAHAIO) is the organizational body that works in this area and provides elaborative definitions of these different terminologies that are used in this field (IAHAIO, 2000). Animal-assisted interventions is the umbrella term that is used to describe all human-animal interactions that are used to enhance the quality of life of clients (Kruger & Serpell, 2010)

Animal-assisted interventions are used by different service providers for educational, motivational, recreational and therapeutic needs (Kruger & Serpell, 2010). For instance, physiotherapists, occupational therapists, psychologists, speech therapists, special educators, doctors, nurses and so on can use AAIs to meet their clients' needs. Different professionals work with animals differently to meet their various treatment requirements in a way that maximizes client satisfaction. For example, a psychologist might involve animal with the client to improve the client's self-esteem, and social skills or reduce stress. An occupational therapist might help clients improve their balance and coordination with the incorporation of animals into different physical exercises. The choice of animals also differs based on this therapeutic or treatment requirement. Thus, every profession will have its own structured and planned animal-assisted intervention with the specific animal that they consider will benefit the client.

Canines or dogs are the most common animals that are seen in the field of AAI. However, various animals such as horses, guinea pigs, rabbits, donkeys, fish, and birds can also be included for these interventions. The terminologies used in AAIs can also be modified and described using these specific animals such as canine-assisted therapy, equine-assisted intervention, dolphin-assisted activities and so on. Being a novel field whose genesis began in 1969 with the term 'pet therapy' by the currently known founding father of animal-assisted therapy, Boris Levinson (Geist, 2011), this field still requires more evidence to prove its efficacy in different populations and settings. The field also lacks a common theoretical framework that can only be achieved with more evidence and theoretical exploration (Narvekar & Ghani, 2022).

The knowledge generated on AAIs is vastly from the US but recently several research papers have been published in Europe, the UK, Australia, Singapore, Japan, India and other countries (Aoki et al., 2012; Hunjan & Reddy, 2023; Muckle & Lasikiewicz, 2017; Stern & Chur-Hansen, 2013). The role of animal-assisted interventions for mental health needs received even more attention during the pandemic (Wong et al., 2021). This shows the increased interest of practitioners and academicians pursuing this area of research for the benefit of humans. More research studies, especially in the eastern world is currently the need of the hour which will help generalize the findings obtained in the western part of the world. Hence, sound and ethical research in AAIs becomes extremely important to achieve this goal.

In such research endeavours like AAI, where animals are also involved in the research as an extra hand or assistant and not as mere subjects, the dynamics of the research environment differs. Here, the participant's welfare is important, and the well-being and protection of the animal involved in the intervention or treatment is also essential. Such a field is different regarding its application to animals in research. Here, the animals are not used but are considered the co-therapists or the medium through which intervention occurs. The field of AAI is only now starting to recognize how crucial animal welfare is to improve the standard of the entire intervention process (Fine & Griffin, 2022). A "one health" perspective currently guides all the human-animal interactions. "One Health" is the outlook that emphasizes wellbeing of both humans and animals in any kind of setting or environment (Pinillos et al., 2016). Thus, we are responsible for the welfare of the animals alongside our human participants.

Academicians and researchers have a lesser understanding of the ethical processes that guide the research activities that involve such significant therapeutic human-animal interactions, unlike other research environments. Since the field of animal-assisted interventions itself is novel, it becomes imperative to enumerate in detail the ethical conduct that guides such scientific interactions. As mentioned earlier, IRB might not be able to guide the ethical conduct of research when human-animal interactions are present. It does not focus on the inclusion of animals in the research study as an additional element or medium of treatment. Therefore, it's recommended in the research forum now that additional implementation of the body such as the Institutional Animal Care and Use Committee (IACUC) should be formed which monitors all the forms of research of which animals form an integral part (Ng, 2021).

IACUC becomes thus an important part of the research that focuses on animal-assisted interventions. Currently, only a few academicians and researchers are aware of this board and the roles and responsibilities that lie with this committee. According to the recent review report, only 10% of the total studies have mentioned the approval from IACUC for their AAI studies (Ng et al., 2019). This number indicates a significant lack of awareness surrounding this topic and also has health hazards because of overlooking both human and animal welfare. The IRB can overlook the welfare of the animal as a sentient being and thus the safety of the human participants, animals and therapists, can be in danger. In addition, being a novel field which lacks formal guidelines that are universally applicable, standardization of interventions becomes impossible leading to differences in the treatment offered (Santaniello et al., 2020). Therefore, it is very possible that institutions overlook the treatment modalities and animals' welfare in the research process.

As evident, it becomes imperative to understand the laws and guidelines concerning the involvement of animals in the research field of animal-assisted interventions. Besides the justification of involving animals in research, additional challenges need to be addressed in academia. This includes but is not limited to housing of animals, veterinary care, standards of hygiene, training of animals, responsibilities of researchers, requirement of expertise on animals, etc. There are also differences in the treatment and attitude towards different animals in different religions and cultures which cannot be ignored in interventional research (Valiyamattam et al., 2018). IACUC, which the chapter focuses upon, can address and supervise all these elements.

INSTITUTIONAL ANIMAL CARE AND USE COMMITTEE (IACUC)

The Institutional Animal Care and Use Committee supervises the ethical use of animals in research, similar to how the IRB oversees the ethical treatment of human participants (Ng et al., 2019). The original IACUC guide was published in the year 1992 and thereby served as an important document that sees the welfare of animals in the animal research community. The current version that is followed in the research community is the 2nd edition of the IACUC which elaborately provides knowledge for the ethical use of animals in different scientific arenas (Institutional Animal Care and Use Committee Guidebook, 2002). This guidebook was developed by the Applied Research Ethics National Association (ARENA), the editorial board of various knowledgeable persons who had experience with IACUC. It talks about all the requirements and the process of getting approvals when the research involves the use of animals such as animal use in laboratories, husbandry, agricultural instruction, veterinary teaching, etc. However, there is no mention of working animals, such as in the case of AAIs. Since many professionals working in the field of AAIs mandate IACUC approval in AAI research, it is now a necessity to learn more about how IACUC works in this direction and how AAI practitioners and researchers can obtain the approval of the same before starting the research.

Moreover, in the field of AAI, we don't use the word 'use' for the involvement of animals, rather the animals are seen as a co-worker who helps humans in delivering intervention services (Evans & Gray, 2012). This is different from the laboratory use of animals, which the guidebook extensively talks about. There is a difference in the way the animals are looked into and hence IACUC, if looking at the proposals in AAIs should approach the research and animal community differently than what is mentioned in the guidebook. Hence, this chapter provides the guidelines that IACUC dictates but are more applicable for research in AAIs to help gain a better understanding of how IACUC can guide AAI studies.

Membership

The IACUC members perform important duties to approve and supervise the treatment of animals in research conduct. Thus, it is important to have a committee comprising different roles and expertise in various arenas. The diversity in the committee helps broaden the perspective and provide the crucial review procedures in more depth (Newcomer & Greer, 2000). It is recommended that the IACUC consists of three to five members with unique roles and responsibilities. The composition can be as follows:

1. Veterinarian: The veterinarian is a trained medical professional who has expertise in animal science and in the context of AAIs, has knowledge about animals and can supervise and monitor the AAI activities.
2. Scientist: The committee should include a scientist who is practicing and has experience in research involving animals.
3. Non-scientist: This includes members who are not scientists such as lawyers, librarians, clergy, etc.
4. Non-affiliated Member: Any person who is not affiliated with the institute where research is conducted or is no immediate family member of the person affiliated with the institute, can take up this position in the committee.
5. Public Member: This member represents the interests of the public and serves as a voice for ethical considerations outside of the research community. In context of AAI research, this can be a person who has love and care for animals and their lives.

These members are oriented and trained to carry out the duties and responsibilities of IACUC with the help of appropriate materials, training by ARENA, adequate resources, and a syllabus that helps to perform the duties consistent with the Animal Welfare Regulations (AWRs). Apart from providing basic information about the responsibilities of being a member, the committee, including the IACUC staff, Chair, etc., trains these members on different areas such as the review process of IACUC, current knowledge about laws, policies, and regulations, and official directives and publications. IACUC needs to have a balance of expertise and diversity among its members to effectively fulfill their responsibilities. It is important to note that no IACUC member should have any potential conflict of interest in the project that seeks approval. If there is, the principal investigator of the research can submit the request for the exclusion of the particular member.

Functions of IACUC

As evident, the major function of IACUC is to ensure that research involving animals is conducted ethically and in compliance with laws and regulations. IACUC performs this function through different tasks such as (a) reviewing protocols of animal use; (b) monitoring programs of animal care and use; and (c) reviewing concerns about animal care and non-compliance issues (Silverman et al., 2014). These tasks are enumerated in detail in the following segment (Figure 1 depicts the same).

Figure 1. Major functions of institutional animal care and use committee

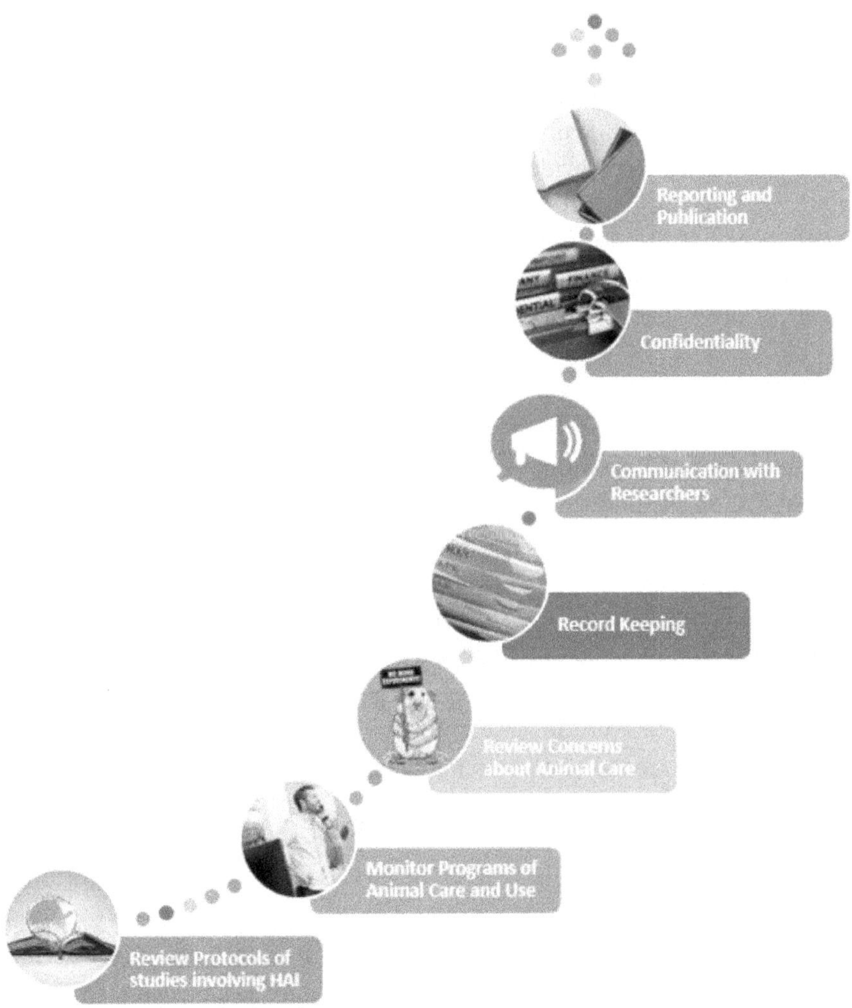

(a) Review Protocols of Animal-assisted Intervention Studies

The first and foremost responsibility of IACUC is evaluating research proposals for all aspects of animal use and care. They need to assess whether the study plan complies with the laws and policies concerning appropriate scientific endeavours. The following are the major areas in the AAI protocols that should be assessed –

1. Rationale of Involving the Animal in the Research

It is known now that animals can help in the improvement of health in humans. Different animals can bring different health outcomes. In the field of AAI, several animals are incorporated to produce physical and mental health benefits. These include but are not limited to dogs, horses, donkeys, cats, guinea pigs, rabbits, elephants, birds, dolphins, and so on. The research proposal should clearly mention the rationale behind involving a particular species of animal in the particular research interest. Enough evidence needs to be provided to ensure that IACUC clearly understands the justification with which the animal is being incorporated for human benefits. On similar lines, the appropriateness of the species needs to be documented with evidence and enough data to convince the committee about the righteous incorporation of animals.

2. Risk Assessment

Risk assessment is essential before conducting a study in any of the scientific fields. IACUC members evaluate potential risks and benefits associated with the use of animals in AAI research and consider both animal and human participants' safety and well-being before providing the investigator with the green signal to initiate the research process. The rule of thumb is that the potential benefits from conducting the study should outweigh the risks or harm that can occur during the research procedure, which can be derived through harm-benefit analysis (Brønstad et al., 2016). In the case of the AAI study, the harm to the animal and the human should be none to minimal compared to the positive outcome that the intervention will produce.

3. AAI Intervention Template

The IACUC analyses the intervention protocol for all the important information that can be potentially deleterious to the animal's welfare. For instance, what will the interaction between humans and animals actually involve? Is it just merely watching animals move and interact or would it involve embracing animals? Whether there will be a handler present during the session? What would be the role of this handler and also the investigator? Are both these roles offered by the same person? Will there be more than one animal at a time? Will the animal work in group settings with multiple contacts at a time? The duration of this contact is also important information to look into. Whether this contact is physical contact and if so whether there was enough care taken to avoid any transfer of diseases should also be analyzed. Most of the studies do not provide this information which is necessary to make an informed decision on the research proposal (Stern & Chur-Hansen, 2013). IACUC sees that all the details are sufficiently provided in the intervention template and gives the results accordingly.

4. Qualifications and Training of the Personnel

The pre-requisite for carrying out the AAI is the qualifications and training of the person carrying out the interventions. Based on the activities or the type of intervention proposed in the research proposal, the IACUC should see to it that the personnel carrying out the intervention have required training and expertise in the area of AAI. According to the universal guidelines provided by IAHAIO, in order to carry out any intervention in this field one needs to have sufficient knowledge about human-animal interactions and the understanding of animal behaviour alongside the handling of animals. A basic degree in healthcare or education is also mandatory for providing interventions. This training and experience should be noted in the research proposal document. Studies have found that a significant proportion of AAI deliveries are provided by those who do not meet the professional requirements (Kerulo et al., 2020). Thus, this step is crucial to safeguard the animals and humans involved in research.

Apart from the investigator or the personnel carrying out AAI activities, other people present and helping with the research tasks should also be trained and provided education regarding proper animal care, handling, and ethical considerations. These people might involve other researchers, staff, students, and interns involved in AAI research. All those who come in contact with animals, in any way in the study environment, should be well-informed about behaving with the animals so that the safety of both animals and humans is not compromised.

5. Training and Certification of the Animal

One major task IACUC performs is checking the credentials of the animals that would be part of AAI research. One requirement of the animals performing AAI activities is that they need to be trained according to their needs. The selection of animals in AAI should be carried out according to the guidelines laid out by organizations such as Delta Society, which establishes standards for the recruitment of animals. This includes explicit animal selection and screening guidelines based on source, temperament, health, and behavior (Palley et al., 2010). Only after successfully completing the examination, after extensive training, the animal is certified or registered as a therapy animal (Warsaw, 2020). IACUC sees that the animal is, thus, licensed to do the work as mentioned in the research proposal. The certifications and the necessary documents should be attached to the proposal in the appendix section so that the IACUC has proof of the same for making informed decisions on the research proposal.

6. Methodology

The IACUC is responsible for ensuring that the methodology and procedure employed in the study do not harm the animals (Mohan & Huneke, 2019). The researcher planning to incorporate animals in the study should submit the research plan enumerating details about the experimental variables, methods of testing, process of procedure employed, study outcomes, and everything that will provide a thorough understanding of the research plan to IACUC. There must be detailed plan descriptions in the research protocol about reducing injuries or harm to animals, if any (Johnson & Besselsen, 2002). In the field of AAI, the studies also employ measures that assess the physiological impact of AAI activities on animals which might need some instruments that can cause some distress or pain. The procedures and techniques should be minimally invasive. The IACUC can take help from Planning Research and Experimental Procedures on Animals: Recommendations for Excellence (PREPARE) to make sure that the research is planned by considering essential criteria for conducting quality animal studies (Smith et al., 2018).

7. Facility Inspection

The site of the data collection is also of paramount importance in the research activities related to AAIs. The selection of the site should have opportunities for data collection while at the same time, there should not be disruptions in a way that affects the work environment. The permission to use this space should also be obtained prior so that the proof of this can be attached to the research protocol in the appendix. This site of the study or facility is also physically inspected as part of facility reviews by the IACUC members. This involves the physical inspection of the rooms and buildings for research purposes. The inspection can be done with the help of a facility inspection checklist provided by the Public Health Service (PHS) Policy on Humane Care and Use of Laboratory Animals available on the website.

AAI researchers can formulate their own checklist for AAI purposes taking help from the existing available checklist.

8. Transportation of Animals

AAI studies might require the transportation of animals to the designated spaces for carrying out the research activities. These spaces might be clinics, hospitals, offices, schools, homes or any other area where the study needs to be conducted. One of the studies focusing on exploring animal welfare in dog-assisted interventions found that transfers of dogs to different facilities for work can increase stress levels in dogs (Marinelli et al., 2009). IACUC needs to ensure that this transportation of the animals has been carefully planned by the investigator so that the safety of the animals is taken care of. This involves the mode of transportation being comfortable and safe for the animals, carriers or crates being appropriate for the size and the species of the animals, vehicles being well-ventilated, and temperature is taken care of. Ensuring that the animals have access to water and food throughout the transportation is important. Regular breaks also need to be provided for bathroom needs.

In addition, trained personnel must be available throughout the journey to take care of emergency requirements, if any. Apart from this, other measures need to be taken to keep the animals stress-free such as avoiding overcrowding, keeping the noise at low, providing familiar bedding and so on. IACUC also needs to inquire whether during transportation the animals will have access to veterinary care in case of any emergencies. Investigators might also require permits for transportation if animals are crossing state or national borders. The research proposal should document these details if any kind of transportation of animal is involved in the research plan.

9. Animal Welfare

In AAI studies, the animal must be stress-free to perform activities in session to maximize the benefits of the session and at the same time, the well-being of the animal is maintained. There are instances where research work can be stressful for the animals at work. Some of the studies have found that therapy dogs had increased salivary cortisol during their workdays as compared to other days where there was no work allotted indicating increased stress response. In addition, with the increased number of therapy sessions, the dogs were found to be more stressed, especially after 3 hours of lengthy sessions (Haubenhofer & Kirchengast, 2007). Therefore, the handler and the investigator should be well-trained to identify their signs of distress. The reviewers should identify whether there are clear provisions in the protocol for enough breaks for the animals and places of rest in case of stressful circumstances during the intervention.

In therapy settings, there might be situations where there occurs harm to the therapy animal. For example, while conducting a session with a child diagnosed with Attention-deficit/hyperactivity Disorder, the child tries to grab the dog's tail and pull it. This disturbs the dog causing fear. In such scenarios, what would be the action taken by the researcher is very important. The plan of action here can be terminating the session protecting the dog from any kind of harm. All respected therapy animal organizations and those who oversee such programs emphasize that the owner/handler of a therapy animal must be observant and function as an advocate for their animal (VonLintel & Bruneau, 2021). IACUC should ensure that animal welfare has been given the most priority in the research plan.

These are some of the major points IACUC looks into while reviewing the investigator's research proposal. This review of the protocol can be of two types: Member Review and the Full Committee Review (Prentice et al., 2014). In member review, each designated member reviews the abstract and the proposal and has the right to approve, appeal modification and request full committee review. In full committee review, all the designated members meet and explicitly review the proposal at the same time. The approval of the proposal is only sought if the majority of the members provide their votes in favour. Once approval is obtained from IACUC, the researcher can go ahead with the study as proposed and described. However, there might be also instances where the research proposal might need modifications and revisions to secure approval. Significant changes would require additional review after the second submission.

(b) Monitor Programs of Animal Care and Use

The work of IACUC does not get completed only with the review of the research protocols but they are vested with power to monitor whether the research activities approved carried out the way it was described in the proposal. This can be done by asking the investigator of the study to report annually on the scientific conduct and any challenges or proposed changes. This may also include inspection of facilities, health surveillance, behavioural assessments, assessing records, monitoring the welfare of the animals and so forth. Some of the major areas IACUC can look at are -

1. Work Environment of the Animal

One crucial task of IACUC is to see that the environment where the animal will be working is adequate and safe. Inadequate working environmental conditions can be a major source of stress for the animals working for assisted interventions. Environmental conditions such as continuous interferences, lack of space, and high temperatures were found to be some of the stressors for animals in the study conducted that reviewed AAIs conducted over 4 years at the Centro Pet Therapy' in Verona (Marinelli et al., 2009). Such environmental concerns should be analysed, if present, by the IACUC members during their inspection.

The risk assessment and the dog welfare plan were developed by the Lincoln Education Assistance with Dogs (LEAD) team in 2018 that ensure that the AAI program is implemented and adhered to in a manner that promotes the welfare of the animals (Figure 1 and 2)

Figure 2. Risk assessment tool for dog-assisted interventions in schools and educational settings

PART B: HAZARD IDENTIFICATION AND CONTROL MEASURES				
Step 1: Identify significant hazards	**Step 2: Identify who might be harmed and how**		**Step 3: Identify precautionary measures already in place**	**Step 4: Identify person/s responsible**
List of significant hazards (anything with potential to cause harm)	**Who might be harmed?**	**Type of harm**	**Existing controls** (Actions already taken to control the risk)	**Name:**
All below:	Pupils, staff	All below:	• Staff should also refer to any internal school policy, if existing, in relation to animals on school premises when organising an animal visit to school.	•
Hygiene	Pupils, staff	Infection, illness	• School infection control procedures to be always followed; • School health and safety procedures to be always followed; • Any significant cuts or abrasions on exposed skin of hands and arms should be covered before contact with the dog; • Hand sanitizer gel and antibacterial wipes are provided for immediate use before and after contact with the dog; • Pupils and adults always wash their hands soon after contact with the dog (or coming into contact with the dog's bedding, water, toys, etc.) and especially before snack/meal times; • Health and safety information, including facts about zoonotic diseases can be accessed through the project website at http://lead.blogs.lincoln.ac.uk	•
Allergies, diseases, parasites	Pupils, staff	Illness, allergic reaction	• Parents/caregivers are asked to identify any pupils known to have allergic reactions to dogs. These pupils may have restricted access to dogs depending on their allergy trigger; • In the rare case that an allergic reaction should occur and does not subside, medical assistance will be sought; • The dogs will have been regularly taken to a veterinarian and have been recently dewormed and treated for fleas; dogs licking participants should be avoided;	•

Figure 3. Dog care plan

Dog Care Plan

Name of dog:____________________ Age:_______________ Gender: __________ Breed: ____________________

Insurance statement of handler with signature:

Indicate type of insurance (e.g., public liability insurance with legally binding indemnity):____________________

for dog (name):____________________ valid (from – to):____________________

to cover any claims made by a setting in case of an incident during an AAI visit.

Name of handler: ____________________ Signature handler: ____________________ Date:________

This Care Plan pertains to the welfare of dogs employed during intervention sessions with participants in any setting.
The Care Plan is part of, and should be read in conjunction with, the settings Risk Assessment document.
Areas relating to safety training with participants and teaching safe behaviour with dogs are detailed in the main risk assessment.

Care and treatment	**Behaviour management**	**Feeding and watering**	**Toileting**	**Enrichment**	**Exercise**
• Visual health-checks carried out by handler before dog begins work in school/other setting; • Dog familiarised with setting; • Dogs will be monitored throughout sessions to ensure their care, cleanliness and treatment is maintained to a high standard; including child/participant behaviour monitoring to ensure the dog is treated with respect; • Dog handlers will have an animal first aid kit in order to administer emergency first aid in circumstances where this may be required; • Dog not to work more than 2 h per day.	• During the intervention sessions, dog handlers and researchers / supervisors trained in dog distress signalling will carry out constant observations of the child/dog interactions to detect signs of stress in the dogs; • If a dog becomes stressed, the dog will be removed from the situation to protect the welfare of all concerned and allow the dog to feel comfortable in their surroundings.	• Dogs will be fed before arrival at the school; • Dogs will have constant access to water; • Preferably no treats to be given; however, handler to decide on treats / rewards throughout sessions as they see fit; • No treats to be given from hands of participants.	• Designated areas for toileting will have been previously agreed between the school / other setting, the researchers / supervisors and dog handlers in advance; • Dogs will be taken outside at regular intervals as the handler sees fit, or as required by the dog during intervention sessions with the child / other participant.	• Dogs may have a toy in the sessions to enable them to display their natural behaviours; • Dogs will have a bed / blanket in a designated space in the room as a rest area away from human interaction, and they should be given the opportunity to use it as they wish; • Children / participants will not be permitted to approach the dog when the dog is in their resting space.	• Designated areas for exercise will have been previously agreed between the school, the researchers and dog handlers in advance; • When toileting, the dog will be given time to exercise outside in order to enable them to display their natural behaviours and also as a break from direct contact with children.

The LEAD Risk Assessment is recommended as a comprehensive measure or toolkit that helps achieve the highest standards of ethical practices in the field of AAIs (Brelsford et al., 2020). This risk assessment can be used with different forms of AAIs, be it AAT, AAA or AAE programs. The tool is useful for risk assessment in relation to managing different AAI sessions in various settings focusing on human contact, animal contact, and protection of animal. Principle Investigators or the IACUC can consider the use of such tools to assess the application of AAI programs.

2. Appropriate Living Conditions of the Animal

Not only the work environment, the living conditions or the housing of these animals should also be looked after. The majority of the time such animals would be apart from their work environment and hence their stay conditions should not be neglected. IACUC would require information on their housing facilities, and they will assess the living conditions in terms of space, ventilation, temperature, lighting, bedding and cleanliness as these are the factors that can affect animal care and health (Pritt & Hammer, 2017). Investigators or the animal's caretakers should ensure that the animal's housing environment allows them for natural behaviour and social interactions while also providing opportunities for privacy and security.

The environmental enrichment cannot be ignored as well. There have to be plenty of opportunities for physical exercise and sensory stimulation for animals. IACUC can also monitor feeding practices to ensure a species-specific balanced diet is provided to the animals and there is no sign of malnutrition or obesity in the animal. Maintaining a high standard of hygiene in the housing facilities of the animal is essential too. Proper waste management, disinfecting the surroundings regularly, and biosecurity are all factors IACUC will look into to monitor animals' living conditions.

3. Emergency, Weekend and Holiday Care

Providing emergency, weekend, and holiday care for animals involved in AAIs is essential to always ensure their well-being and safety. Some of the important things for the staff members is to ensure that they have emergency contact numbers of veterinarians, animal control agencies, and 24/7 veterinary hospitals. The handler or the staff should be trained in first aid and emergency response procedures as well as should be prepared to recognize any distress signals from the animals. Emergency preparedness plans for natural disasters, fires, and medical emergencies are also required.

During weekends and holidays, the facilities should also have backup caregivers in case regular staff members are unavailable. IACUC can conduct these regular check-ins during weekends or holidays to monitor the general well-being of the animals in the facilities. In AAI studies, there might also be incidences of adverse outcomes that need attention. For instance, there might be a death of the animal or a chronic illness that leads to early retirement of the animal (Ng et al., 2019). In such situations, appropriate steps need to be taken, keeping in mind the best possible solution for the situation and the welfare of the animal.

4. Quality Improvement

One of the functions of IACUC is that it should monitor the research activities continuously to assess the effectiveness of animal care programs and identify the different opportunities to improve the quality of care. IACUC can implement quality improvement initiatives based on the monitoring or the feedback from the stakeholders and then further evaluate the outcomes of this process to ensure that improvements are achieved in the direction of increasing the quality care of animals. IACUC can take help from animal wellness guidelines in the AAT program laid out by the American Veterinary Medical Association (AVMA, 2007) for the effective quality improvement of the work. As part of its evaluation of the research program, the IACUC also assists in determining the efficacy of the various training regimens. Any suggestions for improvement are sent in the semi-annual reports to the investigator or the authority (Anderson, 2007).

(c) Review Concerns about Animal Care and Non-compliance Issues

The study review is not completed without evaluating all aspects of research for the potential harm or distress that can be caused by research activities. IACUC needs to assess whether the principle investigator has mentioned and carefully taken all the steps to maintain the animal's well-being. Researchers are also required to report adverse events or deviations from the approved protocols to IACUC. The actions are taken thereby after reviewing these reports. IACUC ensures compliance of the investigator with relevant guidelines, regulations, and institutional policies regarding the involvement of animals in research. There might be instances where there are observations made by the staff of the institution where the research is conducted or the general public or media reporting concerns about the compliance of the investigator. Respondent is the term used in such cases for those who are accused of noncompliance relating to animal use and may include anyone from the scientist to animal caretakers or even IACUC itself (Hansen et al., 2017).

Before classifying an activity as misbehaviour or noncompliance, caution should be exercised because some impressions of noncompliance may stem from misunderstandings or inadequate knowledge. Using fundamentally fair methods to look into and address problems can help to discern between genuine and serious noncompliance and miscommunications or issues of reasonable and valid disagreement. If there is a risk determined for the animal, immediate action is taken to resolve the issue. If the committee determine that the allegation is credible or meets serious noncompliance, a formal inquiry process is initiated. This investigation is conducted in timely manner, with notification to the respondent/s, and coordination (Hansen et al., 2017).

Where significant harm is concluded for the animals involved, prompt interim actions are taken. IACUC can suspend the research projects or take appropriate corrective actions based on the investigation. For instance, it is imperative that prompt action has to be taken to reduce the risk of harm and provide the animal with any necessary remedial care, in cases of such non-compliance (Silverman, 2015). Recommended corrective actions will be based on the severity of the non-compliance incident or responsiveness of the respondent. IACUC can suspend or suggest other corrective actions such as counselling, training, changes in administrative or management processes, appropriate amendment to the animal use protocol, regular reports of the respondent to the IACUC, official letters of reprimand, and direct veterinary or IACUC oversight or monitoring of animal procedures and/or record-keeping (National Institutes of Health, 1986). Moreover, members of the IACUC who are involved in noncompliance discussions need to be unbiased and receptive to any accusations of improper behaviour involving animals. Overall, IACUC is vested with the power to continually assess institutional compliance and determine whether the institution complies with all the relevant regulations, laws, and ethical responsibility of promoting animal welfare.

Apart from these major functions, IACUC also takes care of the following -

1. Record keeping

As any organizational body, IACUC also keeps records of attendance, meetings, inspections, and deliberations. In addition, all the submitted protocols and the modified proposals after review are retained by IACUC till the research work is completed and three years after the research (Laber & Smith, 2014). Those proposals that did not get approvals are also kept for three years to maintain the record. Any adverse events and report findings relevant to IACUC are also retained. On the researcher's side, all the documents containing client's details, intervention sessions, animal inventories, personnel training records, and documentation about emergencies, incidents or veterinary care of the animals should also be maintained and bc up to date so that anytime IACUC want to have look at it, this information should be accessible.

2. Communication with Researchers

Moreover, the IACUC keeps an open communication with the research team which helps the investigator to clarify any concerns they might have related to the conduct of the study. This also makes sure that the researcher carries out the study in accordance with the guidelines and approval of IACUC. Establishing clear channels of communication between the IACUC, researchers, institutional leaders, and other stakeholders involved in animal care and use programs paves the way for the ethical conduct of the research.

3. Confidentiality

IACUC also maintains confidentiality of all the personal and proprietary information submitted by the research team and organization. Inappropriate data disclosure can have a detrimental effect on scientific research and instil mistrust in the IACUC among researchers. There is also the added danger of opposition and conflict from those who are against involving animals in research. Exemptions from disclosure include any information that could negatively affect an individual's competitive position or government interests, such as program effectiveness or compliance, trade secrets, privileged, commercial, or financial information, or personally identifiable information or data that could have an impact on an individual level (Chimes & Sankar, 2014).

4. Reporting and Publication

Similar to academic institutions, sometimes IACUC may choose to review the published work to verify whether the research conduct was governed as the IACUC has approved. They can take help from the Animal Research: Reporting of In Vivo Experiments (ARRIVE) guidelines that consist of a 20-item checklist to promote comprehensive and quality reporting of studies to allow for an accurate critical review of what was done and found through the conducted study. These items are focused on reporting information about the specific characteristics of the animals involved in the research, details of housing, methodology, and methods of analysis (Kilkenny et al., 2014).

5. Evaluating IACUC Performance

Moreover, IACUC is also expected to evaluate their performance in terms of satisfactorily completing its own regulatory functions. A review study conducted by Budda and Pritt (2020) found that research organizations or institutes mainly used qualitative methods like observations, interviews, and surveys or questionnaires to assess the performance of IACUC. The different performance metrics can be useful to track and assess the goals of IACUC which further helps in the quality improvement of the research plans of the investigators. A program of continuing education for IACUC can also help establish standards of quality work in this area.

Recommendations to Advance the Use of IACUC

In order to conduct an ethical research study with the appropriate involvement of animals in animal-assisted interventions, it is evident that IACUC is important. With the humane treatment of animals as its top concern in AAI studies, an efficient IACUC will allow researchers to navigate the regulatory environment through a flexible, science-based program. It can also help the principal investigator, and the research team create and deliver high-quality, repeatable studies. However, sometimes there might be instances where an organization or the researcher do not have access to the already formed IACUC. Therefore, Ng et al. (2019) recommend a partner university or an institution with an IACUC to fulfil the responsibility of reviewing a protocol in the cases where the present institution where the researcher wants to conduct a study does not have an IACUC. This will ensure that all researchers go through IACUC as a mandatory step of such research endeavours without any reason.

At the academic level, the publishers of the journals should explicitly mention the requirement of the IACUC approval in the author guidelines so that this process is followed by researchers worldwide. This will ensure that only those research work will be published that follow the necessary ethical guidelines while conducting the AAI study which further will produce only quality work in this novel field. In addition, animal-assisted organizations have a larger duty in this area as IACUC might have limited expertise with AAI research and its conductions. AAI organizations and regulatory bodies can develop detailed guidelines that help IACUC review research protocols that aim to conduct AAI research.

IMPLICATIONS AND CONCLUSION

This chapter is focused on the unaddressed ethical concern of protecting the rights of the animals in the field of animal-assisted intervention thereby paving the way for more ethical conduction of the research in the field. Enumerating the importance of protecting the welfare of animals in the research world, the chapter elaborately discussed the process of formation of the IACUC, sometimes also called the Animal Ethics Committee or Animal Welfare Committee, and the power vested in it to approve the research proposals. The chapter covered topics such as relevant people who can be a part of this committee, the review process, monitoring of the research, modifications to the proposal, documentation, policies concerning animals, veterinary care during the study tenure, reporting of the adverse outcomes during research and the research findings. The chapter also provided a set of examples of different circumstances that can arise during human-animal interaction research which can cause harm or create unforeseen challenges that can impact the research.

This chapter will benefit the audiences who are interested in conducting animal-assisted interventions as opposed to those who are using animals for research purposes like in the case of laboratory use of animals. As mentioned earlier, the role of animals in this research field is different than the other uses of animals in research activities. From the minimal literature available on this topic, the author tried to provide extensive information and guidelines about this topic. This chapter will be significant in terms of guiding researchers and practitioners on ways of implementing studies with the greatest benefits and minimum harm to humans as well as animals. This chapter has implications for social science researchers, veterinary researchers, as well as public health researchers across the globe. Moreover, efforts to mainstream AAIs will benefit from the research conducted in this area and establishing IACUC and getting approval for the study for the moral scientific conduct of research is deemed necessary for all the evidence that we can count on for human health benefits.

REFERENCES

Anderson, L. C. (2007). Institutional and IACUC responsibilities for animal care and use education and training programs. *ILAR Journal*, 48(2), 90–95. DOI: 10.1093/ilar.48.2.90 PMID: 17420530

Aoki, J., Iwahashi, K., Ishigooka, J., Fukamauchi, F., Numajiri, M., Ohtani, N., & Ohta, M. (2012). Evaluation of cerebral activity in the prefrontal cortex in mood [affective] disorders during animal-assisted therapy (AAT) by near-infrared spectroscopy (NIRS): A pilot study. *International Journal of Psychiatry in Clinical Practice*, 16(3), 205–213. DOI: 10.3109/13651501.2011.644565 PMID: 22486555

AVMA. (2007). Wellness Guidelines for Animals in Animal-Assisted Activity, Animal-Assisted Therapy and Resident Animal Programs. Available online from www.avma.org/issues/policy/animal_assisted_activity.asp

Brelsford, V. L., Dimolareva, M., Gee, N. R., & Meints, K. (2020). Best practice standards in animal-assisted interventions: How the LEAD risk assessment tool can help. *Animals (Basel)*, 10(6), 974. DOI: 10.3390/ani10060974 PMID: 32503309

Brønstad, A., Newcomer, C. E., Decelle, T., Everitt, J. I., Guillen, J., & Laber, K. (2016). Current concepts of harm–benefit analysis of animal experiments–report from the AALAS–FELASA working group on harm–benefit analysis–part 1. *Laboratory Animals*, 50(1, _suppl), 1–20. DOI: 10.1177/0023677216642398 PMID: 27188275

Budda, M. L., & Pritt, S. L. (2020). Evaluating IACUCs: Previous research and future directions. *Journal of the American Association for Laboratory Animal Science; JAALAS*, 59(6), 656–664. DOI: 10.30802/AALAS-JAALAS-20-000077 PMID: 32928341

Chimes, M., & Sankar, P. (2014). Confidential and proprietary information. *The IACUC Handbook*, 524-559. CRC Press.

Evans, N., & Gray, C. (2012). The practice and ethics of animal-assisted therapy with children and young people: Is it enough that we don't eat our co-workers? *British Journal of Social Work*, 42(4), 600–617. DOI: 10.1093/bjsw/bcr091

Fine, A. H., & Griffin, T. C. (2022, January). Protecting animal welfare in animal-assisted intervention: Our ethical obligation. Seminars in Speech and Language (Vol. 43, No. 01, pp. 008-023). Thieme Medical Publishers, Inc.

Geist, T. S. (2011). Conceptual framework for animal assisted therapy. *Child & Adolescent Social Work Journal*, 28(3), 243–256. DOI: 10.1007/s10560-011-0231-3

Hansen, B. C., Gografe, S., Pritt, S., Jen, K. L. C., McWhirter, C. A., Barman, S. M., Comuzzie, A., Greene, M., McNulty, J. A., Michele, D. E., Moaddab, N., Nelson, R. J., Norris, K., Uray, K. D., Banks, R., Westlund, K. N., Yates, B. J., Silverman, J., Hansen, K. D., & Redman, B. (2017). Ensuring due process in the IACUC and animal welfare setting: Considerations in developing noncompliance policies and procedures for institutional animal care and use committees and institutional officials. *The FASEB Journal*, 31(10), 4216–4225. DOI: 10.1096/fj.201601250R PMID: 28821637

Haubenhofer, D. K., & Kirchengast, S. (2007). 'Dog handlers' and dogs' emotional and cortisol secretion responses associated with animal-aassisted therapy sessions. *Society & Animals*, 15(2), 127–150. DOI: 10.1163/156853007X187090

Hunjan, U. G., & Reddy, K. J. (2023). Effect of Animal-Assisted Therapy On Depression, Memory, Attention, And Emotion Regulation. *Human Research in Rehabilitation*, 13(1), 120–129. DOI: 10.21554/hrr.042314

Institutional Animal Care and Use Committee Guidebook. (2002). https://grants .nih.gov/grants/olaw/guidebook.pdf

Johnson, P. D., & Besselsen, D. G. (2002). Practical aspects of experimental design in animal research. *ILAR Journal*, 43(4), 202–206. DOI: 10.1093/ilar.43.4.202 PMID: 12391395

Kerulo, G., Kargas, N., Mills, D. S., Law, G., VanFleet, R., Faa-Thompson, T., & Winkle, M. Y. (2020). Animal-assisted interventions: Relationship between standards and qualifications. *People and Animals : The International Journal of Research and Practice*, 3(1), 4.

Kilkenny, C., Browne, W. J., Cuthill, I. C., Emerson, M., & Altman, D. G. (2014). Improving bioscience research reporting: The arrive guidelines for reporting animal research. *Animals (Basel)*, 4(1), 35–44. DOI: 10.3390/ani4010035

Kruger, K. A., & Serpell, J. A. (2010). Animal-assisted interventions in mental health: Definitions and theoretical foundations. In *Handbook on animal-assisted therapy* (pp. 33–48). Academic Press. DOI: 10.1016/B978-0-12-381453-1.10003-0

Laber, K., & Smith, A. (2014). Submission and Maintenance of IACUC Protocols. *The IACUC Handbook*, 111.

Marinelli, L., Normando, S., Siliprandi, C., Salvadoretti, M., & Mongillo, P. (2009). Dog assisted interventions in a specialized centre and potential concerns for animal welfare. *Veterinary Research Communications*, 33(S1), 93–95. DOI: 10.1007/s11259-009-9256-x PMID: 19578960

Mohan, S., & Huneke, R. (2019). The role of IACUCs in responsible animal research. *ILAR Journal*, 60(1), 43–49. DOI: 10.1093/ilar/ilz016 PMID: 31711237

Muckle, J., & Lasikiewicz, N. (2017). An exploration of the benefits of animal-assisted activities in undergraduate students in Singapore. *Asian Journal of Social Psychology*, 20(2), 75–84. DOI: 10.1111/ajsp.12166

Narvekar, H. N., & Ghani, S. (2022). Animal-Assisted Therapy for the Promotion of Social Competence: a Conceptual Framework. *Human Arenas*, 1-10.

National Institutes of Health (US). Office for Protection from Research Risks. (1986). Public Health Service policy on humane care and use of laboratory animals. Office for Protection from Research Risks (OPRR), National Institutes of Health.

Newcomer, C. E., & Greer, W. (2000). General composition of the IACUC and specific roles of the IACUC members. *The IACUC handbook.*

Ng, Z. (2021). Strategies to assessing and enhancing animal welfare in animal-assisted interventions. In *The welfare of animals in animal-assisted interventions: Foundations and best practice methods* (pp. 123–154). Springer International Publishing. DOI: 10.1007/978-3-030-69587-3_6

Ng, Z., Morse, L., Albright, J., Viera, A., & Souza, M. (2019). Describing the use of animals in animal-assisted intervention research. *Journal of Applied Animal Welfare Science*, 22(4), 364–376. DOI: 10.1080/10888705.2018.1524765 PMID: 30295540

Palley, L. S., O'Rourke, P. P., & Niemi, S. M. (2010). Mainstreaming animal assisted therapy. *ILAR Journal*, 51(3), 199–207. DOI: 10.1093/ilar.51.3.199 PMID: 21131720

Pinillos, R. G., Appleby, M. C., Manteca, X., Scott-Park, F., Smith, C., & Velarde, A. (2016). One Welfare-a platform for improving human and animal welfare. *The Veterinary Record*, 179(16), 412–413. DOI: 10.1136/vr.i5470 PMID: 27770094

Prentice, E. D., Oki, G. S., & Mann, M. D. (2014). General concepts of protocol review. *The IACUC Handbook*, 107-147. CRC Press.

Pritt, S. L., & Hammer, R. E. (2017). The interplay of ethics, animal welfare, and IACUC oversight on the reproducibility of animal studies. *Comparative Medicine*, 67(2), 101–105. PMID: 28381309

Santaniello, A., Dicé, F., Claudia Carratú, R., Amato, A., Fioretti, A., & Menna, L. F. (2020). Methodological and terminological issues in animal-assisted interventions: An umbrella review of systematic reviews. *Animals*, 10(5), 759. DOI: 10.3390/ani10050759 PMID: 32349351

Silverman, J. (2015). The institutional animal care and use committee. In *Research Regulatory Compliance* (pp. 41–78). Academic Press. DOI: 10.1016/B978-0-12-420058-6.00003-4

Silverman, J., Suckow, M. A., & Murthy, S. (2014). *The IACUC handbook.* CRC Press. DOI: 10.1201/b17109

Smith, A. J., Clutton, R. E., Lilley, E., Hansen, K. E. A., & Brattelid, T. (2018). PREPARE: Guidelines for planning animal research and testing. *Laboratory Animals*, 52(2), 135–141. DOI: 10.1177/0023677217724823 PMID: 28771074

Stern, C., & Chur-Hansen, A. (2013). Methodological considerations in designing and evaluating animal-assisted interventions. *Animals (Basel)*, 3(1), 127–141. DOI: 10.3390/ani3010127 PMID: 26487313

Valiyamattam, G., Yamamoto, M., Fanucchi, L., & Wang, F. (2018). Multicultural considerations in animal-assisted intervention. *Human Animal Interactions Bulletin*, hai.2018.0019. DOI: 10.1079/hai.2018.0019

Warsaw. (2020). Personality Assessment For Dogs In AAI (PADA).

Wong, WC, P., Yu WM, R., Chau KY, K., Lee CY, J., & NGAI Tsz-kin, J. (2021). A responding to commentary: "Recognizing the role of animal-assisted therapies in addressing mental health needs during the COVID-19 pandemic". *Asian Journal of Psychiatry*. PMID: 33989958

ADDITIONAL READINGS

ARENA. (2002). Institutional Animal Care and Use Committee Guidebook. https://grants.nih.gov/grants/olaw/guidebook.pdf

AVMA. (2007). *Wellness Guidelines for Animals in Animal-Assisted Activity, Animal-Assisted Therapy and Resident Animal Programs.* www.avma.org/issues/policy/animal_assisted_activity.asp

Fine, A. H. (Ed.). (2010). *Handbook on animal-assisted therapy: Theoretical foundations and guidelines for practice*. Academic Press.

Greene, M. E., Pitts, M. E., & James, M. L. (2007). Training strategies for institutional animal care and use committee (IACUC) members and the institutional official (IO). *ILAR Journal*, 48(2), 131–142. DOI: 10.1093/ilar.48.2.131 PMID: 17420534

Ng, Z. (2021). *The welfare of animals in animal-assisted interventions: Foundations and best practice methods*. Springer International Publishing.

KEY TERMS AND DEFINITIONS

Animal-Assisted Activities: Human-animal interactions that focus on recreational and motivational goals.

Animal-Assisted Coaching/Counselling: Planned interactions between human and animal that focuses on the personal growth of the clients.

Animal-Assisted Education: Structured human-animal activities to improve educational and academic needs.

Animal-Assisted Interventions: Structured interactions between humans and animals that are used to benefit humans.

Animal-Assisted Therapy: Structured and goal-directed human-animal interactions with the help of trained animals to achieve therapeutic outcomes.

Canine-Assisted Therapy: Animal-assisted Therapy conducted with the help of canine or a dog.

Complementary and Alternative Medicine: Treatment practices that do not fall under mainstream medicinal practices or evidence-based treatments.

Co-Therapist: An additional therapist that helps the main therapist to facilitate the therapeutic process.

Dolphin-Assisted Activities: Animal-assisted activities carried out with the help of dolphins.

Equine-Assisted Intervention: Animal-assisted Intervention carried out with the help of equine or horses.

Handler: A person who trains or has the responsibility of the animal.

Human-Animal Interactions: Different forms of relationships and interactions between human and animal

Humane Treatment: treating others with respect, compassion, and sympathy in a way that do not incur harm.

Laboratory Animals: Animals that are used in the laboratory for research purposes in the field of animal science and medicine.

Principle Investigator: A researcher who is the primary responsible person for the investigation or conduct of the research.

Research Ethics: It is a set of principles and guidelines that help investigators conduct their studies with respect, integrity, and adherence to human rights.

Subjects: Living beings participating in research about whom the research investigator is studying a concept.

Working Animals: Domesticated animals that are used by humans for meeting different needs in society such as farming, transportation or rehabilitation.

Chapter 8
Empowering Research Through Strategic Administration:
A Historical and Forward-Looking Analysis

Goran Trajkovski
https://orcid.org/0000-0002-3745-3009
Western Governors University, USA

ABSTRACT

This chapter traces the remarkable evolution of research administration from its modest origins to its current strategic role in advancing scientific research. It provides a comprehensive overview of how the field has adapted to the changing needs of the scientific community and the increasing complexity of the research enterprise. The impact of key policy developments, such as the Bayh-Dole Act and the Belmont Report, on the professionalization of research administration is discussed, along with an examination of the current landscape, including the benefits and challenges of technological integration, the rise of specialized roles, and the increasing emphasis on compliance and accountability. The potential influence of emerging technologies, such as artificial intelligence, automation, and advanced analytics, on research administration is also investigated. Ultimately, the chapter underscores the vital role of research administrators as strategic partners in the advancement of scientific knowledge and societal progress.

DOI: 10.4018/979-8-3693-3848-3.ch008

INTRODUCTION: THE VITAL ROLE OF RESEARCH ADMINISTRATION

Research administration is the backbone of scientific advancement, providing the essential support and guidance that enables researchers to focus on their groundbreaking work. It encompasses a wide range of activities, from submitting funding proposals and managing budgets to ensuring regulatory compliance, protecting intellectual property, and upholding research ethics (Vallejo et al., 2020). As research projects continue to grow in scope and complexity, the role of skilled research administrators has become increasingly crucial in navigating the intricate landscape of grants, contracts, and institutional policies.

In today's competitive research environment, administrators play a vital role in assisting institutions to secure and retain funding while helping researchers navigate the complexities of the research process (Kayla et al., 2023). They serve as the bridge between the scientific community and the administrative world, ensuring that projects adhere to legal, ethical, and institutional frameworks while facilitating the efficient operation of research activities. Without their expertise and dedication, many groundbreaking discoveries and innovations would not be possible.

However, despite their critical contributions, the strategic significance and multifaceted responsibilities of research administrators often remain little known beyond the walls of academia. This lack of visibility can lead to a limited understanding of the profession's true value and the challenges administrators face in an ever-evolving research landscape.

EXPLORING THE EVOLUTION AND FUTURE OF RESEARCH ADMINISTRATION

This chapter aims to shed light on the dynamic evolution of research administration, tracing its journey from modest beginnings to its current role as an indispensable partner in scientific progress. By examining the historical context, current practices, and future trends, we seek to provide a comprehensive understanding of how research administration has adapted to the changing needs of the scientific community and the increasing complexity of the research enterprise.

Through an in-depth analysis of key policy milestones, technological advancements, and the shifting roles of research administrators, we will explore how the profession has transformed over the past 70 years. From the early days of informal support for individual scientists to the highly specialized and technology-driven practices of today, research administration has undergone a remarkable metamorphosis.

As we delve into the current landscape of research administration, we will examine the benefits and challenges of technological integration, the rise of specialized roles, and the increasing emphasis on compliance and accountability (Trajkovski, 2023). We will also explore how research administrators are adapting to global trends such as sustainability, open science, and data privacy regulations, highlighting the need for continuous learning and adaptability in the face of change (Shykhnenko & Nozhovnik, 2023).

While this chapter primarily focuses on the research administration landscape in the United States, it also draws on global trends and practices to provide a more comprehensive understanding of the field. By incorporating international perspectives, we aim to highlight the commonalities and differences in research administration practices across countries and foster a more inclusive understanding of the discipline.

Looking to the future, we will investigate the potential impact of emerging technologies such as artificial intelligence, automation, and advanced analytics on research administration. While these innovations hold great promise for streamlining processes and enhancing decision-making, they also raise important questions about the changing role of research administrators and the skills they will need to succeed in an increasingly data-driven and automated world (Kayla et al., 2023).

Ultimately, this chapter aims to underscore the vital role of research administrators as strategic partners in the advancement of scientific knowledge. By providing a comprehensive overview of the profession's past, present, and future, we hope to foster a greater appreciation for the critical work of research administrators and inspire continued growth and innovation in the field.

The Past: Modest Origins and Post-War Growth

Research administration has its roots in the early 20th century when universities and research institutions began to receive external funding for scientific projects. In these early days, the management of research grants and contracts was often handled informally by individual researchers or department heads, with little centralized oversight or support. This decentralized approach allowed for flexibility and autonomy but also led to inconsistencies in practices and a lack of accountability, particularly in areas such as financial management and research ethics.

The landscape of research administration began to change dramatically in the post-World War II era, as the United States government significantly increased its investment in scientific research and development. The influx of federal funding, driven by the desire to maintain technological superiority and address pressing societal challenges, transformed the scale and complexity of research projects. Universities and research institutions found themselves struggling to keep pace with the growing administrative demands of these large-scale, government-funded initiatives (Link & Wagner, 2021).

In response to this rapid expansion, institutions began to establish more formalized research administration structures and processes. The 1950s and 1960s saw the emergence of dedicated research administration offices and the professionalization of the field, as institutions recognized the need for specialized expertise in managing grants, contracts, and compliance issues. This period also marked the beginning of a shift towards greater centralization and standardization of research administration practices, as institutions sought to ensure consistency and accountability across their growing research portfolios.

The post-war period also saw significant developments in research administration in other countries. For example, in the United Kingdom, the Research Councils were established to support and coordinate research across various disciplines. In Europe, research management practices began to emerge as universities and research institutions sought to navigate the increasing complexity of research funding and regulation.

Key Policy Milestones Propel Professionalization

The evolution of research administration during this period was shaped by several key policy milestones that would have far-reaching impacts on the field. One of the most significant was the passage of the Bayh-Dole Act in 1980, which fundamentally reshaped the relationship between academic research and industry. The Act allowed universities and other non-profit institutions to retain ownership of intellectual property rights for federally funded research, paving the way for increased technology transfer and commercialization activities (Link & Wagner, 2021). This development added a new dimension to the responsibilities of research administrators, who now had to navigate the complex terrain of patents, licenses, and industry partnerships.

Another pivotal moment came with the publication of the Belmont Report in 1979, which established the ethical principles and guidelines for the protection of human subjects in research. The report's emphasis on respect for persons, beneficence, and justice laid the groundwork for the development of Institutional Review Boards (IRBs) and the strengthening of research ethics oversight (Office for Human Research Protections, 2022; Znamirowski, 2021). Research administrators played a critical role in implementing these new ethical frameworks and ensuring compliance with evolving regulations and standards (Ainembabazi et al., 2023).

The passage of the Bayh-Dole Act had a profound impact on the research landscape, unleashing a wave of academic entrepreneurship and university-industry collaboration. By granting universities and other non-profit institutions ownership of intellectual property rights for federally funded research, the Act created powerful incentives for the commercialization of research discoveries. This shift towards a more entrepreneurial model of academic research was reflected in the dramatic increase in university patenting and licensing activities in the years following the Act's passage.

While the Bayh-Dole Act focused on the commercialization of research, another significant development in the late 1970s aimed to protect the rights and welfare of human subjects involved in research. The Belmont Report, published in 1979, established the ethical principles and guidelines for the protection of human subjects in research. The report's emphasis on respect for persons, beneficence, and justice laid the groundwork for the development of Institutional Review Boards (IRBs) and the strengthening of research ethics oversight (Office for Human Research Protections, 2022).

The Belmont Report, while less frequently discussed in the context of research administration, had an equally profound impact on the field. The report's articulation of the ethical principles and guidelines for human subjects research helped to establish a new framework for research oversight and accountability. The creation of Institutional Review Boards (IRBs) and the strengthening of informed consent procedures placed new responsibilities on research administrators to ensure that their institutions were complying with these evolving ethical standards.

Despite these significant policy developments and the growing professionalization of research administration, the field still faced challenges in keeping pace with the rapid advancement of scientific research and the increasing complexity of the funding landscape. Many institutions struggled with inconsistent practices, inadequate staffing, and a lack of standardized processes for managing grants and contracts (White, 2020). These challenges highlighted the need for further professionalization and the development of best practices in research administration.

Addressing Shortfalls Leads to Centralization

In response to these challenges, many institutions began to centralize their research administration functions in the 1980s and 1990s, moving away from the decentralized, department-level approach of earlier years. The goal of centralization was to improve efficiency, consistency, and compliance by establishing a single, institutional-level office responsible for managing all aspects of sponsored research. This shift towards centralization was accompanied by a growing emphasis on professional development and the creation of specialized roles within research administration, such as pre-award and post-award management, compliance, and technology transfer.

Centralization offered numerous benefits for institutions, including improved coordination and communication across the research enterprise, greater consistency in policies and procedures, and enhanced oversight of compliance and financial management issues. By consolidating research administration functions under a single, centralized office, institutions were better able to navigate the complex regulatory landscape and ensure that their research activities were being conducted in accordance with federal, state, and institutional requirements (Shykhnenko & Nozhovnik, 2023; Galvão, & de Carvalho, 2020).

Centralization also brought its own set of challenges, particularly in terms of maintaining flexibility and responsiveness to the needs of individual researchers and departments. Some critics argued that centralized research administration could lead to a "one-size-fits-all" approach that failed to account for the unique needs and circumstances of different research projects and disciplines. Others raised concerns about the potential for centralized offices to become overly bureaucratic and focused on compliance at the expense of facilitating research and innovation.

Despite these challenges, the trend towards centralization continued throughout the 1990s and early 2000s, as institutions sought to balance the competing demands of research support and regulatory compliance. The establishment of centralized research administration offices represented a significant milestone in the professionalization of the field, as institutions recognized the need for specialized expertise and dedicated resources to manage the increasingly complex world of sponsored research.

As research administration has continued to evolve in the 21st century, the field has been shaped by a growing emphasis on technology, data-driven decision-making, and collaboration across institutional and disciplinary boundaries. While the core functions of research administration remain focused on facilitating research and ensuring compliance, the tools and strategies used to achieve these goals have become increasingly sophisticated and data driven.

The Present: Specialization and Technological Integration

In the current research administration landscape, the field has evolved to encompass a wide range of specialized roles and functions, each designed to support specific aspects of the research enterprise. This specialization reflects the increasing complexity of the research funding and regulatory environment, as well as the growing recognition of the strategic importance of research administration in advancing institutional goals and priorities.

The increasing specialization of research administration roles is a global trend, with many countries recognizing the need for dedicated professionals to support various aspects of the research enterprise. For instance, in Australia, universities have established specialized units to manage research ethics, grants, and commercialization activities (White, 2020). International initiatives, such as the development of comprehensive frameworks for research management and administration, have also emerged to promote standardization and best practices in research administration worldwide.

Pre-Award and Post-Award Management

One of the most significant developments in modern research administration has been the emergence of pre-award and post-award management as distinct areas of specialization. Pre-award administrators are responsible for supporting researchers in the development and submission of grant proposals, ensuring that all application requirements are met, and budgets are accurately prepared. They also play a critical role in identifying funding opportunities and providing guidance on sponsor guidelines and regulations.

Post-award administrators, on the other hand, are responsible for managing the financial and administrative aspects of sponsored projects once funding has been secured. This includes tasks such as setting up grant accounts, monitoring expenditures, preparing financial reports, and ensuring compliance with sponsor requirements and institutional policies. Post-award administrators also play a key role in managing subcontracts, overseeing project closeout, and facilitating audits and other sponsor reviews (Crane et al., 2023; Bloch, 2020).

Specialized Roles in Research Administration

Research administration has also seen the development of specialized roles in areas such as contracts and grants management, research compliance, technology transfer, and research development. Contracts and grants managers are responsible for negotiating and administering complex research agreements, while compliance officers ensure that research activities are conducted in accordance with ethical, legal, and regulatory standards. Technology transfer professionals work to facilitate the commercialization of research discoveries and innovations, and research development officers help to identify and pursue new funding opportunities and strategic partnerships.

The increasing specialization of research administration roles reflects the growing recognition of the diverse skills and expertise required to support the modern research enterprise effectively. By allowing professionals to focus on specific aspects of research management, institutions can ensure that each function is performed with the highest level of expertise and efficiency (Shykhnenko & Nozhovnik, 2023).

Technology Integration and Its Impact

Another defining feature of the current research administration landscape is the growing integration of technology into all aspects of the research management process. From electronic grant submission and award management systems to financial management and compliance tracking tools, technology has become an essential component of modern research administration.

The adoption of electronic research administration (eRA) systems has been a particularly significant development in recent years. These systems provide a centralized platform for managing the entire grant lifecycle, from proposal development and submission to award management and closeout. By automating many of the routine tasks associated with research administration, eRA systems have the potential to greatly enhance efficiency, reduce errors, and improve communication and collaboration among research team members.

Benefits and Challenges of Technology Integration

Research administrators are also increasingly leveraging data analytics and business intelligence tools to inform decision-making and support strategic planning. By analyzing data on research funding, publications, and other metrics, administrators can gain valuable insights into institutional strengths and weaknesses, identify emerging trends and opportunities, and make data-driven decisions about resource allocation and strategic investments. Figure 1 illustrates the various technologies involved and their interconnections within the context of modern research administration.

While the integration of technology has brought many benefits, it has also presented new challenges and complexities. Implementing and maintaining complex eRA systems can be costly and time-consuming, requiring significant investments in hardware, software, and personnel. There is also a risk that an overreliance on technology can lead to a loss of flexibility and adaptability, as well as a disconnect between the systems used by administrators and the needs of researchers (Giuffrida & Hall, 2023; Atabek, 2020).

Figure 1. Technological ecosystem in research administration

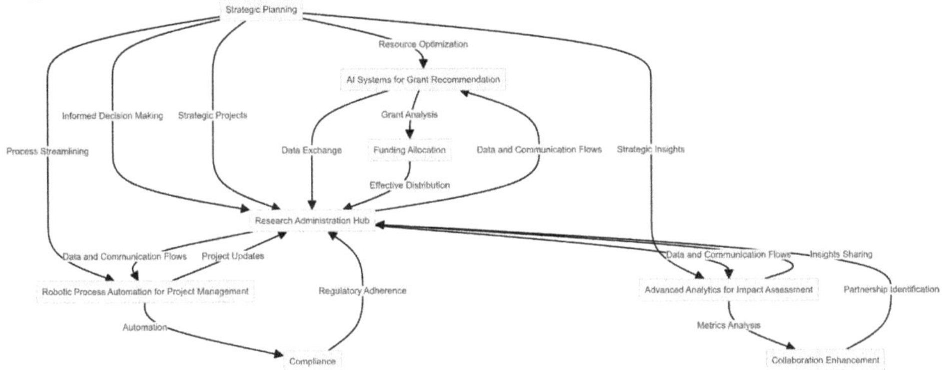

This diagram illustrates the complex interplay of various technological systems and processes within research administration. Centralized around the 'Research Administration Hub,' it highlights how data flows from multiple sources—such as AI systems for grant recommendation, project updates, and compliance data—into this hub, which facilitates strategic planning, process streamlining, and informed decision-making. The diagram shows the interconnected nature of resourcc optimization, data exchange, and advanced analytics, emphasizing their roles in enhancing efficiency, compliance, and collaboration. This ecosystem supports the comprehensive management of research projects by integrating technology to drive strategic insights and optimize resource allocation, ultimately contributing to the successful execution and oversight of research activities

Another challenge is the need for ongoing training and support to ensure that research administrators have the skills and knowledge needed to use these tools effectively. As the pace of technological change continues to accelerate, it is becoming increasingly important for research administrators to engage in continuous learning and professional development to stay current with the latest tools and best practices in the field (Shykhnenko & Nozhovnik, 2023).

Navigating the Current Landscape

The current research administration landscape is characterized by a complex interplay of specialization, technology integration, and an ever-evolving regulatory and funding environment. To navigate this landscape effectively, research administrators must be skilled at balancing the competing demands of compliance and flexibility, while also staying attuned to the needs and priorities of researchers and institutional stakeholders. A key challenge facing research administrators today is the need to keep pace with the rapid pace of change in the research enterprise. From new funding mechanisms and regulatory requirements to emerging technologies and scientific advances, this landscape is constantly evolving, requiring administrators to be agile and adaptable in their approach to research management (Shykhnenko & Nozhovnik, 2023).

Many institutions are investing in the professional development and training of their research administration staff, as well as in the development of more collaborative and cross-functional teams. By fostering a culture of continuous learning and improvement, institutions can ensure that their research administration practices remain aligned with best practices and responsive to the changing needs of the research community.

Another trend in the current research administration landscape is the growing emphasis on research impact and accountability. As funding agencies and other stakeholders increasingly demand evidence of the social, economic, and scientific impact of research investments, research administrators are being called upon to develop more robust systems for tracking and reporting on research outcomes and impacts.

This emphasis on impact and accountability is driving a shift towards more data-driven and evidence-based approaches to research management, as well as a greater focus on stakeholder engagement and communication. By working closely with researchers, funders, and other stakeholders to demonstrate the value and impact of research, administrators can help to build support for continued investment in scientific discovery and innovation (Giuffrida & Hall, 2023).

The Technological Future: AI, Automation, and Analytics

As we look to the future of research administration, emerging technologies such as artificial intelligence (AI), automation, and advanced analytics will play an increasingly important role in shaping the field. These technologies hold the potential to revolutionize the way research is managed and supported, offering new opportunities for efficiency, insight, and innovation (Trajkovski, 2024).

AI and Automation: Streamlining Processes and Enhancing Decision-Making

One of the most significant ways in which AI and automation are likely to impact research administration is through the streamlining of routine tasks and processes. By automating activities such as data entry, document management, and basic compliance checks, AI-powered tools can help to reduce administrative burdens and free up research administrators to focus on more strategic and value-added activities.

Examples of AI and automation in research administration may include a university research office implementing an AI-driven grant recommendation system that analyzes previous grant application data and current research trends to recommend potential funding opportunities to researchers. A federal research funding agency employing advanced analytics and data visualization tools to assess and visualize the impact of funded projects across various scientific fields. A research institution deploying Robotic Process Automation (RPA) for streamlining the administrative aspects of research project management, such as budget tracking, compliance monitoring, and report generation (Bhima et al., 2023).

Figure 2. Data collection processes in research administration.

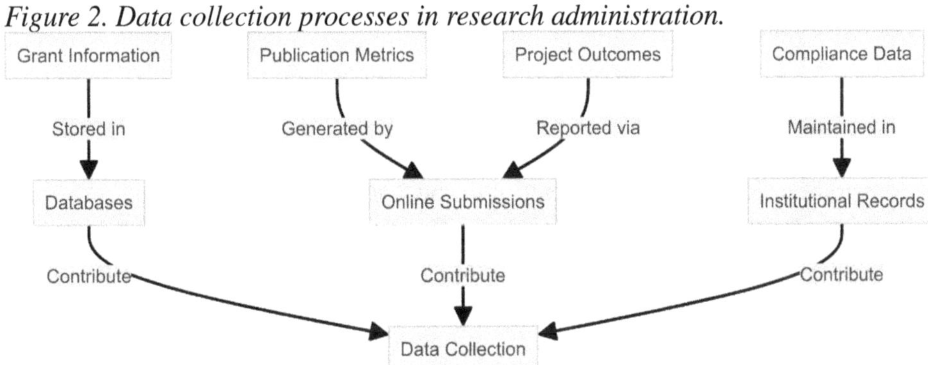

This diagram illustrates the various pathways through which data is collected in research administration. The chart showcases how grant information, publication metrics, project outcomes, and compliance data are gathered from multiple sources, including databases, online submissions, and institutional records. These data sources contribute to a centralized data collection process that is crucial for comprehensive research management. By integrating information from different domains, this system supports efficient monitoring, evaluation, and reporting of research activities, thereby enhancing the overall effectiveness of research administration

Figure 3. Data processing and analysis in research administration.

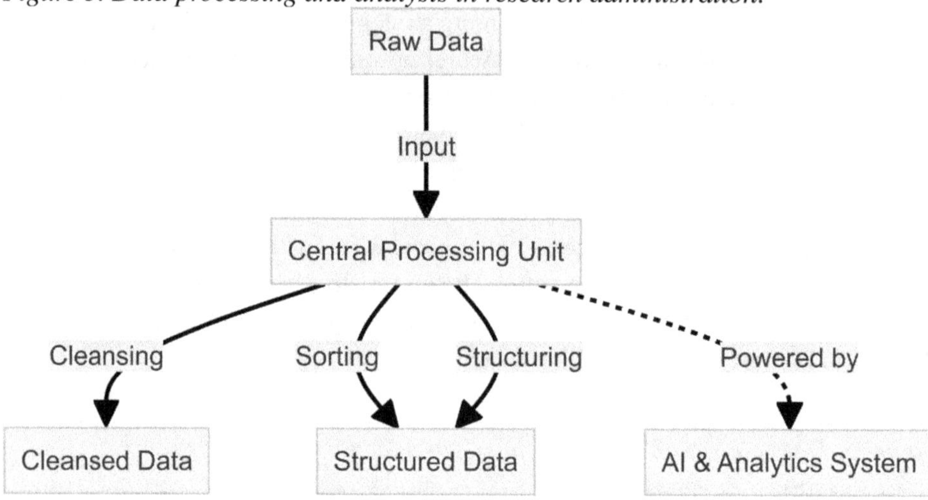

This diagram depicts the workflow of data processing and analysis in research administration. Raw data is funneled into the 'Central Processing Unit,' where it undergoes essential transformation steps. The data is cleansed, removing errors and inconsistencies, resulting in 'Cleansed Data.' It is then sorted and structured, creating 'Structured Data' that is ready for analysis. Powered by an 'AI & Analytics System,' this processed data is used for in-depth analysis and to generate valuable insights. The flowchart highlights the critical role of data processing in converting raw data into meaningful, actionable information that supports research management, compliance, and decision-making

Figure 4. Analytics and insight generation in research administration.

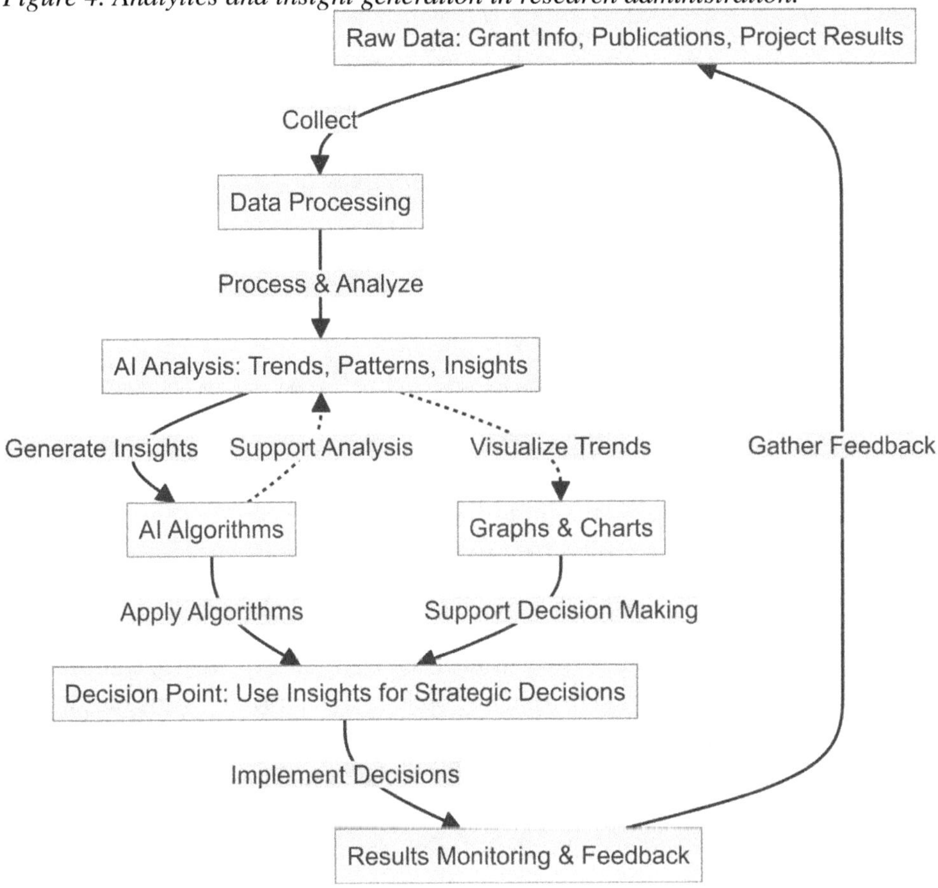

This diagram illustrates the comprehensive process of generating insights from raw data, encompassing grant information, publications, and project results. The workflow begins with data collection, followed by processing and analysis. AI algorithms are applied to detect trends, patterns, and insights, which are then visualized using graphs and charts to support decision-making. Insights generated are used at the decision point to guide strategic decisions, which are implemented and subsequently monitored for results. Feedback gathered from this process loops back to inform and refine data collection, creating a continuous cycle of improvement and strategic enhancement in research administration

Figure 5. Decision-making framework in research administration.

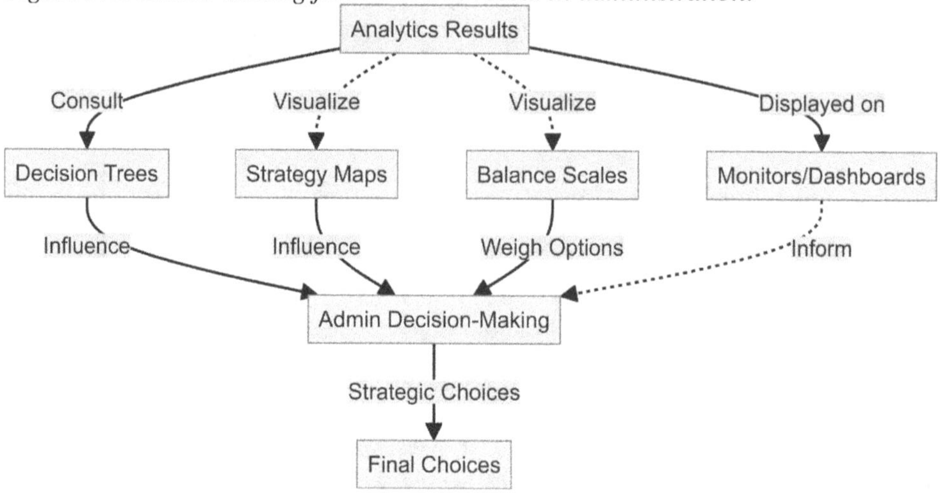

This diagram visualizes the integration of analytics results into administrative decision-making processes. It illustrates how analytics results are consulted and visualized through various tools such as decision trees, strategy maps, and balance scales. These visualizations aid in weighing options and informing decisions, which are displayed on monitors or dashboards for real-time analysis. The insights gathered from these tools influence administrative decision-making, leading to strategic choices that culminate in final decisions. This framework highlights the role of visual tools in simplifying complex data to enhance decision-making processes in research administration, ensuring that decisions are informed by comprehensive and accurate data analysis

Figure 6. Implementation and continuous improvement cycle in research administration.

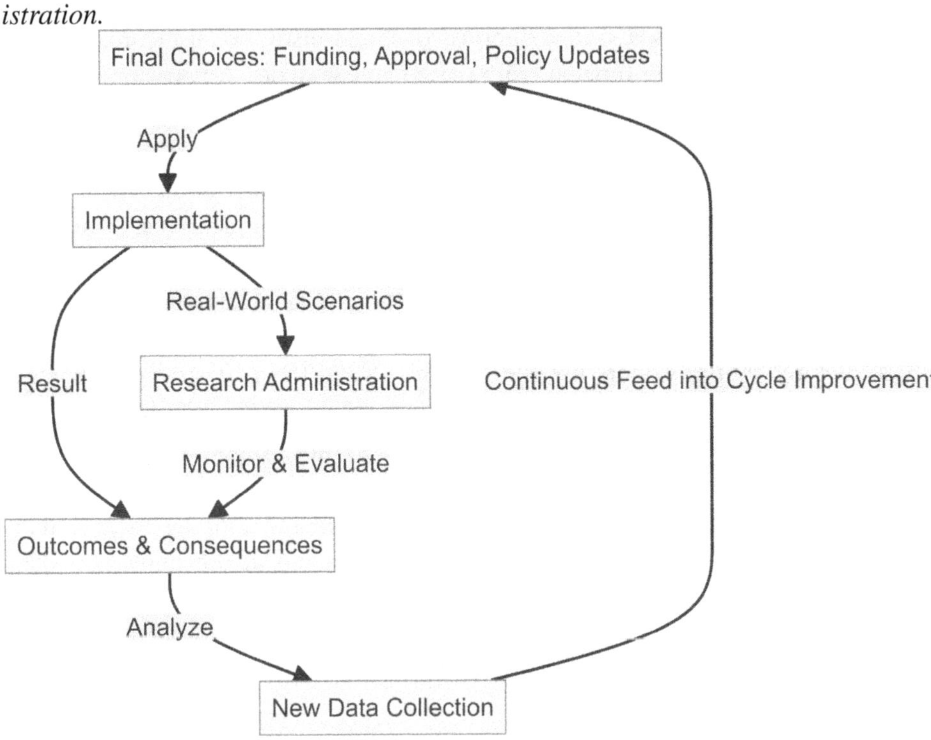

This flowchart depicts the cyclical process of implementing final decisions, such as funding allocations, project approvals, and policy updates, and how these decisions are applied in real-world scenarios. It illustrates the steps from the initial implementation to monitoring and evaluating outcomes and consequences. These results feed into new data collection, which is analyzed to inform further improvements, creating a continuous loop of feedback and refinement. This cycle emphasizes the importance of ongoing assessment and adaptation in research administration, ensuring that practices and policies remain effective and responsive to evolving needs and challenges

In addition to streamlining routine tasks, AI and machine learning algorithms could also be used to enhance decision-making and support more proactive research management. By analyzing large volumes of data on research funding, publications, and other metrics, AI tools could help administrators to identify patterns and trends, predict outcomes, and make more informed decisions about resource allocation and strategic investments.

Figure 2 shows different streams of data entering the process, such as grant information, publication metrics, project outcomes, and compliance data from different data sources like databases, online submissions, and institutional records. Figure 3 illustrates the data being funneled into the AI and analytics system, highlighting the AI's role in cleansing, sorting, and structuring the data. Figure 4 summarizes the data-driven decision-making process and outlines how advanced analytics and AI influence the decision-making process in research administration. The process starts with the collection of raw data (grant information, publications, project results), which is processed and analyzed by AI systems to identify trends, patterns, and insights. Research administrators use these insights to make informed strategic decisions. Figure 5 leads the flow from analytics to the decision-making phase, where research administrators use the insights gained to make strategic choices. Administrators at this stage are consulting the analytics results highlighting the direct impact of AI-generated insights on their decisions. Figure 6 illustrates the implementation of decisions into real-world research administration scenarios, such as funding allocation, project approval, or policy updates, looping back to data collection, indicating that the outcomes of these decisions also become new data points, feeding into a continuous cycle of improvement and refinement.

Challenges and Implications of Emerging Technologies

While these examples demonstrate the potential benefits of emerging technologies in research administration, it is important to recognize that the implementation of these tools also presents challenges and practical implications. One key challenge is the need for significant investment in infrastructure, hardware, and software, as well as the development of new skills and capabilities among research administration staff. Institutions must carefully consider the costs and benefits of implementing these technologies and ensure that they are aligned with overall strategic goals and priorities.

Another practical implication is the need for effective change management and communication strategies to support the adoption of new technologies and processes. Research administrators may face resistance or skepticism from researchers or other stakeholders who are accustomed to traditional ways of working. It is essential to engage these stakeholders early in the process, provide clear communication and training, and demonstrate the value and benefits of the new technologies.

The implementation of emerging technologies in research administration raises important questions about data privacy, security, and ethics. Institutions must ensure that they have robust policies and procedures in place to protect sensitive data, comply with relevant regulations and standards, and ensure that the use of these technologies is aligned with institutional values and principles.

By exploring these practical implications and challenges in more depth, research administrators can gain a more comprehensive understanding of what is involved in successfully implementing emerging technologies in their work. They can also learn from the experiences of other institutions and organizations that have already embarked on this journey, and identify best practices and lessons learned to inform their own technology adoption strategies.

ADVANCED ANALYTICS: DRIVING INSIGHT AND IMPACT

In addition to AI and automation, advanced analytics and data visualization tools are also likely to play an increasingly important role in the future of research administration. By leveraging large volumes of structured and unstructured data from a wide range of sources, these tools can help administrators to gain new insights into the research enterprise and to communicate the impact of research in more compelling and accessible ways.

Figure 7. Collaboration and communication in research administration

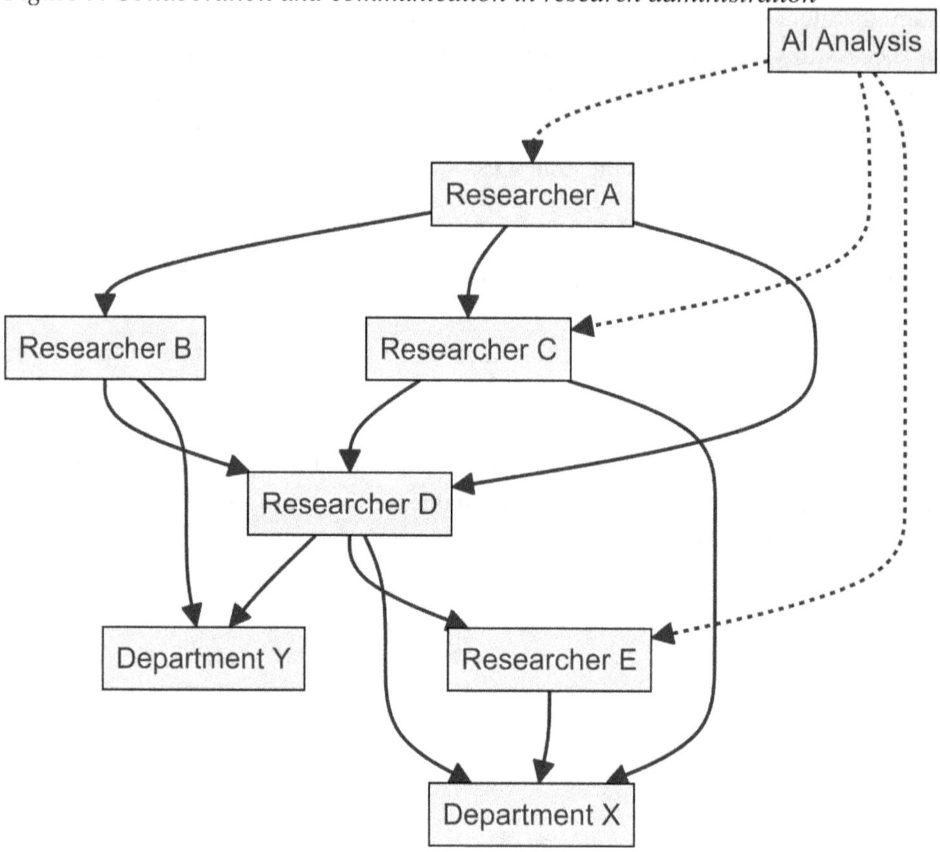

This diagram illustrates the network of interactions among researchers and departments within a research institution. It shows how AI analysis enhances collaboration by linking various researchers (A through E) and departments (X and Y). The arrows represent the flow of information and collaboration, highlighting direct and indirect connections facilitated by AI. The solid lines depict direct communication paths, while the dashed lines represent the influence of AI in fostering data-driven insights that inform research and departmental strategies. This network model emphasizes the importance of integrated communication and collaboration, supported by AI, in enhancing research administration's effectiveness and fostering interdisciplinary teamwork

Network analysis tools, for example, could be used to map the collaborative relationships among researchers within and across institutions, identifying key players and potential opportunities for new partnerships and collaborations. Figure 7 illustrates how network analysis tools can map the collaborative relationships among researchers within and across institutions, highlighting key players, potential areas for new partnerships, and how AI tools can uncover these insights (Trajkovski, 2023). Bibliometric analysis tools could be used to track the citation impact of research outputs and to identify emerging areas of scientific excellence and opportunity. Data visualization tools, such as dashboards and infographics, could also be used to communicate the results and impact of research to a wide range of stakeholders, from funders and policymakers to the general public. By presenting complex data in more engaging and accessible formats, these tools can help to build support for research and to demonstrate the value and relevance of scientific inquiry to society as a whole.

As with AI and automation, the effective use of advanced analytics in research administration will require a combination of technical skills and domain expertise. Research administrators will need to be able to work closely with data scientists and other experts to identify relevant data sources, develop appropriate analytical models, and interpret and communicate the results in meaningful ways. They will also need to be skilled at translating analytical insights into actionable recommendations and strategies for improving research management and support (Battle & Scheidegger, 2020). This may require the development of new frameworks and methodologies for integrating data-driven insights into decision-making processes and for aligning analytical activities with institutional goals and priorities.

Preparing for the Future: Skills and Strategies

As research administration continues to evolve in the face of AI, automation, and advanced analytics, professionals in the field will need to adapt and develop a diverse set of skills to thrive in this rapidly changing landscape (Figure 8).

Figure 8. Skills and competencies for future readiness in research administration

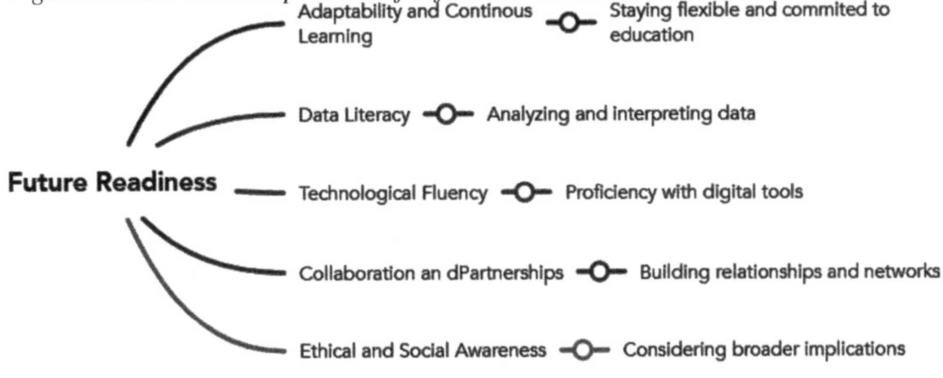

This mind map outlines the essential skills and competencies required for future readiness in research administration. At the center is 'Future Readiness,' which is supported by various interconnected attributes such as 'Data Literacy' for analyzing and interpreting data, 'Technological Fluency' for proficiency with digital tools, and 'Ethical and Social Awareness' for considering broader implications. Other critical skills include 'Collaboration and Partnership' for building relationships and networks, and 'Adaptability and Continuous Learning' for staying flexible and committed to education. This diagram highlights the multifaceted nature of future readiness, emphasizing the need for a diverse set of skills to navigate the evolving landscape of research administration effectively

Data Literacy

Data literacy will become increasingly crucial, as research administrators must be comfortable working with large volumes of data and capable of understanding and interpreting complex analytical results. They will also need to effectively communicate data-driven insights to a wide range of stakeholders, from researchers and funders to institutional leaders and the general public, ensuring that everyone is well-informed and engaged in the decision-making process (Berndsen, 2021; Shykhnenko & Nozhovnik, 2023).

Technological Fluency

Technological fluency will be a key competency for research administrators. As the research enterprise becomes increasingly digitized and data-driven, administrators will need to be comfortable with a wide range of technologies, from eRA systems and data management platforms to AI and machine learning tools. They must be able to assess the potential impact of new technologies on research management and develop strategies for integrating these tools into existing workflows and processes, ensuring a smooth and efficient transition.

Ethical and Social Awareness

The adoption of advanced technologies in research administration also raises important ethical and social considerations. Research administrators will need to have a deep understanding of the ethical and social implications of these tools and be skilled at identifying and mitigating potential biases and risks. They will also be responsible for developing governance frameworks and policies that ensure the responsible and equitable use of these technologies, promoting transparency, accountability, and fairness in the research enterprise (Bezuidenhout et al., 2020).

Collaboration and Partnership

Collaboration and partnership will also be essential skills for research administrators in the future. As research becomes increasingly interdisciplinary and cross-functional, administrators will need to work closely with a wide range of stakeholders both within and beyond their institutions. Building and managing complex partnerships, fostering a culture of collaboration and knowledge-sharing across disciplinary and organizational boundaries, and facilitating effective communication and coordination will be key to success in this evolving landscape.

Adaptability and Continuous Learning

Given the rapid pace of change in the research enterprise, adaptability and a commitment to continuous learning will be critical for research administrators. They will need to be proactive in seeking out new skills and knowledge, staying up to date with the latest trends and best practices in the field, and embracing a mindset of lifelong learning. This may involve attending conferences and workshops, participating in online learning programs, and engaging in mentoring and coaching activities to continually enhance their skills and expertise.

Supporting the Development of Essential Skills and Competencies

To support the development of these essential skills and competencies, institutions will need to invest in the ongoing training and professional development of their research administration staff. This may include providing opportunities for staff to attend relevant events, access online learning resources, and engage in peer-to-peer learning and knowledge-sharing (Shykhnenko & Nozhovnik, 2023). Institutions may also need to rethink their organizational structures and processes to better support collaboration and cross-functional work, creating new roles and teams focused on data management, analytics, and technology integration, and fostering a culture of experimentation and innovation.

Ultimately, the future of research administration will require a delicate balance between the strategic use of advanced technologies and the human skills and judgment that are essential for effective research management and support.

THE TECHNOLOGICAL FUTURE: TRANSFORMING RESEARCH ADMINISTRATION

The advent of artificial intelligence, automation, and advanced analytics is poised to transform the landscape of research administration in profound ways. As these technologies continue to evolve and mature, they offer the potential to streamline processes, enhance decision-making, and drive new insights and innovations in research management and support (Figure 9). However, realizing the full potential of these technologies will require more than just technical implementation. It will require a fundamental shift in the way research administrators approach their work, and in the skills and competencies they bring to bear on the challenges and opportunities of the research enterprise.

Figure 9. Transformative elements and organizational shifts in research administration

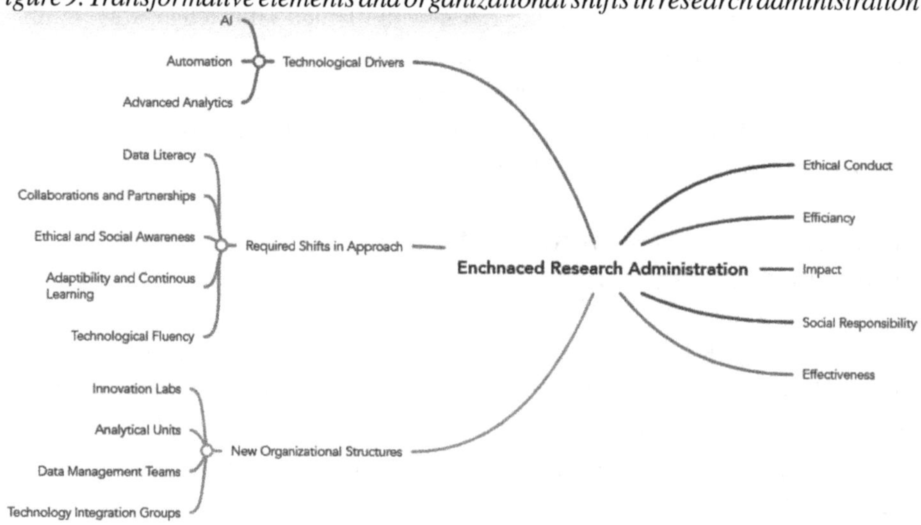

This comprehensive diagram illustrates the interplay of various factors driving transformation in research administration. At the core is 'Enhanced Research Administration,' which is influenced by 'New Organizational Structures' like technology integration groups, data management teams, and innovation labs. Key elements such as 'Ethical Conduct,' 'Social Responsibility,' and 'Technological Drivers' including AI and automation are highlighted as essential for advancing the field. The diagram also underscores the importance of 'Required Shifts in Approach,' emphasizing the need for adaptability, continuous learning, technological fluency, and collaboration. It presents a holistic view of how integrating advanced technologies and ethical practices can lead to impactful and efficient research administration

To thrive in this new technological landscape, research administrators will need to become more data-literate, technologically fluent, and ethically aware. They will need to be skilled at collaborating across disciplinary and organizational boundaries, and at adapting to the rapid pace of change in the research enterprise. At the same time, institutions will need to rethink their organizational structures and processes to better support the integration of advanced technologies into research management and support. This may require the creation of new roles and teams focused on data management, analytics, and technology integration, as well as a culture of experimentation and innovation.

The goal of leveraging advanced technologies in research administration should be to enhance the efficiency, effectiveness, and impact of research, while also ensuring that it is conducted in an ethical and socially responsible manner. By combining the power of these tools with the human expertise and judgment of skilled research administrators, institutions can position themselves for success in a rapidly evolving research landscape and contribute to the advancement of knowledge and the betterment of society.

The Role Shift: From Administrators to Strategists

As the research landscape continues to evolve and become increasingly complex, the role of research administrators is undergoing a significant shift. No longer solely focused on administrative tasks and regulatory compliance, research administrators are now being called upon to serve as strategic partners and advisors, working closely with researchers and institutional leaders to advance the research enterprise. This shift towards a more strategic role is being driven by a number of factors, including the increasing complexity of research funding and regulatory environments, the growing emphasis on interdisciplinary and collaborative research, and the need for institutions to demonstrate the impact and value of their research activities (Bandarian, 2020).

From Compliance Officers to Strategic Advisors

Traditionally, research administrators have been viewed primarily as compliance officers, responsible for ensuring that research activities are conducted in accordance with institutional policies and regulatory requirements. While compliance remains a critical function of research administration, the scope of the role is expanding to include a broader range of strategic activities.

Internationally, research administrators are also taking on more strategic roles and contributing to institutional decision-making. In the United Kingdom, the Association of Research Managers and Administrators (ARMA) has emphasized the importance of research administrators as strategic partners in driving research excellence and impact. Global networks, such as the International Network of Research Management Societies (INORMS) and the European Association of Research Managers and Administrators (EARMA), provide platforms for knowledge sharing and professional development among research administrators worldwide (Shykhnenko & Nozhovnik, 2023).

In this new paradigm, research administrators are being called upon to serve as strategic advisors, working closely with researchers to identify funding opportunities, develop competitive proposals, and build collaborative partnerships. They are also being asked to provide strategic guidance to institutional leaders on issues such as research priorities, resource allocation, and research infrastructure development. To fulfill this strategic role, research administrators must have a deep understanding of the research landscape, including funding trends, emerging areas of scientific opportunity, and the priorities of key stakeholders such as funding agencies, industry partners, and policymakers. They must also be skilled at navigating complex organizational dynamics and building relationships with a wide range of internal and external stakeholders.

Decentralizing Support: Embedding Administrators in Research Units

One approach to facilitating this strategic shift is to decentralize research administration support by embedding administrators directly within research units and departments. This approach allows administrators to work more closely with researchers, gaining a deeper understanding of their needs and priorities, and providing tailored support and guidance throughout the research lifecycle. Embedded research administrators can help to bridge the gap between the administrative and scientific functions of the institution, facilitating communication and collaboration between researchers and central administrative offices. They can also serve as a valuable resource for researchers, providing guidance on funding opportunities, proposal development, and regulatory compliance (Bazhin et al., 2021).

By decentralizing research administration support, institutions can create a more agile and responsive research support infrastructure that is better equipped to meet the evolving needs of researchers and the research enterprise. However, this approach also requires careful planning and coordination to ensure that research administrators have the necessary skills, resources, and support to be effective in their roles.

Developing New Skills and Competencies

As research administrators transition from their traditional roles as compliance officers to strategic advisors, they will need to develop a diverse set of skills and competencies (Figure 10) to effectively navigate this new landscape.

Figure 10. The evolving role of research administrators: key skills and competencies

This mind map delineates the multifaceted role of research administrators and the critical skills required for their evolving responsibilities. At the center is the 'Evolving Role of Research Administrators,' highlighting core competencies such as 'Leadership and Change Management' for providing guidance and vision, and 'Strategic Planning' for aligning actions with institutional goals. Essential skills include 'Financial and Business Acumen,' encompassing financial management capabilities, and 'Effective Communication and Influence,' necessary for disseminating and receiving information. The diagram also emphasizes the importance of 'Collaboration and Partnership' for fostering partnerships and collaborative efforts. This comprehensive view underscores the diverse and strategic role of research administrators in modern research environments

Strategic planning will be at the forefront of these new responsibilities, as research administrators will be expected to develop and implement plans that align with institutional research priorities and goals. This will require a keen ability to conduct environmental scans, analyze data on research performance and impact, and engage with a wide range of stakeholders to build consensus around strategic priorities. By honing these skills, research administrators can help guide their institutions towards a more focused and impactful research agenda (Shykhnenko & Nozhovnik, 2023).

Collaboration and partnership-building will also be crucial in this new strategic role. As research becomes increasingly interdisciplinary and collaborative, administrators will need to be adept at building and managing complex partnerships both within and beyond their institutions. This may involve developing collaborative agreements, managing conflicts and tensions, and fostering a culture of trust and mutual respect among partners. By cultivating strong relationships and facilitating effective communication, research administrators can help break down silos and promote a more integrated and collaborative research environment.

Effective communication and the ability to influence others will be essential skills for research administrators in their strategic roles. They will need to be able to articulate the value and impact of research to a wide range of stakeholders, from researchers and funders to institutional leaders and the broader public. This will require not only strong communication skills but also the ability to persuade and influence others, whether it be to secure funding, build support for research priorities, or navigate complex organizational dynamics. By developing a compelling narrative and engaging stakeholders at all levels, research administrators can help build a shared vision for the future of research at their institutions.

As the research funding landscape becomes increasingly competitive and complex, financial and business acumen will be critical competencies for research administrators. They will need to have a strong understanding of financial management and business principles, enabling them to develop and manage complex budgets, negotiate contracts and agreements, and identify new sources of funding and revenue. By bringing a strategic and entrepreneurial mindset to research administration, administrators can help their institutions stay ahead of the curve and secure the resources needed to support cutting-edge research.

Finally, leadership and changing management skills will be essential for research administrators as they guide their institutions through the evolving research landscape. They will need to be effective leaders and change agents, able to develop and implement strategies that drive innovation and adaptability. This may involve building coalitions of support, fostering a culture of experimentation and risk-taking, and navigating the complex challenges and opportunities that arise in the face of change. By embracing a leadership role and driving positive change, research administrators can help position their institutions for success in an increasingly dynamic and competitive research environment.

Supporting the Development of Critical Skills and Competencies

To support the development of these critical skills and competencies, institutions will need to invest in the ongoing training and professional development of their research administration staff. This may include providing opportunities for formal education and training, such as workshops, seminars, and certification programs, as well as mentoring and coaching initiatives that allow administrators to learn from more experienced colleagues (Shykhnenko & Nozhovnik, 2023). By creating a culture of continuous learning and growth, institutions can ensure that their research administration staff are equipped with the skills and knowledge needed to succeed in their new strategic roles.

In addition to investing in training and development, institutions may also need to rethink their recruitment and hiring strategies to attract individuals with the diverse skills and experiences needed to thrive in these more strategic positions. This may involve looking beyond traditional research administration backgrounds and considering candidates with experience in areas such as business development, strategic planning, and partnership management. By bringing a fresh perspective and a broader range of skills to the table, these individuals can help drive innovation and change within the research administration field.

Embracing the Strategic Shift

Ultimately, the shift towards a more strategic role for research administrators presents a significant opportunity for institutions to enhance the efficiency, effectiveness, and impact of their research activities. By empowering research administrators to serve as strategic partners and advisors, institutions can create a more agile, responsive, and forward-thinking research support infrastructure that is better equipped to meet the evolving needs of researchers and the research enterprise (Shykhnenko & Nozhovnik, 2023). As the research landscape continues to change and grow more complex, the strategic role of research administrators will only become more critical, making it essential for institutions to invest in the skills, competencies, and talent needed to succeed in this new era of research administration.

Embracing this strategic shift will require a significant investment in the development of new skills, competencies, and organizational structures. Institutions will need to provide ongoing training and professional development opportunities for their research administration staff, as well as create new roles and career paths that reflect the growing strategic importance of the function. At the same time, research administrators themselves will need to be proactive in seeking out new opportunities for learning and growth, and in building the relationships and networks necessary to succeed in their new strategic roles. They will need to be comfortable working in a fast-paced and constantly evolving environment and be willing to take on new challenges and responsibilities as the needs of the research enterprise evolve.

The success of this strategic shift will depend on the ability of institutions and research administrators to work together in a spirit of collaboration and partnership. By fostering a culture of innovation, adaptability, and continuous learning, institutions can create a research support infrastructure that is truly responsive to the needs of researchers and the research enterprise.

KEY RECOMMENDATIONS AND CONCLUDING PERSPECTIVES

As we reflect on the remarkable journey of research administration over the past 70 years, it is evident that the field has undergone a profound transformation. From its humble beginnings to its current role as a vital strategic partner in the research enterprise, research administration has continuously evolved to meet the changing needs of the scientific community. Looking ahead, it is clear that the pace of change will only continue to accelerate. To successfully navigate this dynamic landscape and continue to support the advancement of research, institutions and research administrators must embrace a proactive and adaptive approach (Shykhnenko & Nozhovnik, 2023).

First and foremost, it is essential to recognize and embrace the strategic shift that is occurring within research administration. Institutions must empower research administrators to serve as true partners and advisors in the research enterprise, providing them with the necessary skills, competencies, and organizational structures to succeed in this new role. This may require a willingness to rethink traditional roles and responsibilities, and to invest in the ongoing development and growth of research administration professionals (Palad, 2023).

Fostering a culture of collaboration will also be critical to the future success of research administration. As the research landscape becomes increasingly interdisciplinary and complex, close partnerships between research administrators, researchers, and other key stakeholders will be essential. Institutions should actively work to break down silos between administrative and scientific functions, and to create an environment that encourages knowledge-sharing, communication, and collaboration across all levels of the organization.

Technology and data will play an increasingly important role in the future of research administration, and institutions must be prepared to invest in the development of robust data management and analytics capabilities. By integrating these tools into their research support infrastructure, institutions can enhance efficiency, gain valuable insights, and drive greater impact in their research activities. It will be important to approach the use of technology and data in a thoughtful and strategic manner, ensuring that these tools are deployed in a way that aligns with institutional goals and priorities (Borde et al., 2022).

As the use of advanced technologies such as artificial intelligence becomes more prevalent in research administration, it will be crucial to prioritize ethics and social responsibility. Institutions must develop clear governance frameworks and policies around the use of these technologies, and work to ensure that they are being deployed in a manner that promotes transparency, fairness, and accountability. By taking a proactive and responsible approach to the use of technology, institutions can help to build trust and confidence in the research enterprise and ensure that the benefits of these tools are realized in an equitable and sustainable way (Nazarenko, 2023).

Cultivating a culture of continuous learning and professional development will be essential for research administrators to thrive in the rapidly changing research landscape. Institutions should provide ongoing opportunities for training and education and encourage research administrators to actively seek out new knowledge and skills. By fostering a mindset of lifelong learning and growth, institutions can ensure that their research administration staff are equipped with the tools and capabilities needed to succeed in an increasingly complex and dynamic environment (Shykhnenko & Nozhovnik, 2023).

Overcoming Resistance to Change

As research administration continues to evolve towards a more strategic and technology-driven approach, it is crucial to acknowledge and address the potential resistance to change that may arise within the field. The shift from traditional roles and practices to a more proactive, data-informed, and collaborative model may face internal challenges, as some research administrators and stakeholders may be hesitant to embrace new ways of working. To overcome resistance to change, institutions and research administration leaders should consider a multifaceted approach that focuses on communication, culture, stakeholder engagement, support, and recognition (Figure 11). First and foremost, it is crucial to clearly communicate the vision and benefits behind the strategic shift in research administration. By articulating the rationale and potential advantages, such as increased efficiency, better decision-making, and enhanced impact, leaders can help stakeholders understand the necessity and value of the proposed changes (Bodnar & Horishna, 2021).

Figure 11. Managing resistance to change: key strategies and elements

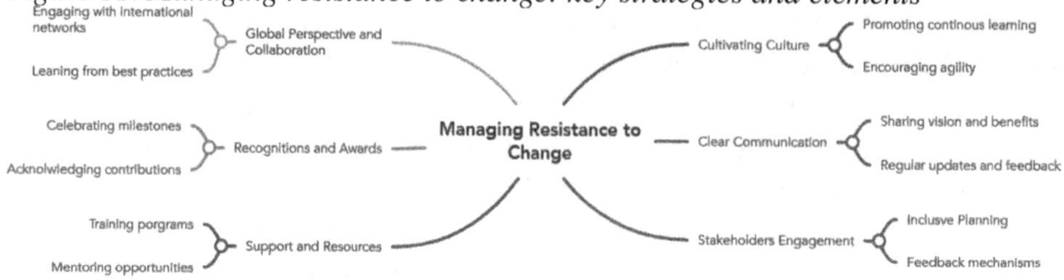

This mind map explores the critical strategies and components for effectively managing resistance to change in research administration. At the center is 'Managing Resistance to Change,' surrounded by various interconnected strategies. 'Global Perspective and Collaboration' emphasizes engaging with international networks and learning from global best practices. 'Cultivating Culture' and 'Clear Communication' stress the importance of promoting continuous learning, agility, and sharing vision and benefits. 'Stakeholder Engagement' highlights inclusive planning sessions and feedback mechanisms, while 'Support and Resources' focus on providing training programs and mentoring opportunities. 'Recognition and Rewards' underscore celebrating milestones and acknowledging contributions. This comprehensive approach ensures a supportive and communicative environment that fosters acceptance and adaptability in the face of change

Equally important is the cultivation of a culture of adaptability and learning within research administration. Encouraging a mindset of continuous growth and development among staff and emphasizing the importance of staying agile and responsive to the evolving research landscape will create an environment that embraces change. Providing opportunities for professional development, training, and knowledge-sharing will equip research administrators with the necessary skills and competencies to thrive in their new roles and contribute to the success of the strategic initiatives.

Involving stakeholders in the change process is another critical strategy for overcoming resistance. By engaging research administrators, researchers, and other key stakeholders in the planning and implementation of strategic initiatives, institutions can foster a sense of ownership and collaboration. Seeking input, listening to feedback, and incorporating insights from diverse perspectives will help build trust and buy-in for the proposed changes. This participatory approach ensures that stakeholders feel heard and valued, reducing the likelihood of resistance, and promoting a shared commitment to the new direction.

As research administration continues to evolve globally, it is necessary for professionals in the field to engage in international collaboration and knowledge sharing. By actively participating in global networks, such as INORMS and EARMA, administrators can gain valuable insights into best practices, emerging trends, and shared challenges across different countries. Embracing a global perspective will enable research administrators to adapt to the changing landscape and effectively support research excellence worldwide.

The Vital Role of Research Administrators in Advancing Science and Society

At its core, the research enterprise is about pushing the boundaries of human knowledge and understanding and using that knowledge to solve real-world problems and improve people's lives. Whether it is developing new treatments for disease, creating more sustainable sources of energy, or exploring the mysteries of the universe, research has the power to transform our world in profound ways. But none of this would be possible without the dedicated efforts of research administrators, who work tirelessly behind the scenes to support and enable the work of researchers. From navigating complex funding landscapes and regulatory requirements to building collaborative partnerships and managing vast amounts of data, research administrators play a critical role in ensuring that research can move forward efficiently and effectively.

Figure 12. Research administrators at work: core responsibilities and impact.

This diagram illustrates the multifaceted role of research administrators and their significant impact on scientific and societal progress. At the center is 'Research Administrators at Work,' surrounded by their key responsibilities and areas of influence. 'Key Responsibilities' include navigating funding landscapes, ensuring regulatory compliance, building partnerships, and managing data. 'Research Support' highlights critical tasks such as project planning, resource allocation, ethical oversight, and collaboration facilitation. The 'Scientific and Societal Impact' section showcases the broad outcomes of their work, including medical breakthroughs, environmental solutions, technological innovations, and social improvements. Finally, 'Global Transformation' links their efforts to enhanced quality of life through improved health, sustainable living, and advanced technology, underscoring their vital contribution to societal advancement

As we have seen throughout this chapter, the role of research administrators has evolved significantly over time and will continue to do so in the years ahead. Through it all, the core mission remains the same: to support the advancement of research and the pursuit of knowledge for the betterment of society (Figure 12). Ultimately, the success of research administration will be measured not just by the efficiency and effectiveness of research support functions, but by the impact that research has on the world around us. By working closely with researchers and other key stakeholders to enable groundbreaking discoveries and innovations, research administrators can help to create a brighter, more sustainable, and more equitable future for all.

CONCLUDING INSIGHTS AND FUTURE DIRECTIONS

This chapter significantly advances the field of research administration by providing a detailed analysis of both historical and current trends, with a particular focus on the integration of advanced technologies such as AI and automation. The chapter introduces innovative frameworks for understanding the evolution of research administration practices and highlights strategic approaches for effective management and compliance.

The chapter's focus on the United States limits the scope of international applicability, and the reliance on secondary data sources necessitates empirical validation to confirm some of the proposed frameworks and trends. Future studies should address these limitations by exploring diverse international contexts and conducting primary research to assess the frameworks' validity.

Future research should investigate the application of the discussed frameworks across various international contexts to evaluate their effectiveness in different institutional environments. Empirical studies are also needed to assess the impact of AI and automation in research administration, focusing on efficiency and ethical considerations. Longitudinal studies could provide valuable insights into the evolving roles of research administrators and the long-term effects of technological integration on research productivity and compliance.

REFERENCES

Ainembabazi, P., Castelnuovo, B., Okoboi, S., Arinaitwe, W., Sebuyira, L., Parkes-Ratanshi, R., & Byakika-Kibwika, P. (2023). Building Capacity for Research Ethics Committees in Uganda. *Journal of Empirical Research on Human Research Ethics*, 18(3), 91–98. DOI: 10.1177/15562646231177410 PMID: 37272030

Atabek, O. (2020). Experienced educators' suggestions for solutions to the challenges to technology integration. *Education and Information Technologies*, 25(6), 1–17. DOI: 10.1007/s10639-020-10243-y

Bandarian, R. (2020). Explaining the evolution stages of strategic roles of research and technology management. *International Journal of Research, Innovation and Commercialisation*. .DOI: 10.1504/IJRIC.2020.10031470

Battle, L., & Scheidegger, C. (2020). A Structured Review of Data Management Technology for Interactive Visualization and Analysis. *IEEE Transactions on Visualization and Computer Graphics*, 27(2), 1128–1138. DOI: 10.1109/TVCG.2020.3028891 PMID: 33031039

Bazhin, A., Taverna, M., Lee, S., Mayer, B., & Werner, J. (2021). Development and assessment of a decentralized management system for doctoral student administration. *International Journal of Educational Administration, Management, and Leadership*. .DOI: 10.51629/ijeamal.v2i2.44

Berndsen, C. (2021). Teaching data management and literacy to support course-embedded research projects. *The FASEB Journal*, 35(S1). Advance online publication. DOI: 10.1096/fasebj.2021.35.S1.03442

Bezuidenhout, L., Quick, R., & Shanahan, H. (2020). "Ethics When You Least Expect It": A Modular Approach to Short Course Data Ethics Instruction. *Science and Engineering Ethics*, 26(4), 2189–2213. Advance online publication. DOI: 10.1007/s11948-020-00197-2 PMID: 32067185

Bhima, B., Zahra, A., & Nurtino, T. (2023). *Enhancing Organizational Efficiency through the Integration of Artificial Intelligence in Management Information Systems. APTISI Transactions on Management*. ATM., DOI: 10.33050/atm.v7i3.2146

Bloch, C. (2020). Heterogeneous impacts of research grant funding. *Research Evaluation*. Advance online publication. DOI: 10.1093/reseval/rvaa025

Bodnar, O., & Horishna, O. (2021). Mechanisms for overcoming resistance to changes in general secondary education institutions. *Educational Horizons*, (2), 180–189. DOI: 10.25128/2415-3605.20.2.24

Borde, P., Arora, R., & Kakoty, S. (2022). Transformational health-care leaders in collaborative entrepreneurial model to achieve UNSDG: a qualitative study. *Leadership in health services*, ahead-of-print ahead-of-print. https://doi.org/.DOI: 10.1108/LHS-03-2022-0032

Crane, K., Blatch-Jones, A., & Fackrell, K. (2023). The post-award effort of managing and reporting on funded research: a scoping review. *F1000Research*, 12. https://doi.org/DOI: 10.12688/f1000research.133263.2

Giuffrida, I., & Hall, A. (2023). Technology integration in higher education and student privacy beyond learning environments - A comparison of the UK and US perspective. *British Journal of Educational Technology*, 54(6), 1587–1603. DOI: 10.1111/bjet.13375

Kayla, C., Jennifer, P., Sandra, B., & Johnston, K. (2023). Building our research administrator workforce as our clinical and translational research programs become increasingly complex. *Frontiers in Pharmacology*, 14, 1295255. Advance online publication. DOI: 10.3389/fphar.2023.1295255 PMID: 37942484

Link, A., & Wagner, C. (2021). The publicness of publicly funded research. *Science & Public Policy*, 48(5), 757–762. Advance online publication. DOI: 10.1093/scipol/scab050

Nazarenko, T. (2023). Strategic resources for Digital transformation of Economic activity and Public administration in the field of Landscaping. *Artificial societies*. .DOI: 10.18254/S207751800025960-5

Office for Human Research Protections. (2022, October 17). *The Belmont Report: Ethical principles and guidelines for the protection of human subjects of research*. U.S. Department of Health & Human Services. https://www.hhs.gov/ohrp/regulations-and-policy/belmont-report/index.html

Palad, J. (2023). Strategies for Improving Organizational Efficiency, Productivity, and Performance through Technology Adoption. *Journal of Management and Administration Provision*. .DOI: 10.55885/jmap.v2i3.230

Shykhnenko, K., & Nozhovnik, O. (2023). Innovations in professional training of research administrators in the USA and Ukraine: A technological approach. *Education. Innovation.Practice*, 11(9), 71–78. Advance online publication. DOI: 10.31110/2616-650X-vol11i9-010

Trajkovski, G. (2023). *A more beautiful prompt: The art of human-centered prompt engineering* (1st ed., Kindle ed.). The Art of Living Large. ASIN: B0CJQZ68TM. https://beautifulprompt.expert

Trajkovski, G. (2024). Bridging the public administration-AI divide: A skills perspective. *Public Administration and Development*, 1–15. DOI: 10.1002/pad.2061

Vallejo, M., Ramirez, M., Gurumendi, I., & Perdomo, T. (2020). Importance of research and drafting in the scientific production of the health area. *International Journal of Research*, 8(1), 83–89. DOI: 10.29121/granthaalayah.v8.i1.2020.251

White, M. (2020). Why Human Subjects Research Protection Is Important. *The Ochsner Journal*, 20(1), 16–33. DOI: 10.31486/toj.20.5012 PMID: 32284679

Znamirowski, B. (2021). *Getting a Grip on Research Data Management. Bulletin - Association of Canadian Map Libraries and Archives*. ACMLA., DOI: 10.15353/acmla.n169.4739

KEY TERMS AND DEFINITIONS

Research Administration: The management and support of scientific and academic research projects, involving tasks such as proposal submission, budget management, regulatory compliance, intellectual property protection, and ethics oversight. It serves as the foundational infrastructure that enables researchers to focus effectively on their work.

Bayh-Dole Act (1980): A pivotal US legislation that allowed universities, small businesses, and non-profits to retain intellectual property rights of inventions developed from federally funded research, significantly changing the landscape of academic research and industry collaboration.

Belmont Report (1979): A report that established the ethical principles and guidelines for the protection of human subjects in research, emphasizing respect for persons, beneficence, and justice. This led to the development of Institutional Review Boards (IRBs) to oversee research ethics.

Centralization: The process of consolidating research administration functions within an institution to improve efficiency, consistency, and compliance, moving away from a decentralized approach where tasks were managed individually by researchers or departments.

Electronic Research Administration (eRA): The use of electronic systems and tools to manage the grant lifecycle, from proposal submission to project closeout, which aims to streamline administrative processes and enhance efficiency and communication in research management.

Data Literacy: The ability to read, understand, create, and communicate data as information. In the context of research administration, it includes the capacity to work with large volumes of research data and to interpret and convey data-driven insights effectively.

Technology Integration: The incorporation of advanced technologies such as AI, automation, and analytics into research administration to improve processes, decision-making, and the overall management of research projects.

Strategic Planning: In research administration, this refers to developing and implementing plans that align with institutional research goals and priorities, requiring skills in environmental scanning, data analysis, and stakeholder engagement.

Compliance: Adhering to the laws, regulations, guidelines, and specifications relevant to research activities. In research administration, this includes ensuring that research projects comply with ethical standards, safety regulations, and funding requirements.

Technology Transfer: The process of transferring scientific findings from one organization to another for further development and commercialization. In academia, it involves the transfer of discoveries made through research to the private sector.

Chapter 9
Data Ethics Policy Trends for Human Subjects Protections in the Era of the AI Bill of Rights

Robin Throne
https://orcid.org/0000-0002-3015-9587
University of the Cumberlands, USA

ABSTRACT

For institutional review boards (IRBs) and human research protection programs (HRPPs) specifically, the ubiquity of artificial intelligence (AI), machine learning (ML), and data mining (DM) within research today have created various and significant challenges for protections of human subjects and human subjects' data, including personally identifiable information, in research regulated by U.S. federal agencies. This chapter presents an analysis of the Blueprint for the AI Bill of Rights released in September 2022 by the White House Office of Science and Technology Policy in the context of human subjects' research and PII to address the mounting data ethical issues surrounding AI, ML, and DM, and the implications for IRBs and HRPPs for the protection of human subjects' data.

INTRODUCTION

> *"Doing no harm is the first principle of an ethical AI system."*
> *~ Sri Amit Ray*[1]

DOI: 10.4018/979-8-3693-3848-3.ch009

The data and research ethics surrounding artificial intelligence (AI), machine learning (ML), and data mining (DM) have been widely discussed within scholarships and among regulatory bodies. In recent research, much has been published in the recent scholarship about the implications of the institutional review board (IRB) and human research protection program (HRPP) for the protection of human subject data for AI, ML, and DM. For IRBs, more specifically, the emergence of AI, ML, and DM within research has created various challenges for human subject's research (Bernstein et al., 2021). At the same time, the number of IRBs has continued to increase with the increases in the amount of human subjects research[2] overall (Nwokeji & Iwanenko, 2021).

Further, "AI and ML research involving human data challenges the federal human subjects guidelines stemming from the difficulty in defining 'about whom' the data is being collected" (Eto, 2022, p. 2). For example, Doerr and Meeder (2022) have noted the group harms that may result from AI and ML used for data science research in the health sciences, and the authors called for an ethical review of these forms of research to consider whether the current human research federal regulations protect human subjects and public policy. These trends, along with the lack of uniformity in the review of research by IRBs, contribute to a complex issue for the protection of human subjects' data and personally identifiable information (PII) by the entities charged with this protection (Eto, 2022; Nwokeji & Iwanenko, 2021).

In addition, Gurwitz (2020) raised concerns over prescription data mining during the COVID-19 pandemic, whereby IRBs were asked to relax restrictions under the global emergency and said that post-pandemic data privacy concerns still needed mediation. While withholding health data from the research community may be a breach that escalates further threats to society, health data sharing requires "careful anonymization of identifiers" to assure individual privacy (Gurwitz, 2020, p. 779). As Doerr and Meeder (2022) stressed, this is further complicated by technological advances, such as "when new approaches such as AI and ML for big health data research emerge, it becomes necessary to determine how or if the current regulatory framework allows IRBs to fulfill their role" (p. 36). This void in oversight may be further complicated by IRB members who do not understand the algorithmic methods used within AI, ML, and DM applications (Greene et al., 2021).

Finally, Serpico (2021) called out HRPPs as an added ethics steward for institutional officials and IRBs to ensure compliance for research conducted. To address the myriad of concerns and IRB/HRPP implications, this chapter presents an analysis of the *Blueprint for the AI Bill of Rights* (White House Office of Science and Technology Policy [OSTP]. 2022) in the context of these data ethics and IRB implications for research involving AI, ML, and DM for all forms of human subjects research, not solely healthcare research.

BACKGROUND

Many scholars have called out IRBs and institutional ethics committees for not keeping pace with the data ethics issues fostered by technological advancements in AI, ML, and DM (Kimmons & Veletsianos, 2018). Large public data sets may result in group harm, especially for public health and other disciplines, and IRBs may determine AI, ML, or DM research as not human subjects research and, therefore, not requiring informed consent (Kimmons & Veletsianos, 2018). Publicly available data sets are exempt under the Common Rule and as such, not subject to IRB oversight (Hirsch et al., 2021; Pike, 2019) and several scholars have called out this gap as contributing to the risk for human subjects who share various identifiers within publicly available data, from political affiliations to sexual orientation to immigration status, which may impose risk yet remain directly or indirectly available in these datasets (Panicker & Huh-Yoo, 2022; Pike, 2019).

Still, other fields of research, such as business, law, medicine, and computer science often lack the ethical and responsible research standards and review processes long held in other research areas (Bernstein et al., 2021). Social media is one of the most significant contributors to open-source human subjects data, intentional or unintentional (Marmo, 2021). Social media users may unintentionally share personally identifiable information and often indirectly consent to use the data for research purposes (Panicker & Huh-Yoo, 2022). Scandals over the past five years have demonstrated the "unfortunate reminders of the vulnerability of individuals to, and the relative ease of, the large-scale misuse of personally identifiable data that were detailed enough to create psychographic profiles of individuals" (Martinez-Martin et al., 2018, p. 2).

Personally identifiable information (PII) as defined by the U.S. General Services Administration states

> *"The term "PII," as defined in OMB Memorandum M-07-1616 refers to information that can be used to distinguish or trace an individual's identity, either alone or when combined with other personal or identifying information linked or linkable to a specific individual. The definition of PII is not anchored to any single category of information or technology. Rather, it requires a case-by-case assessment of the specific risk that an individual can be identified. In performing this assessment, an agency needs to recognize that non-PII can become PII whenever additional information is made publicly available - in any medium and from any source - that, when combined with other available information, could be used to identify an individual." (OMB, 2007)*

OMB M-10-22 provided additional commentary on the definition of PII previously defined in OMB M-07-16 by stating

"The definition of PII is not anchored to any single category of information or technology. Rather, it demands a case-by-case assessment of the specific risk that an individual can be identified. In performing this assessment, it is important for an agency to recognize that non-PII can become PII whenever additional information is made publicly available — in any medium and from any source — that, when combined with other available information, could be used to identify an individual." (OMB, 2019)

Likewise, the National Institute of Standards and Technology defined PII as "any information about an individual maintained by an agency, including (1) any information that can be used to distinguish or trace an individual's identity, such as name, social security number, date and place of birth, mother's maiden name, or biometric records; and (2) any other information that is linked or linkable to an individual, such as medical, educational, financial, and employment information" (McCallister et al., 2010, p. ES-1). This definition includes "personal characteristics, including photographic image (especially of the face or other identifying characteristic), fingerprints, handwriting, or other biometric data (e.g., retina scan, voice signature, facial geometry)" (p. ES-1). Finally, The Health Insurance Portability and Accountability Act of 1996 (HIPAA) distinguishes personal health information (PHI) from PII and indicates 18 specific types of personal identifiers that define PHI (Alder, 2022) as well as a deidentification standard to remove these identifiers (HIPAA Privacy Rule, 2013).

These various identifiers of what constitutes PII allow IRBs to interpret what constitutes the Revised Common Rule definition of personal information as "private information for which the identity of the subject is or may readily be ascertained by the investigator or associated with the information" (Protection of Human Subjects, 2018). Eto (2022) noted that distinctions were added to expand consideration for what constitutes data and PII within research as "using, studying, or analyzing individuals' information or biospecimens or generating identifiable private information or identifiable biospecimens" (p. 6). Whether PII is sufficiently protected within research involving AI, ML, or DM continues to be widely debated among scholars and government officials. Eto (2022) recommended a modification of the OHRP decision tree for IRB reviewers[3] to better evaluate the protections for "human focused data" as a stepwise process to where and how the technology is used within the research whether it be labeled as AI, ML, DM, natural language, neural networks, predictive modeling, or other forms (p. 8). However, in the end, current rules may allow data science research to be conducted well outside requirements for human subject identifiers (Greene et al., 2021).

In the meantime, IRBs utilize these existing federal definitions to determine when data involves human subjects identifiers and PII, and thereby may be determined as human subjects research requiring IRB oversight. The deidentification of datasets gathered via AI, ML, or DM can be complicated or covert, making ethical oversight for the research challenging, especially when regulatory informed consent does not apply (Eto, 2022; Hirsch et al., 2021; Panicker & Huh-Yoo, 2022). Thus, data or text mining can disrupt the traditional oversight over participant consent as individual consent for the use of the data may be unclear, undistinguishable within application consents, or otherwise understated, which may, in turn, adversely affect populations and communities (Panicker & Huh-Yoo, 2022; Skorburg & Friesen, 2020). In worst cases, individuals may be unaware and uninformed as to the use and abuse of their PII without informed consent (Doerr & Meeder, 2022).

Thus, an enduring question continues as to where are the bounds that AI, ML, and DM researchers must remain cognizant of to ensure ethical protections of data that involves PII and other human subject identifiers. With the technological advancement, and increased use of big data research involving AI, ML, and DM, in addition to the personal information security concerns among intelligent technologies, the meaning of data privacy for individuals continues to evolve as privacy is no longer a binary value (Office for Human Research Protections [OHRP], 2019). For example, Throne and Hendon (2024) noted

The issue with open-source tools is that the IRB cannot run the algorithm. Can the process be holistic to assist the IRB reviewer in understanding that the AI can transparently run in the algorithm in a research population without noting the structure of the predictive algorithm? As stated within the discloser of the toolkit, the IBM Developer Staff (2018) noted, "It can be problematic, however, to base predictions, directly or indirectly, on attributes protected through social norms, organizational policies, or legal regulations" (para. 5). There is a growing need to justify an understanding of HRP policy and increased transparency so as to mitigate HSR privacy and confidentiality. Thus, increasing the possibility of bias and reidentification of the information may not address future harms" (p. 11)

Free the Data but Not the Human Subjects Identifiers

The "Free the Data" movement found momentum in the 2010s and arrived in the 2020s to face the competing momentum of a rising individual agency for data privacy.[4] Since mid-2022, a wide-ranging set of discussions and debates surrounding personally identifiable information (PII) and data ethics for artificial intelligence (AI), machine learning (ML), and data mining (DM) has advanced a new era of considerations for data privacy and PII protection. Over recent months, many regulatory bodies and other governmental groups globally have issued new guidance, regulations, and restrictions for PII and data privacy for these continually evolving technologies. Thus, it remains essential in this time of momentous change that leadership of institutional review boards (IRBs), human research protection programs

(HRPP), and other research ethics committees remain attentive to these emergent policy events in the context of human subjects research (HSR) as discussions and regulatory considerations have rapidly advanced throughout late 2022.

In September 2022, the White House Office of Science and Technology Policy released the *Blueprint for the AI Bill of Rights* presents "five principles that should guide the design, use, and deployment of automated systems to protect the American public in the age of artificial intelligence". The following month, Public Responsibility in Medicine and Research (PRIM&R) hosted a webinar to explore the ongoing challenges of research protections for IRBs and HRPPs in this digital age amid the expansion of these technologies.[5] The webinar presenters referenced Tamiko Eto's White Paper from August 2022 that offered an AI decision tree for human subjects research (HSR) as adapted from the Office for Human Research Protection (OHRP) post-2018 IRB review decision trees and intended to further the discussion of human subjects protections within AI for IRB capacity building for research that involves these advancing technologies. Although not overtly noted, the digital landscape involving AI should address ML and DM as similar risks to PII continue. Eto had previously defined AI human subjects research (AI HSR) with Canca in 2020 as "research conducted to develop AI tools involving human subjects" as framed within the U.S. federal regulatory definitions for human subjects and research."[6] Thus, consistency in definitions and categories for these technologies in the context of human subjects research is helpful to policy monitoring.

Further in July 2022, The Secretary's Advisory Committee on Human Research Protections (SACHRP) was charged to provide advisory guidance to OHRP to address this evolving nature of AI and ML and the risks and implications to human research protections.

> *"SACHRP recommends that the Secretary consider establishing fora and mechanisms to facilitate dialogue, and ultimately, regulatory guidance, about how the interests of groups predictably affected by AI research might be considered and protected, consistent with maintaining scientific integrity. Further, SACHRP recommends that, based on such opportunities for dialogue, the Secretary establish formal guidance to ensure that anticipated benefits as well as risks of harm of research to affected groups, particularly of research outside the biomedical domain, are considered when HHS considers funding research projects that use AI or that refine AI methods and algorithms, when such group benefits and harms may predictably be at stake."* [7]

With this initial guidance provided, the discussion continued to the October 2022 SACHRP meeting to reach a consensus on guidance for AI research that was grouped as AI/ML and Big Data (BD), as well as data derived from intelligent technologies under the umbrella of the Internet-ofThings (IoT). SACHRP guided the 10 questions raised by the Secretary and stressed the need for the Secretary to continue to consider the Common Rule's protections of human subjects data identifiability while concurrently noting the scientific value of these technologies to advance knowledge:

"AI/ML and BD research expose the limits of the traditional concept of identifiability that serves as the basis for privacy protections under the Common Rule. The explicit ability to identify an individual from a specific dataset is a characteristic that was appropriate when data were analyzed in isolation, when data collection occurred primarily in the context of well-defined research studies (i.e., before the widespread use of electronic health records and ubiquitous data collection outside of healthcare), and before the routine collection and use of genomic data, which are arguably intrinsically identifying. SACHRP urges the Secretary to follow through on the Common Rule's commitment to regularly reexamine the meaning of identifiability in response to evolving technology and research practices."[8]

Correspondingly, in November 2022, 31 U.S. state attorneys general submitted a letter to the U.S. Federal Trade Commission calling for mandated data collection limits,[9] and in the letter, characterized the dangers of PII sold by data brokers: "Data brokers profile consumers by scouring social media profiles, internet browsing history, online and offline purchase history, credit card information, and government records, including driver's license/motor vehicle records, census data, birth certificates, marriage licenses, and voter registration information. Data brokers also use this information to categorize consumers into audiences based on susceptibility to certain advertising or likelihood to buy certain products…" and thereby the failures of traditional policy to protect consumers in the use of current data privacy requirements.[9] Similar concerns about indirect identifiers used for market research and other commercial purposes have been raised by IRBs, HRPPs, and health science researchers for some time.[10]

Concurrent with U.S. policy considerations for PII and data involving human subjects identifiers, by late 2022, legislation changes have also been in process globally. In October 2022, the European Data Protection Supervisor (EDPS), the European Union's (EU) independent data protection authority, opened negotiations[11] for a Council of Europe convention on artificial intelligence, human rights, democracy, and the rule of law to complement the European Commission's proposed Artificial Intelligence Act[12] to strengthen the protection of individual fundamental rights. By November 2022, EU members continued to refine the AI Act.[13]

Leaders of IRBs and HRPPs must continue to be attentive to this fluctuating digital landscape whereby the contronym of sanction applies while IRBs, HRPPs, and regulatory organizations grapple with the refinement of policy and procedures amid new guidance and best practices for PII used within research that involves these technologies, including AI, ML, DM, BD, and IoT. The social good and advances to scientific knowledge from these technologies have been widely recognized across scholarship, and yet, the risks to human subject identifiers and PII remain.

Thus, it is critically essential in 2024 for IRB, HRPP, and other research ethics leadership to continue monitoring the debate, guidance, and new thoughts that arise as these diverse groups grapple with the need for standards and boundaries for individual protections and data privacy within HSR. Leaders must continue to gain best practices and establish standards for protecting human subject identifiers and PII in digital spaces not overtly addressed within the Common Rule and follow these concurrent regulatory discussions for what changes are needed to protect human subjects data in the digital era.

Data Privacy and the AI Bill Of Rights

The Secretary's Advisory Committee on Human Research Protections (SACHRP) was charged with providing advisory guidance to OHRP to address the evolving nature of AI and ML and, thereby, the implications to human research protections. Initial guidance was provided, and discussion continued to the October 2022 SACHRP meeting to further reach a consensus on AI research as well as social media research implications.

> *"SACHRP recommends that the Secretary consider establishing fora and mechanisms to facilitate dialogue, and ultimately, regulatory guidance, about how the interests of groups predictably affected by AI research might be considered and protected, consistent with maintaining scientific integrity. Further, SACHRP recommends that, based on such opportunities for dialogue, the Secretary establish formal guidance to ensure that anticipated benefits as well as risks of harm of research to affected groups, particularly of research outside the biomedical domain, are considered when HHS considers funding research projects that use AI or that refine AI methods and algorithms, when such group benefits and harms may predictably be at stake." (SACHRP, 2022)*

At the October 2022 session, SACHRP members noted that the *Blueprint for the AI Bill of Rights* was released by the White House Office of Science and Technology Policy in September 2022 and guidance should remain in context with the framework and other "existing frameworks or tools that funding agencies, investigators, HRPP staff, and IRBs can use to illuminate and mitigate ethical concerns with human-focused AI research and development." The Committee furthered the calls for the potential of group harms and inequities within AI research that the Common Rule does not address and that these harms and inequities should be addressed at the federal level.

Hosseini et al. (2022) stressed that the biases embedded in data sets and third-party technologies, and those resulting from researchers' prejudices," elevate risk and ethical challenges for these technologies (p. 29). Kurfess et al. (2022) noted that these types of research must go beyond consideration for harms and ethical protections to the need for just outcomes and reduce the reproduction of "systematic and historic inequities for those on the margins" (p. 1). In tandem, OSTP's release of the *AI Bill of Rights* also calls out the potential for inequities, biases, and potential discrimination in the overall use of these technologies:

> *"Information about or algorithmically-determined assumptions related to identity should be carefully limited if used to target or guide surveillance systems in order to avoid algorithmic discrimination; such identity-related information includes group characteristics or affiliations, geographic designations, location-based and association-based inferences, social networks, and biometrics." (OSTP, 2022, p. 34).*

Further, under a principle for "Data Privacy," the *AI Bill of Rights* calls for individual agency over data and how it is to be used while concurrently noting that the U.S. lacks a "comprehensive statutory or regulatory framework governing the rights of the public when it comes to personal data" (OSTP, 2022, p. 31). As federal agencies continue to seek guidance to inform federal regulations in these areas, it is of the utmost importance for policymakers to work collectively to formulate the best ethical standards for the collection of PII across these technological applications, regardless of typology. These issuances and ongoing discussions illustrate the importance of IRBs and HRPPs continually reassessing best practices in policies and procedures for reviewing research involving AI, ML, and DM.

Ethical Issues for Personally Identifiable Data and Human Subject Protections

For more than a decade, research administrators and data ethicists have called for a better understanding of the challenges emerging technologies bring to data involving PII and the resulting ethical issues for human subjects protections (Eto, 2022; Martinez-Martin et al., 2018; Narayanan & Felten, 2014; OHRP, 2019; Skorburg & Friesen, 2020). Others have noted that this evolution has been ongoing for two decades or more. For example, Pater et al. (2022) reported, "Our computational capacity has greatly increased in the last 20 years, but new capabilities in fields like AI and machine learning are changing the nature of research and leaving significant gaps in both ethical norms and oversight" (p. 1) and Rossi et al. (2022) indicated social media data scraping has gone on unchecked throughout its evolution. Likewise,

Hosseini et al. (2022) noted that IRBs, HRPPs, and research ethics committees are not adequately equipped to address these issues across social science research. Others have also noted the gaps that exist within the regulations as research involving AI, ML, or DM may utilize datasets that fall outside of definitions for human subject research, and thereby lack of oversight may result in ethical challenges (Hosseini et al., 2022; Rossi et al., 2022). As outlined in the *AI Bill of Rights*, without explicit consent by data owners and the individuals whose data are to be used, the research has inherent risks and ethical concerns for research that falls within the gap of IRB oversight (Eto, 2022; Hosseini et al., 2022; Panicker & Huh-Yoo, 2022).

Still, others stress that the burden of ethical oversight for the employment of these technologies must also be placed upon the researcher as much cybersecurity research falls outside of IRB regulations and is thus not subject to this oversight (Rossi et al., 2022). Others recommend the solution may lie in consideration of technological applications that identify datasets as a potential solution to PII and human subjects' data protection (Nwokeji & Iwanenko, 2021; Rossi et al., 2022).

CONCLUSION

The current scholarship has presented many cases where further policy and future research considerations are needed for ethical assurances for the protection of human subjects in AI, ML, and DM. Still, others have called out the limitations of current regulatory language that does not encompass these challenges nor provide sufficient guidance that limits interpretation for IRBs that review this research. In addition, OSTP's (2022) *AI Framework* may further these discussions and has been a positive development as regulatory bodies begin to reference these principles in ongoing discussions of needed regulatory change and implications for IRBs and HRPPs. While numerous scholars, regulators, research associations, and government officials have called for the continuing debate and establishment of clear standards to address human subjects' protections within research involving AI, ML, and DM, more work is needed. Doerr and Meeder (2022) called for policymakers to consider the fitness of the Common Rule and in better context with The Belmont Report[14], for necessary regulation changes and further guidance for interpretation and implementation in relation to this research.

Hosseini et al. (2022) recommended that IRBs and HRPPs establish research ethics and data use frameworks to "provide guidance and best practices on how to minimize biases embedded in data sets and third-party technologies and those resulting from researchers' personal prejudices" (p. 29). Leadership for IRBs, HRPPs, and other research ethics committees should remain attentive to the ongoing debate and discussion and, in so doing, must consider how and where to establish best practices for this research within the organization or institution for which it has oversight. Finally, future research is needed for bias and risk assessment and/or effectiveness testing in the use of data ethics frameworks and procedures for protecting human subjects data and PII within research involving AI, ML, and DM.

REFERENCES

Alder, S. (2022). What is considered PHI under HIPAA? *HIPAA Journal*. https://www.hipaajournal.com/considered-phi-hipaa

Bernstein, M. S., Levi, M., Magnus, D., Rajala, B., Satz, D., & Waeiss, C. (2021). ESR: Ethics and society review of artificial intelligence research. Preprint. https://doi.org//arXiv.2106.11521DOI: 10.48550

Doerr, M., & Meeder, S. (2022). Big health data research and group harm: The scope of IRB review. *Ethics & Human Research*, 44(4), 34–38. DOI: 10.1002/eahr.500130 PMID: 35802789

Eto, T. (2022). Conducting an effective IRB review of artificial intelligence human subjects research (AI HSR). https://techinhsr.com/wp-content/uploads/2022/08/AI-HSR-WHITE-PAPER-TechInHSR-08.2022-1.pdf

Friesen, P., Douglas-Jones, R., Marks, M., Pierce, R., Fletcher, K., Mishra, A., Lorimer, J., Véliz, C., Hallowell, N., Graham, M., Chan, M. S., Davies, H., & Sallamuddin, T. (2021). Governing AI-driven health research: Are IRBs up to the task? *Ethics & Human Research*, 43(2), 35–42. DOI: 10.1002/eahr.500085 PMID: 33683015

Greene, T., Martens, D., & Shmueli, G. (2021). Barriers for academic data science research in the new realm of behavior modification by digital platforms. https://papers.ssrn.com/sol3/ papers.cfm?abstract_id=3946116

Gurwitz, D. (2020). Repurposing current therapeutics for treating COVID-19: A vital role of prescription records data mining. *Drug Development Research*, 81(7), 777–781. DOI: 10.1002/ddr.21689 PMID: 32420637

HIPAA Privacy Rule, 45 CFR §164.514(a) (2013).

Hirsch, D. D., Bartley, T., Chandrasekaran, A., Norris, D., Parthasarathy, S., & Turner, P. N. (2021). Business data ethics: Emerging trends in the governance of advanced analytics and AI. https://papers.ssrn.com/sol3/papers.cfm?abstract_id=3828239

Hosseini, M., Wieczorek, M., & Gordijn, B. (2022). Ethical issues in social science research employing big data. *Science and Engineering Ethics*, 28(3), 1–21. DOI: 10.1007/s11948-022-00380-7 PMID: 35705883

Kimmons, R., & Veletsianos, G. (2018). Public internet data mining methods in instructional design, educational technology, and online learning research. *TechTrends*, 62(5), 492–500. DOI: 10.1007/s11528-018-0307-4

Kurfess, F., Vasilaky, K., Cheuk, T., Jenkins, R., Nolan, G., Hajrasouliha, A., & St John, E. (2022, August). Assessment of ethics and social justice aspects in data science and artificial intelligence. In *2022 ASEE Annual Conference & Exposition*. DOI: 10.18260/1-2--41550

Marmo, R. (2021). Social media mining. In *Encyclopedia of organizational knowledge, administration, and technology* (pp. 2153-2165). DOI: 10.4018/978-1-7998-3473-1.ch149

Martinez-Martin, N., Insel, T. R., Dagum, P., Greely, H. T., & Cho, M. K. (2018). Data mining for health: Staking out the ethical territory of digital phenotyping. *NPJ Digital Medicine*, 1(1), 1–5. DOI: 10.1038/s41746-018-0075-8 PMID: 31211249

McCallister, E., Grance, T., & Scarfone, K. (2010). Guide to protecting the confidentiality of personally identifiable information (PII). Special Publication (NIST SP), National Institute of Standards and Technology. https://tsapps.nist.gov/publication/get_pdf.cfm?pub_id=904990

Narayanan, A., & Felten, E. W. (2014). No silver bullet: De-identification still doesn't work. *White Paper, 8*. https://www.cs.princeton.edu/~arvindn/publications/no-silver-bullet-de-identification.pdf

Nucera, S., Tartarisco, G., Epasto, A., Smeriglio, D., Mazzeo, A., Pioggia, G., & Anastasi, A. (2018). Ubiquitous, wearable, mobile: Paradigm shifts in e-learning and diffusion of knowledge. In *Handbook of research on mobile devices and smart gadgets in K-12 education* (pp. 286–307). IGI Global., DOI: 10.4018/978-1-5225-2706-0.ch018

Nwokeji, J., & Iwanenko, W. (2021). Requirements analysis and design of IRB system. *American Journal of Operations Management and Information Systems*, 6(4), 75–85. DOI: 10.11648/j.ajomis.20210604.12

Office for Human Research Protections (OHRP). (2019, September 19). Privacy and health research in a data-driven world [Workshop]. https://www.hhs.gov/ohrp/sites/default/files/report-privacy-and-health-10-31-19.pdf

Panicker, S., & Huh-Yoo, J. (2022, October 18). Research ethics and oversight in the digital age. PRIM&R Webinar. https://www.pathlms.com/primr/courses/47358

Pater, J., Fiesler, C., & Zimmer, M. (2022). No humans here: Ethical speculation on public data, unintended consequences, and the limits of institutional review. *Proceedings of the ACM on Human-Computer Interaction, 6*(GROUP), 1-13. DOI: 10.1145/3492857

Pike, E. R. (2019). Defending data: Toward ethical protections and comprehensive data governance. *Emory Law Journal*, 69, 687.

Protection of Human Subjects, 45 CFR §46.102(e)(5) (2018).

Rossi, A., Arenas, M. P., Kocyigit, E., & Hani, M. (2022, June). Challenges of protecting confidentiality in social media data and their ethical import. In *2022 IEEE European Symposium on Security and Privacy Workshops (EuroS&PW)* (pp. 554-561). IEEE. DOI: 10.1109/EuroSPW55150.2022.00066

Serpico, K. (2021). Making metrics meaningful: How human research protection programs can efficiently and effectively use their data. *Ethics & Human Research*, 43(5), 26–35. DOI: 10.1002/eahr.500102 PMID: 34496157

Shin, S. Y. (2021). Privacy protection and data utilization. *Healthcare Informatics Research*, 27(1), 1–2. DOI: 10.4258/hir.2021.27.1.1 PMID: 33611870

Skorburg, J., & Friesen, P. (2020). Ethical issues in text mining for mental health. In *The atlas of language analysis in psychology* (in press). Guilford.

The Secretary's Advisory Committee on Human Research Protections (SACHRP). (2022, July 21). Considerations for IRB review of research involving artificial intelligence. https://www.hhs.gov/ohrp/sachrp-committee/recommendations/attachment-e-july-25-2022-letter/index.html

Throne, R., & Hendon, M. (2024). Belmont 2.0: Ethical research expectations for PII in AI, ML, and data mining/scraping. In *Methodologies and ethics for social sciences research* (pp. 1-19). IGI Global.

Throne, R., Hendon, M., & Kozinski, J. (2023). Graduate student investigator: Best practices for human research protections within online graduate research. In *Dynamic curriculum development and design strategies for effective online learning in higher education* (pp. 1–15). IGI Global. DOI: 10.4018/978-1-6684-8646-7.ch001

United States Office of Management and Budget (OMB). (2007). *OMB circular M-07-1616 compliance supplement* [Washington, D.C.]. Executive Office of the President, Office of Management and Budget.

United States Office of Management and Budget (OMB). (2019). *OMB circular M-10-23 compliance supplement* [Washington, D.C.]. Executive Office of the President, Office of Management and Budget.

White House Office of Science and Technology Policy (OSTP). (2022). A blueprint for an AI bill of rights. https://www.whitehouse.gov/ostp/ai-bill-of-rights/

ADDITIONAL READING

Brown, N. (2023). Research ethics in a changing social sciences landscape. *Research Ethics*, 19(2), 157–165. DOI: 10.1177/17470161221141011

Coeckelbergh, M. (2020). *AI ethics*. MIT Press. DOI: 10.7551/mitpress/12549.001.0001

Kumar, R., Joshi, A., Sharan, H. O., Peng, S. L., & Dudhagara, C. R. (Eds.). (2024). *The ethical frontier of AI and data analysis*. IGI Global. DOI: 10.4018/979-8-3693-2964-1

Mazilu, M. C. (2022). Web scraping and ethics in automated data collection. In *Education, research and business technologies* (pp. 285–294). Springer. DOI: 10.1007/978-981-16-8866-9_24

Rahman, R. U., Wadhwa, D., Bali, A., & Tomar, D. S. (2020). The emerging threats of web scraping to web applications security and their defense mechanism. In *Encyclopedia of criminal activities and the deep web* (pp. 788–809). IGI Global. DOI: 10.4018/978-1-5225-9715-5.ch053

Vasiliu-Feltes, I., & Thomason, J. (Eds.). (2021). *Applied ethics in a digital world*. IGI Global. DOI: 10.4018/978-1-7998-8467-5

Xiao, G. (2022). Data misappropriation: A trade secret cause of action for data scraping and a new paradigm for database protection. *The Columbia Science and Technology Law Review*, 24(1), 125 172. DOI: 10.52214/stlr.v24i1.10456

KEY TERMS AND DEFINITIONS

Data Mining/Scraping: While data mining and data scraping are often used synonymously across disciplines, the distinction between mining versus scraping typically refers to data mining as the gathering of large datasets for analysis, while data scraping involves the process of gathering large datasets. Due to the overlap in meaning, for the purposes of this chapter, the terms are concomitant (Throne & Hendon, 2024).

Ubiquitous Digital Technologies: Ubiquitous digital technologies refer to all everyday objects embedding intelligent microprocessors that can create architectures that underpin ubiquitous computing, such as networking, artificial intelligence, and wireless computing. Ubiquitous technologies employ sensitive human-computer interaction without spatial or temporal constraints (Nucera et al., 2018; Throne et al., 2023).

ENDNOTES

1 Ray, S. (2022). *Ethical AI systems: Frameworks, principles, and advanced practices*. Inner Light Publishers.

2 Protection of Human Subjects 45 CFR §46 (2018). https://www.ecfr.gov/on/2018-07-19/title-45/subtitle-A/subchapter-A/part-46

3 Tamiko Eto's White Paper from August 2022 that offered an AI decision tree for human subjects research (HSR) was adapted from the Office for Human Research Protection (OHRP) post-2018 IRB review decision trees and intended to further the discussion of human subjects protections within AI for IRB capacity building. See https://techinhsr.com/wp-content/uploads/2022/08/AI-HSR-WHITE-PAPER-TechInHSR-08.2022-1.pdf

4 Sklar, E. B., "Be Careful Where You Spit: Do HIPAA-Covered Genetic Tests Actually Provide Greater Privacy Protection to Consumers?" *Seton Hall Legislative Journal* 44, no. 1 (2019): 7; Wachter, S., & Mittelstadt. B., "A Right to Reasonable Inferences: Re-thinking Data Protection Law in the Age of Big Data and AI." *Colum. Bus. L. Rev.* (2019): 494.

5 Panicker, S., & Huh-Yoo, J., "Research Ethics and Oversight in the Digital Age," PRIM&R Webinar, October 18, 2022, https://www.pathlms.com/primr/courses/47358

6 Canca, C., & Eto, T., "Artificial Intelligence (AI) and Human Subject Protections," The Collaborative Institutional Training Initiative (CITI Program), June 1, 2020, https://about.citiprogram.org/news/newwebinar-artificial-intelligence-ai-and-human-subject-protections/

7 The Secretary's Advisory Committee on Human Research Protections (SACHRP), "Considerations for IRB Review of Research Involving Artificial Intelligence," July 21, 2022, https://www.hhs.gov/ohrp/sachrp-committee/recommendations/attachment-e-july-25-2022letter/index.html

8 The Secretary's Advisory Committee on Human Research Protections (SACHRP), "IRB Considerations on the Use of Artificial Intelligence in Human Subjects Research," October 19, 2022, https://www.hhs.gov/ohrp/sachrp-committee/recommendations/irb-considerations-use-artificialintelligence-human-subjects-research/index.html

9 Federal Trade Commission, p. 4 [See Citation 9].

10 Go Fair, "Fair Principles", https://www.go-fair.org/fair-principles/; Johns Hopkins Libraries Data Services, "Protecting Human Subject Identifiers", https://guides.library.jhu.edu/protecting_identifiers; US HHS Health Information Privacy, "Guidance Regarding Methods for De-identification of Protected Health Information in Accordance with the Health Insurance Portability and Accountability Act (HIPAA) Privacy Rule", October 25, 2022, https://www.hhs.gov/hipaa/for-professionals/privacy/special-topics/deidentification/index.html#guidancedetermination

11 European Data Protection Supervisor, "AI Convention: Stronger Protection of Fundamental Rights is Necessary," https://edps.europa.eu/press-publications/press-news/press-releases/2022/ai-conventionstronger-protection-fundamental-rights-necessary_en

12 European Union, "The Artificial Intelligence Act," https://artificialintelligenceact.eu/

13 EURACTIV, "Last-minute Changes to EU Council's AI Act Text Ahead of General Approach," https://www.euractiv.com/section/digital/news/last-minute-changes-to-eu-councils-ai-act-text-ahead-ofgeneral-approach/

14 US National Commission for the Protection of Human Subjects of Biomedical and Behavioral Research. (1978). *The Belmont Report: Ethical principles and guidelines for the protection of human subjects of research* (Vol. 1): Department of Health, Education, and Welfare, National Commission for the Protection of Human Subjects of Biomedical and Behavioral Research.

Chapter 10
The Belmont Report: Guiding Ethical Principles in Human Research

Muhammad Usman Tariq
https://orcid.org/0000-0002-7605-3040
Abu Dhabi University, UAE & University College Cork, Ireland

ABSTRACT

This chapter explores the Belmont Report, a cornerstone in the ethics of human subject research. It details the report's historical development, emphasizing the three pivotal principles: Respect for People, Beneficence, and Justice. These principles advocate for autonomy through informed consent, the minimization of harm coupled with the maximization of benefits, and fairness in research burden distribution. The chapter discusses the operational challenges in applying these principles across modern research settings and examines the evolving role of Institutional Review Boards (IRBs) in enforcing these standards. Additionally, it addresses the limitations of the Belmont Report in contemporary issues like digital data privacy and the globalization of research practices. The chapter concludes by emphasizing the necessity for dynamic ethical frameworks that adapt to ongoing technological and societal shifts, urging continued ethical vigilance in human subject research.

INTRODUCTION

The Belmont Report arose from U.S. historical research events, ethical concerns, and regulatory needs in the mid-20th century. During this time, several high-profile ethical violations marred the landscape of biomedical and behavioural research, eventually leading to demands for strict guidelines to protect human subjects. The

DOI: 10.4018/979-8-3693-3848-3.ch010

most famous of these ethical violations was the Tuskegee Syphilis Study, which began in 1932 and continued until 1972.

This study, conducted by the US Public Health Service, aimed to examine the natural progression of untreated syphilis in African American men under the banner of free treatment. The men were not informed of their diagnosis and refused treatment, although penicillin became the standard treatment in the 1940s. The study's publication in 1972 caused an outcry, lawsuit, and national scandal, underscoring the urgent need for ethical oversight in human research. Other influential cases include hepatitis studies at Willowbrook State School and the Stanford Prison Experiment. In Willowbrook, children with developmental disabilities were intentionally infected with hepatitis to study disease progression and possible treatments. Although parental consent was obtained, the coercive conditions under which consent was given – offering access to an overcrowded facility only if parents agreed to the study – raised significant ethical concerns. Psychologist Philip Zimbardo's 1971 Stanford Prison Experiment included college students role-playing prisoners and guards in a simulated prison environment. The experiment quickly disappeared, and the participants behaved in extreme and violent ways. This study highlighted ethical issues related to informed consent and the psychological harm research can cause participants.

Following these and other controversies, it has been increasingly recognized that current ethical standards and control mechanisms are inadequate. This recognition led to the National Commission for the Protection of Human Subjects in Biomedical and Behavioural Research in 1974. The Commission's mission was to identify critical ethical principles supporting biomedical and behavioural research involving humans and developing guidelines. Such studies should be conducted using the following principles: the role of the National Commission in the Protection of Human Subjects in Biomedical and Behavioural Research.

The National Commission for the Protection of Subjects, Used in Biomedical and Behavioural Research, was created by the National Research Act of 1974. Mandate of Commission The aim was to address the ethical issues of research involving human subjects, which the Commission took seriously and thoroughly. The Commission comprises diverse experts, including medical professionals, ethical researchers, legal scholars, and laypeople, who ensure various perspectives. This diversity is crucial to addressing today's complex ethical issues and ensuring that the guidelines produced are comprehensive and universally applicable.

One of the Commission's main tasks was identifying and formulating the basic ethical principles guiding human research. To this end, the commission extensively reviewed historical and contemporary research cases, studied existing ethical rules and guidelines, and participated in public discussions. This process culminated in the publication of the Belmont Report in 1979. The Belmont Report identified three primary ethical principles: respect for people, benevolence, and justice. These principles were derived from the broader ethical traditions of Western philosophy and adapted to the specific context of human subject research.

BACKGROUND

Respect for People

This principle contains two ethical beliefs: individuals should be treated as autonomous agents, and individuals with limited autonomy are entitled to protection. The committee emphasized the importance of informed consent, which ensured that participants were fully aware of the study procedures, risks, benefits, and their right to withdraw at any time. The concept of respect for a person also includes the requirement to provide additional protection to vulnerable populations such as children, prisoners, and people with impaired judgment.

Benefit

This principle requires researchers to maximize potential benefits and minimize potential participant harm. The commission interpreted charity as a moral obligation to do good and a necessary avoidance of harm. Applying this principle requires a rigorous assessment of the risk-benefit ratio to ensure that the potential benefits justify the potential risks.

Equity

This principle concerns the fair distribution of the benefits and burdens of research. The commission was particularly concerned about the exploitation of vulnerable populations, as demonstrated in the Tuskegee Syphilis Study. Fairness requires that the selection of research subjects be carefully scrutinized so that no group is unduly burdened or excluded from the benefits of the research. This principle also requires equal treatment of all research participants and avoidance of bias in the design and conduct of the research.

In addition to articulating these principles, the committee also provided practical guidelines for their application in research. These guidelines include the establishment of Institutional Review Boards (IRBs) that oversee research involving human subjects and ensure that ethical standards are followed throughout the research process. IRBs are tasked with reviewing research proposals, evaluating their ethical implications, and overseeing ongoing research to protect the rights and well-being of the participants.

The committee also addressed the need for ongoing research ethics training for researchers and IRB members. This emphasis on education reflects the view that ethical research practices must be continually strengthened and updated in response to new developments and challenges in this field. The Belmont Report and the National Commission's impact was profound and lasting. The principles and guidelines expressed in the report were incorporated into federal regulations governing human subject research, known as the Common Rule. These regulations provide a framework for the ethical conduct of research and have been adopted by several federal agencies to ensure a consistent and comprehensive approach to human protection.

The legacy of the Belmont Report extends beyond the United States and influences the international guidelines and standards for research ethics. Documents such as the Declaration of Helsinki, the CIOMS Guidelines, and the International Ethical Guidelines for Biomedical Research on Humans reflect the principles expressed in the Belmont Report and adapt them to a global context. In short, the Belmont Report is a critical milestone in the history of research ethics. This arose from conducting research involving humans under intense ethical scrutiny and humane care. The National Commission for the Protection of Subjects in Biomedical and Behavioural Research played a crucial role in identifying and articulating such research's ethical principles. The principles of respect for people, benevolence, and justice, as well as the practical instructions given by the commission, had a lasting impact on the ethical conduct of research, the protection of people, and the integrity of the research enterprise.

Core Ethical Principles

Respect for the Person: The Right to Self-Determination and the Requirement of Informed Consent

The principle of respect for a person is the basis of research ethics, and it emphasizes the inherent value of the person and the right to make independent decisions. This principle consists of two main parts: recognition of autonomy and protection of people with reduced autonomy. Recognition of autonomy means recognizing and respecting the ability of individuals to make informed and voluntary decisions regarding their participation in research. This respect is achieved through the informed consent process, which assumes that potential research participants receive all the information they need to make informed decisions about their participation. This information includes details about the purpose of the study, procedures, risks, benefits, and the right to withdraw at any time without penalties. For example, in a clinical trial testing a new drug, researchers must inform participants about the drug's known side effects, likelihood of receiving a placebo, duration of the trial, and potential benefits.

The consent process must ensure that the participants fully understand this information and voluntarily agree to participate. Researchers often use consent forms and verbal explanations to facilitate understanding and should be prepared to answer any questions that participants may have. Another aspect of respect for people involves providing additional protection to people with disabilities, such as children, the cognitively impaired, and prisoners. These groups may be more vulnerable to coercion or may not fully understand the consequences of their participation. For example, when researching with children, obtaining informed consent requires permission from parents or guardians and consent from children so that they can receive information appropriate to their age and understanding.

The case of Henrietta Lacks is an example of a violation of the principle of respect for a person. Her cancer cells were taken without her knowledge or consent in the 1950s, leading to major scientific breakthroughs but raising severe ethical issues. This case underlines the need for informed consent and respect for autonomy because it highlights the consequences of ignoring these ethical obligations.

Benefit

The obligation to minimize harm and maximize benefits. The principle of use includes the ethical obligation to maximize potential benefits and minimize potential harm to the research participants. This principle is based on the Hippocratic principle of "do no harm" and goes beyond simply avoiding harm to promote participants' well-being actively.

In practice, the utility requires that researchers conduct a thorough risk-benefit analysis before beginning the research. To do this, it is necessary to identify all possible risks for participants, assess the likelihood and severity of these risks, and compare them with the expected benefits of the study. The goal is to ensure that the potential benefits justify the potential risks and that those risks are minimized as much as possible. For example, in a study investigating a new surgical technique, researchers must consider the physical risks associated with the procedure, such as infection or complications, and weigh the risks against potential benefits, such as improved health outcomes or reduced recovery. Risk reduction measures included strict sterilization protocols, thorough preoperative screening, and careful postoperative monitoring.

The Tuskegee Syphilis Study illustrates a violation of the principle of beneficence. In this study, researchers deliberately withheld effective treatment from participants with syphilis, which caused unnecessary suffering and death. The study design prioritized scientific curiosity over the well-being of participants, ignoring their right to appropriate medical care. This case highlights the importance of prioritizing the well-being of research participants and ensuring that the search for knowledge does not come at the expense of human health and dignity.

Another example is the field of behavioral research, such as studies on psychological interventions. Researchers must consider the possible emotional and psychological effects of their interventions and take measures to support them. For example, in studies examining the effects of therapy on trauma survivors, researchers should provide access to counseling services and develop stress management protocols to ensure that participants receive the support they need. The benefits also extend to the dissemination of the research findings. Scientists are ethically obligated to share their findings with the scientific community and the public, contributing to collective knowledge and benefiting society. Dissemination must be performed responsibly to ensure the results are reported accurately and without exaggeration or omission.

Justice: justice in the distribution of the benefits and burdens of research. The principle of justice refers to the fair distribution of the benefits and burdens of research. It deals with ethical issues related to the choice of research subjects and equal treatment of different social groups. Fairness requires that no group of individuals be unnecessarily burdened to participate or systematically excluded from the research. This principle aims to prevent the exploitation of vulnerable populations and ensure that selecting research subjects is fair and equitable. For example, the exploitation of marginalized populations in research has a long history, as evidenced by the Tuskegee syphilis study, in which poor African Americans were treated and injured. Such exploitation violates the principle of fairness, emphasizing the need to carefully consider who is chosen for study and why.

In current research, the principle of fairness can be applied to clinical trials of new drugs. Researchers must ensure that the participants represent a diverse cross-section of the drug-using population. This means including women, minorities, and different age groups in the research rather than focusing only on a homogenous group that may not reflect the larger population. For example, studies of heart disease, which historically focused mainly on men, have also shifted to women because the disease manifests differently between the sexes, and the effectiveness of treatment can vary. In addition, the law requires that research benefits be available to all, especially those who have borne the burden of participation. This includes ensuring that successful treatments or interventions developed through research are available to the study population. For example, if research conducted in an impoverished community leads to a breakthrough in medicine, efforts should be made to make treatment affordable and accessible to that community.

Another aspect of justice is the fair distribution of resources and research opportunities. Funding, support, and recognition are equally shared between researchers and institutions. It also includes addressing systemic biases that may disadvantage certain groups, such as women or minority researchers, and ensuring they have equal opportunities to participate and benefit from scientific development. The principle of justice also intersects with the ethical aspects of post-investigation duties. Researchers are responsible for considering the long-term effects of their work on the communities involved and addressing possible negative consequences. This may include providing follow-up care or support to participants who experience adverse effects from participation.

Basic ethical principles such as respect for people, benevolence, and justice form the basis of ethical research and ensure research participants' protection and fair treatment. These principles, expressed in the Belmont Report, provide a comprehensive framework for addressing the complex ethical issues that arise in research on human subjects. Respect for people emphasizes the importance of autonomy and informed consent, which ensures that individuals have the necessary information and freedom to make voluntary decisions about their participation. The utility principle obliges researchers to maximize potential benefits and minimize potential harm, conduct thorough risk-benefit analyses, and implement measures to protect the well-being of participants. The principle of justice ensures a fair distribution of the benefits and burdens of research, prevents exploitation of vulnerable population groups, and promotes equal access to the fruits of research. These principles guide researchers to conduct ethical and responsible research that promotes trust and integrity within the scientific community. By following these principles, researchers can ensure that their work advances knowledge and respects and protects those involved with human dignity and rights.

APPLICATION OF THE BELMONT PRINCIPLES IN MODERN RESEARCH

Informed consent is the cornerstone of ethical research, ensuring that participants are fully aware of what their participation entails and voluntarily agree to participate. In contemporary research, applying informed consent to different settings presents unique challenges that require careful consideration and adaptation. In clinical trials, informed consent usually involves a detailed process in which participants are thoroughly informed of the study's purpose, procedures, risks, benefits, and rights. Written consent forms and verbal explanations often facilitate this. For example, in multisite clinical trials of a new cancer treatment, participants received detailed brochures, spoke with medical professionals, and asked many questions before agreeing to participate.

Informed consent becomes more complex in situations in which literacy levels vary significantly. For example, in a global health study conducted in a resource-poor setting, participants may have had limited literacy or exposure to scientific concepts. Researchers must adapt consent processes to these situations, often using visual aids, simplified languages, and verbal consent procedures. In a study investigating malaria interventions in rural Africa, researchers may use pictorial consent forms and community meetings to ensure the participants understand the study.

Digital and online research also bring new challenges and opportunities for informed consent. With the proliferation of online surveys, apps, and social media surveys, there is a need for clear, concise, and easily accessible digital forms of informed consent. Researchers must ensure that participants understand how their data will be used and that their privacy is protected. For example, an online study of social media behavior might use interactive consent forms with pop-up explanations and consent checkboxes to engage participants effectively.

Informed consent for research on vulnerable populations requires additional safeguards. For example, researchers must obtain consent from parents or guardians and the children in research involving children. This includes explaining the study in an age-appropriate language and ensuring children understand that they can withdraw anytime. In studies involving people with cognitive disabilities such as dementia, researchers may involve legal guardians in the consent process and use simplified explanations to facilitate understanding.

The principle of respect for people also requires that participants withdraw from the study at any time without penalty. Researchers must explain this right to ensure a straightforward withdrawal procedure. In longitudinal studies, where participants are followed over a long period, researchers must confirm their consent regularly to ensure they understand and accept their participation. Assessment of the risk-benefit ratio when planning and conducting research. The benefit principle requires that researchers maximize benefits and minimize harm, which requires a thorough assessment of the risk-benefit ratio when planning and conducting research. This process is critical to ensure that the research is ethically sound, and that the well-being of the participants comes first.

In clinical trials, risk-benefit assessment begins with a detailed analysis of the potential risks and benefits. Researchers must consider the severity and likelihood of risks ranging from physical harm, such as drug side effects, to psychological stress, such as that caused by invasive procedures. Benefits may include potential medical advances, better treatment, and personal health benefits for the participants. For example, when testing a new HIV vaccine, researchers must weigh the potential side effects against the significant benefits of preventing an HIV infection.

Innovative methods are often used to reduce risk. In drug research, Phase I trials usually involve small groups of healthy volunteers to assess safety and dosage. Phase II and III trials are conducted with larger groups of patients to assess efficacy and monitor side effects. Adaptive trial designs allow researchers to modify protocols in response to the interim results, improving safety and efficacy. For example, if an unexpected side effect occurs, the study design can be changed by reducing the dose or introducing additional monitoring.

In behavioral and social science research, risks may include emotional distress, privacy violations, or social consequences. Researchers should implement strategies to reduce these risks, such as providing counseling resources, ensuring confidentiality, and using anonymized data. For example, trauma recovery studies may include on-site counselors who support participants experiencing stress and use coded data to protect their identities. Ethical review committees, such as Institutional Review Boards (IRBs), are crucial when assessing the risk-benefit ratio. They reviewed research proposals to ensure risks were minimized and justified by their potential benefits. Investigators must provide detailed protocols, risk assessments, and plans for monitoring and managing adverse events. For example, an IRB that reviews research on a new surgical technique reviews the procedure's safety, the surgeon's expertise, and the procedures used to manage complications.

In modern research, the development of technology has brought new dimensions to risk-benefit assessment. For example, genomic research offers the potential for major medical breakthroughs but raises concerns about genetic privacy and discrimination. Researchers must address these risks by implementing strong data protection measures and ensuring participants fully know their potential consequences. The principle of futility extends to the dissemination of the research results. Researchers are ethically obligated to report results accurately and publicly to advance scientific knowledge and public health. This includes publishing negative or inconclusive results to avoid unnecessary research and to guide future research. For example, a failed drug trial must be reported to avoid repeating the same mistakes and inform ongoing activities in the field.

The principle of justice requires that the benefits and burdens of research be fairly shared. This requires ensuring that the selection of research subjects is fair and that no group is overburdened or excluded from the potential benefits. Historically, research has often used marginalized populations for little benefit. The Tuskegee Syphilis Study is an egregious example in which poor African American men were defrauded and refused treatment. Modern research ethics requires vigilance to prevent such exploitation and ensure that research subjects are selected for fair and transparent reasons.

Fair selection means including diverse populations in clinical trials to ensure the research findings are generalizable and applicable. Recruiting participants of different genders, ages, ethnicities, and socioeconomic backgrounds is necessary. For example, cardiovascular research must include women and minority groups that have been historically underrepresented to understand how treatments work in different populations. Researchers often use community engagement strategies to promote a fair selection. This requires working with community leaders, stakeholders, and local organizations to build trust and encourage participation. For example, in a study on diabetes management in Spanish communities, researchers could work with local health clinics and cultural organizations to recruit participants and ensure that the research was culturally sensitive.

Ensuring equal choices also means removing potential barriers to participation. Logistical challenges such as transportation and leisure time may prevent participation in poor or rural areas. Researchers can offer transportation stipends, flexible schedules, and compensation for lost wages to facilitate participation. For example, studies on rural health interventions could offer mobile clinic visits and financial incentives to ensure diverse participation. Vulnerable populations such as prisoners, children, and people with cognitive disabilities require special ethical considerations to avoid exploitation. Research on these groups should be reviewed to ensure that participation is voluntary and that additional safeguards are in place. For example, researchers must show that participation is mandatory in studies involving prisoners and that the benefits outweigh the risks. This may include providing independent consent advocates and ensuring participation does not affect the probation.

The principle of justice also extends to the distribution of research profits. Researchers have an ethical obligation to ensure that the study population benefits from the results. This may include access to successful treatments or interventions developed through the research. For example, suppose research on new tuberculosis treatment is being conducted in an impoverished community. In that case, efforts must be made to ensure the community can access treatment after the trial. Ensuring equity in global health research requires addressing health disparities and promoting equitable outcomes. Researchers must consider the broader implications of their work on global health inequality and work to reduce inequalities. For example, malaria prevention studies may involve collaboration with local health authorities to implement and maintain effective interventions throughout the study period.

The principle of justice also requires transparency and accountability during the investigative process. It requires reporting research findings in a manner that is accessible to all stakeholders, including participants, decision-makers, and the general public. Researchers must avoid selective reporting or misrepresenting results and ensure that the information produced benefits society.

Applying Belmont Principles to modern research ensures ethical research and human protection. By providing informed consent, researchers respected participants' autonomy and provided them with the information and support they needed to make voluntary decisions. Assessment of risk-benefit ratios ensures that potential harm is minimized and justified by the expected benefits, contributing to the well-being of participants. Ensuring a fair selection of research topics promotes equality and inclusion, prevents exploitation, and ensures that research benefits are widely shared.

As expressed in the Belmont Report, these principles continue to guide ethical research practices and promote trust and integrity in the scientific community. As research methods and contexts evolve, continued commitment to these principles will ensure that participants' human dignity and rights are respected and that the social benefits of research are maximized. By following these ethical principles, researchers can navigate the complexities of modern research and advance their knowledge while maintaining respect, benevolence, and justice.

The Belmont Report and IRB: An Evolving Relationship

With a particular emphasis on institutional review boards (IRBs) roles, the Belmont Report provides the underpinnings of ethical principles in research. These boards conduct casual trials and procedures involving human subjects following generally accepted ethical principles. IRBs are responsible for supervising the assessment and meeting of the Belmont principles regarding respect for people, beneficence, and justice in a research study to spare the risk to subjects' rights and well-being.

Therefore, the first principle of respect for people is implemented through the prism of the IRB's comprehensive and detailed consideration of informed consent mechanisms. IRBs assess the information disclosure level that participants may give to test conclusive, elaborate, and easily understandable consent forms. They ensured that the researchers obtained informed consent, emphasizing those with restricted capacity to consent. They are children, prisoners, and mentally challenged individuals. For instance, when an IRB committee considers a study of Alzheimer's disease, the committee may demand that the researchers translate the consent document to the patient level and ensure that the patient's legal representatives appreciate the implications of the research (Faden & Beauchamp, 1986).

Beneficence is a significant concept, and its management is prescribed by the various rules and regulations set out by the IRBs, along with thorough risk-benefit assessments. IRBs critically review proposals to ensure that every risk is controlled to the highest possible level and that the benefits of the research proposition are harnessed to the optimum degree possible. They evaluated the plans suggested by the researchers regarding the safety aspects of the modality and the events that were tracked during the research. For example, when an IRB assesses a new chemotherapy trial, the committee will examine the probable side effects and the available precautions to reduce them to ensure that the chance of recovery justifies the risks involved (Emanuel et al., 2000).

Here, IRBs conform to the principle of justice in conducting research through a fair selection of subjects. To avoid using individuals who may be coerced to participate in research or those who appear to have no agency over their lives, IRBs scrutinize recruitment strategies. They ensure that participants are not unfairly excluded from a site where they may benefit others or that certain groups are systematically exposed to costs that others are not exposed to. For instance, in research on cardiovascular health, an IRB may deem it necessary to have a mix of both male and female participants from various ethnic groups to have an idea of what is universal and what is fair for all, without discriminating the gender or ethnic group of a participant (Kass et al., 2007).

These case descriptions demonstrate the specific considerations for ethical subjects. They depict how the ethical principles of the Belmont Report are implemented in the IRBs' decision-making process. These examples clearly show the importance of IRBs in protecting participants and encouraging ethical practices in different research areas.

Case Study 1: HIV Prevention Trials: An Update in Developing Countries

However, in 1996 and 1997, several clinical trials were conducted to observe the efficacy of short-course antiretroviral drugs in PMTCT in developing countries. These trials raised improper ethical concerns, and the principle of justice was violated. The 'development' trials have become controversial. Ethicists have claimed they were unfair to the participants because some were given placebos. At the same time, there are existing disease treatments in developed nations.

The responsibilities of the IRBs that supervised these trials involved facing challenges that arose from the need to ensure that scientific research standards were met without compromising the rights of human subjects. They had to ensure that the research was beneficial to the general health of the public, specifically to the patients, but at the same time, they had to respect the rights of the patients who were part of the research. Finally, through extensive regulation of placebo-controlled trials, the IRBs made researchers explain why placebo control had to be used and made sure that their participants underwent effective informed consent, which included making them aware of the dangers involved and the availability of other medical treatments (Angell, 1997)

Case Study 2: The SUPPORT trial

In the context of the trial context, the SUPPORT Trial is also known as the Mortality for Neonates with Hypoxic-Ischemic Encephalopathy-Related Clinical Trials: Supplement to Randomized Trial; Intrapartum "Fetal Oxygen Obsession." The SUPPORT study conducted between 2004-2009 aimed to determine the best oxygen concentration levels among premature infants. The goal of the trial was to determine the optimal concentration of oxygen that is less hazardous to patients so that they are unlikely to suffer from blindness or any other related condition. However, this study has been criticized for its poorly informed procedures (Kimmelman, 2020).

The university's Institutional Review Board (IRB) approved the day-of-the-week effect hypothesis, considering the general healthcare benefits derived from the study and the overall scientific standard employed in conducting the study proposal. Subsequently, even the U.S. Department of Health and Human Services and Office for Human Research Protections (OHRP) investigated and informed the parents that the consent forms did not sufficiently describe the risk of different oxygen levels to the parents. As a result, the IRB was required to reconsider the way that consent is given for future experiments, the establishment of which must be more apparent in terms of risks and benefits; in doing so, the principle of Respect for People was reaffirmed (OHRP, 2013).

Case Study 3: We Focused on Gene Therapy or Treatment for Hemophilia, a Major Genetic Disorder

British researchers attempted to use gene therapy to treat hemophilia in the early 2000s by undertaking a study to correct the gene that resulted in the disorder. The trial contained certain risks, such as immune response and insertional mutagenesis. The IRB's responsibility was to weigh these risks against the possible benefits that would accrue from a treatment that would have a permanent fix for hemophilia.

Due to the concerns of the IRB, several safeguards to protect the participants were established, including stopping rules for any adverse events reported by the participants and constant supervision of the participants during their activities. Patients were made aware of all the risks involved, including the fact that this was a trial for therapy, and they were unsure of the long-term side effects. This rigorous oversight protected the principle of beneficence, as it insisted on maximizing benefits over risks and the entire disclosure rights of participants to the risks involved (Sheridan, 2003).

Case Study 4: This Study Focuses on the Case of Henrietta Lacks and Research Involving Biospecimens

In the case of Henrietta Lacks, whose cells were taken without her knowledge to develop the first "immortal" cell line, it is clear that respect for individuals and informed consent to participate in collecting and using biospecimens should be a fundamental principle. In translational research. Because of the ethical breach in the Lacks case, many changes have been made to how IRBs review research involving human tissue samples (Kimmelman, 2020). Advanced IRBs require informed consent from analysts to use samples when tests cannot be identified. They ensure that members are informed of the possible uses of their tissue, including commercial applications, and of their right to withdraw their consent. This case highlights the importance of IRBs in dealing with today's ethical challenges and strengthening the standards of the Belmont Report (Skloot, 2010).

Case Ponder 5: The Stanford Prison Experiment

In a 1971 investigation of the Stanford prison, college students acted as prisoners and bell ringers. The meditation ended early because of extraordinary mental problems among the members. Although IRBs were not as common at that time, this case has taught IRBs in mental health research ever since. High-level IRBs review the research design for potential psychological harm and require strong safeguards, rigorous research, and members' mental well-being. The testament of the experiment states the importance of interest and the requirement of the IRB to ensure that the risks of emotional harm are minimized and that the well-being of participants is prioritized (Arduino, 2007).

Case Study 6: Antibody Trials Amid the COVID-19 Widespread

The rapid development of antibodies against COVID-19 has presented particular moral challenges to IRBs, including reconciling critical and open welfare with moral standards of goodness and justice. The IRB ensured vaccinations were administered quickly and safely through a thorough risk-benefit assessment and impartial member allocation. IRBs require vaccinations to cover different populations, ensuring the findings are relevant across different statistical groups. They also ordered precise and thoroughly educated consent forms that clearly explained the litmus test of natural hazards and the unclear long-term effects of vaccination. This ensured that the study met the standards of the Belmont Report for addressing the global welfare crisis (World Health Organization, 2020).

Case Study 7: Mental Inquire About Injury Recuperation

Research examining the effectiveness of various treatments for trauma survivors requires careful ethical consideration because of the potential for traumatization. The IRB overseeing the study required extensive support systems, including access to counselors and regular check-ins, to monitor participants' mental health. A rigorous IRB review process ensured that the study adhered to the principle of utility, minimized harm, and provided appropriate support to participants (Schnurr et al., 2007).

Challenges and Critiques

The Belmont Report, distributed in 1979, provides foundational moral rules for investigating human subjects. As investigative strategies and settings have advanced, a few restrictions have been developed, especially regarding modern issues such as computerized information security and worldwide investigative hones (Vollmer & Howard, 2010).

One critical impediment is the Belmont Report's need for direction in advanced information protection. With computerized innovations and enormous amounts of information, analysts presently collect and analyze large amounts of individual data through online stages, social media, and wearable gadgets. However, the standards of regard for people, advantages, and equity require reinterpretation to address the complexities of advanced information collection and utilization. Occasionally, educated assent within the computerized age faces challenges such as guaranteeing that members completely understand how their information will be utilized, put away, and shared. In numerous cases, advanced information is collected latently without expressing assent, raising moral concerns about protection and independence (Metcalf & Crawford, 2016).

In addition, the Belmont report does not address information security issues or potential information security breaches. In an era where cyber-attacks and data breaches are becoming increasingly common, analysts must implement strong data assurance measures to protect participant data (Vollmer & Howard, 2010).

A helpful guideline that emphasizes harm minimization should include considerations for additional harm, such as identity breaches and unauthorized information. This has led to the demand for a review of ethical rules, especially regarding data and information security in computational research (Mittelstadt & Floridi, 2016). Global questions about mills are another challenge in the Belmont Report. These standards were created in the context of the United States and can only partially explain the social contrasts and moral standards of other countries. For example, informed consent forms that are feasible in the Western world may not be suitable for societies with different communication styles and structures.

Analysts working in different social contexts must adapt their approach to ensure informed consent and be informed by nearby traditions. This may involve exploiting community consensus or shutting down nearby pioneers to promote understanding and belief (Bhutta, 2004). Equity guidelines, which focus on an impartial selection of research topics, also face challenges in global research. Indeed, state-building studies have sometimes abused vulnerable populations without corresponding benefits (Kimmelman, 2020).

The equality goal of the Belmont Report should be expanded to address misconduct and ensure that the benefits of the investigations are shared equally. This involves effective mediation and ensuring nearby communities benefit from research in their areas (Benatar, 2002). Furthermore, the rapid pace of innovative development often exceeds the ability of moral rules—new advances such as artificial intelligence (AI) and machine learning offer previously untapped moral situations. AI survey frameworks can inadvertently propagate biases in training data, leading to unjustified results (Kimmelman, 2020).

Belmont Standards should be reviewed and updated to guide the ethical use of AI in surveys, ensuring that these innovations are created and, updated). communicated in a way that makes sense, is simple, and is human rights friendly (Florido et al., 2018Speaking about the clarification and implementation of the standards. The translation and implementation of the standards of The Belmont Report have been the subject of an ongoing debate. These debates often focus on how the three central standards are changed and adapted to inquiries related to the contemporary environment (Vollmer & Howard, 2010).

Analysts must adapt by providing comprehensive information and ensuring the acceptance concept is manageable. In complex clinical trials, consent forms can be long and full of technical language, making it difficult for members to understand fully. This raises questions regarding how consent forms can be distinguished when providing relevant information. Some have suggested multi-level consent forms, where members get an overview of the platform and can request information on a point-by-point basis if desired (Manson & O'Neill, 2007).

Another area of discussion is guidance on utility and how harm and benefit can be accurately assessed and adjusted. In cutting-edge research, such as quality control and stem cell therapy, long-term risks are regularly obscured, making it challenging to ensure that the benefits outweigh potential harm. This vulnerability can lead to conflict among analysts, ethicists, and governing bodies regarding the moral acceptability of such issues. For example, the use of inventions to change the quality of people has been discussed, as well as the potential for unexpected results compared with the treatment of inherited diseases (Baylis & Robert, 2004). Equity guidelines, which require common sense to conduct surveys of benefits and burdens, are particularly inconvenient for global well-being research.

Experts argue that more prosperous countries or subjects regularly benefit, even if considered morally, while the burden falls on poorer and more vulnerable populations (Vollmer & Howard, 2010). For example, clinical trials are typically conducted during drug trials. In poorer countries, future medicines are regularly estimated to be beyond the reach of the trial population. This has led to demands for stricter ethical codes to ensure that research conducted in impoverished settings directly benefits these communities (Emanuel, 2008). Disagreements arise regarding the relative weights of Belmont's principles. In some cases, the standards may conflict, forcing analysts and national audit authorities to prioritize one guideline over another (Rafi & Snyder, 2015). For example, to investigate the circumstances in a crisis, such as amid a plague, the requirement for rapid information gathering and intercession (utility) may conflict with the requirement for carefully trained consent (respect for people).

IRBs must carefully examine these conflicts, often case-by-case, to decide the most critical moral course of action (Levine, 2008). Applying the Belmont Report to social and behavioural research presents additional challenges. Unlike biomedical research, social and behavioural research often attempts to negotiate more subjective and variable wonders, making it difficult to apply standard moral rules. For events, securing informed consent in ethnographic research, in which intuition and data collection can be fluid and progressive, requires an adaptive and context-specific approach.

Analysts must be permanently enclosed between members, and consent forms must be adapted to the energetic nature of the research environment (Guillemin & Gillam, 2004). In addition, the increasingly widespread use of conspiracy research, which combines techniques from specific fields, complicates the application of Belmont Standards. Moral rules that work well for one teacher may not be easily transferred to another. For example, combining therapeutic and social science data raises complex data security, informed consent, and randomization issues. Analysts must explore these exciting challenges by creating customized moral systems that satisfy the core standards and appeal to the specific perspectives of each field (Resnik, 2012).

In response to these challenges and controversies, some have called for a revised and expanded moral system based on the Belmont Report while dealing with contemporary issues. It appears to include additional standards, such as those dealing with community engagement and social responsibility, to reflect the complexity of advanced research better. For example, the concept of "benefit sharing" has been proposed to ensure that all partners, including participants and communities, receive a fair return from preliminary research ((CIOMS, 2016).

Thus, while the Belmont Report introduced significant moral research, its standards require constant reinterpretation and adaptation to meet contemporary challenges. Contradictions in the translation and implementation of norms highlight the need for an adaptive and context-sensitive approach to studying morality. By continuously advancing and fostering moral rules, the scientific community can protect its members' rights and well-being in the ever-changing arena of logical inquiry.

The Future of Ethical Guidelines in Human Research

The moral rules overseeing human inquiry have advanced, reflecting rapid progress in innovation, societal changes, and a developing understanding of the moral suggestions of logical requests. As we set out in this new era of logical disclosure, it is essential to consider developing moral contemplations within the computerized age and the requirement for energetic and ethical frameworks that can adjust to the ever-changing scene of innovative and societal headways.

One of the foremost critical rising moral contemplations within the computerized age is the issue of information protection and security. With the multiplication of advanced advances, tremendous amounts of individual and delicate information are collected, removed, and analyzed without satisfactory security. This raises concerns about the potential abuse of information, such as character burglary, surveillance, and control (Vitak & Ashktorab, 2016). In addition, the expanding dependence on manufactured insights (AI) and machine learning calculations raises questions about responsibility, straightforwardness, and reasonableness in decision-making. As analysts create and utilize these innovations, it is fundamental to establish vigorous moral rules that guarantee information security and promote responsible use. (Vollmer & Howard, 2010).

Another primary concern is the morals of manufactured insights (AI) in healthcare. AI has the potential to revolutionize healthcare by providing personalized medications, moving forward with demonstrative exactness, and streamlining healthcare administration. However, there are concerns about predisposition, security, and responsibility in AI decision-making. AI frameworks may propagate existing predispositions and segregate them against specific bunches based on their characteristics (Vollmer & Howard, 2010). Additionally, AI frameworks may not be straightforward or responsible for their choices, leading to unintended results. To address these concerns, analysts must create moral systems prioritizing straightforwardness, decency, and accountability for AI advancement.

To expand these developing moral contemplations, there is a need for energetic moral systems that can react to mechanical and societal changes. Conventional moral systems frequently center on particular issues or applications that can become obsolete as innovation propels. To address this challenge, analysts must create moral systems that are versatile, adaptable, and responsive to changing circumstances. This may include creating layered systems that can suit distinctive levels of hazard or complexity or setting up a preparation for routinely updating and revising moral rules.

Moreover, there is a developing recognition of the significance of partner engagement and support in moral decision-making. Historically, moral committees frequently depended on master's suppositions and scholastic investigations to illuminate their choices. In any case, this approach may only partially capture different partners' perspectives and concerns. Analysts must cooperate with different partners to address this challenge through inquiries about preparing and counting members, community leaders, and policymakers. This may help guarantee that research is socially mindful, socially delicate, and responsive to the desires and values of diverse communities (Vitak & Ashktorab, 2016).

Another essential consideration is the need for an intriguing ethical approach. Historically, morals have been viewed as teaching in the scholarly community. However, numerous moral issues have emerged at the intersection of numerous disciplines, such as law, medicine, reasoning, and human science. Analysts must develop intriguing approaches to coordinating knowledge from different areas to address these complex issues. This could include collaborating with specialists from different disciplines or creating courses integrating ethics with other areas.

In expanding these emerging ethical contemplations and the need for energetic moral systems, there is a growing acknowledgment of the significance of moral instruction in investigating and preparing programs. Generally, inquiries about preparation programs have focused primarily on specialized aptitudes and methodological skills. Moral instruction is now recognized as an essential component of inquiry into preparing programs (Vitak & Ashktorab, 2016). This includes joining moral instruction in enquiring about preparing programs from the early stages and giving continuous back and assets to analysts as they explore complex moral issues.

In addition, there is a developing acknowledgment of the significance of ethics in enquiring about arrangement advancement. Enquiring about approaches often provides far-reaching suggestions for investigating hones and outcomes. To address this challenge, analysts must lock in with policymakers through the approach advancement handle to guarantee that approaches are informed by moral contemplations and promote responsible investigations.

CONCLUSION

In conclusion, the long-term moral rules in human inquiry are complex and multifaceted. Rising moral contemplations within the computerized age require energetic moral systems that can react to mechanical and societal changes. To address these challenges successfully, analysts must create layered systems that suit distinctive levels of hazard or complexity, engage with differing partners through investigative preparation, embrace interdisciplinary approaches to morals, provide ethics education in inquiry about preparing programs, and lock in with policymakers to investigate approach improvement (Vitak & Ashktorab, 2016). Doing so guarantees that human investigations are conducted capably, ethically, and about human nobility.

The Belmont Report's compelling testimony to the development of moral research reinforces the importance of moral reflection in human research. Since its publication in 1979, the report has helped researchers and educational and administrative organizations ensure that human subjects are protected from harm and that research has been conducted ethically and competently. Reporting standards for people, interests, and equality have been widely adopted and continue to influence research.

The significance of the Belmont Report goes beyond its immediate impact on the scientific community. It has significantly improved investigative ethics approaches and rules, both widely and globally. The report's emphasis on the importance of informed consent, risk-benefit analysis, and the safety of vulnerable populations has influenced the design of moral life research and treating human subjects with respect and dignity (Kimmelman, 2020).

Despite the criticism, progress is still needed in the ethical conduct of research, ongoing care, and ensuring adaptation. As research progresses, untapped challenges and ethical situations emerge, which require analysts, educators, and administrators to be alert and responsive to emerging questions. Therefore, we call for moral vigilance and adaptability in human research. This includes:

- Ensuring that research members are adequately educated about the risks and benefits of collaboration for which their consent is voluntary.
- Constant monitoring and evaluation of the morale of unused quests for strategies and progress.
- We respond to the needs of vulnerable people, such as children, older adults, and people with disabilities, who may be at greater risk of abuse or harm
- Increasing directness and responsibility in questions provoke details of conflicting cases and events.
- Preparation and guidance of analysts, members of the control committee of the organization, and other partners in ethical investigations.
- Develop global cooperation and coordination on moral issues to ensure regional coherence and consistency.

By understanding these standards and codes, we can ensure that human research can advance our understanding of the world while protecting the rights and dignity of those interested.

REFERENCES

Adams, D. P., & Miles, T. P. (2013). The application of Belmont Report principles to policy development. *Journal of Gerontological Nursing*, 39(12), 16–21. DOI: 10.3928/00989134-20131028-07 PMID: 24219075

Adashi, E. Y., Walters, L. B., & Menikoff, J. A. (2018). The Belmont Report at 40: Reckoning with time. *American Journal of Public Health*, 108(10), 1345–1348. DOI: 10.2105/AJPH.2018.304580 PMID: 30138058

Barrow, J. M., Brannan, G. D., & Khandhar, P. B. (2017). *Research ethics*. Europe PMC.

Blood, B. (2003). The Belmont Report: Ethical principles and guidelines for the protection of human subjects research. *Bracketing Interview, 58*, 115. *Foundations for Research*, 59, 421.

Brothers, K. B., Rivera, S. M., Cadigan, R. J., Sharp, R. R., & Goldenberg, A. J. (2019). A Belmont reboot: Building a normative foundation for human research in the 21st century. *The Journal of Law, Medicine & Ethics*, 47(1), 165–172. DOI: 10.1177/1073110519840497 PMID: 30994072

Cox, D. J., Suarez, V. D., & Marya, V. (2023). Ethical principles and values guiding modern scientific research. In *Research ethics in behavior analysis* (pp. 35–61). Academic Press. DOI: 10.1016/B978-0-323-90969-3.00008-6

Emanuel, E. J., Wendler, D., & Grady, C. (2000). What makes clinical research ethical? *Journal of the American Medical Association*, 283(20), 2701–2711. DOI: 10.1001/jama.283.20.2701 PMID: 10819955

Emanuel, E. J., Wendler, D., & Grady, C. (2008). An ethical framework for biomedical research. *The Oxford textbook of clinical research ethics*, 123-135. Oxford University Press.

Faden, R. R., & Beauchamp, T. L. (1986). *A history and theory of informed consent*. Oxford University Press.

Ferdowsian, H., Johnson, L. S. M., Johnson, J., Fenton, A., Shriver, A., & Gluck, J. (2020). A Belmont Report for animals? *Cambridge Quarterly of Healthcare Ethics*, 29(1), 19–37. DOI: 10.1017/S0963180119000732 PMID: 31581963

Friesen, P., Kearns, L., Redman, B., & Caplan, A. L. (2017). Rethinking the Belmont Report? *The American Journal of Bioethics*, 17(7), 15–21. DOI: 10.1080/15265161.2017.1329482 PMID: 28661753

Guillemin, M., & Gillam, L. (2004). Ethics, reflexivity, and "ethically important moments" in research. *Qualitative Inquiry*, 10(2), 261–280. DOI: 10.1177/1077800403262360

Kimmelman, J. (2020). What is human research for? reflections on the omission of scientific integrity from the Belmont report. *Perspectives in Biology and Medicine*, 63(2), 63. DOI: 10.1353/pbm.2020.0017 PMID: 33416651

Lantos, J. D. (2020). The Belmont Report and innovative clinical research. *Perspectives in Biology and Medicine*, 63(2), 389–400. DOI: 10.1353/pbm.2020.0026 PMID: 33416660

Levine, G. (2008). *Realism, ethics and secularism: essays on Victorian literature and science*. Cambridge University Press. DOI: 10.1017/CBO9780511484872

Manton, A., Wolf, L. A., Baker, K. M., Carman, M. J., Clark, P. R., Henderson, D., & Zavotsky, K. E. (2014). Ethical considerations in human subjects' research. *Journal of Emergency Nursing: JEN*, 40(1), 92–94. DOI: 10.1016/j.jen.2013.11.001 PMID: 24384483

Miller, R. B. (2003). How the Belmont report fails. *Essays in Philosophy*, 4(2), 119–134. DOI: 10.5840/eip2003423

Nagai, H., Nakazawa, E., & Akabayashi, A. (2022). The creation of the Belmont Report and its effect on ethical principles: A historical study. *Monash Bioethics Review*, 40(2), 157–170. DOI: 10.1007/s40592-022-00165-5 PMID: 36357708

Nicolaides, A. (2016). Bioethical considerations, the common good approach and some shortfalls of the Belmont Report. *Medical Technology SA*, 30(1), 15–24.

Perlman, D. (2006). Putting the" ethics" back into research ethics: A process for ethical reflection for human research protection. *The Journal of Research Administration*, 37(1/2), 13.

Rafi, N., & Snyder, B. (2015). *Ethics in research: How to collect data ethically. Doing Research in Emergency and Acute Care: Making Order Out of Chaos*. Wiley.

Resnik, D. B. (2012). *Environmental health ethics*. Cambridge University Press. DOI: 10.1017/CBO9781139161848

Schupmann, W., & Moreno, J. D. (2020). Belmont in context. *Perspectives in Biology and Medicine*, 63(2), 220–239. DOI: 10.1353/pbm.2020.0028 PMID: 33416649

Shore, N. (2006). Re-conceptualizing the Belmont Report: A community-based participatory research perspective. *Journal of Community Practice*, 14(4), 5–26. DOI: 10.1300/J125v14n04_02

Siddiqui, W., & Sharp, R. R. (2021). Beyond the Belmont Report. *The American Journal of Bioethics*, 21(10), 1–4. DOI: 10.1080/15265161.2021.1972649 PMID: 34554060

Vitak, J., Shilton, K., & Ashktorab, Z. (2016, February). Beyond the Belmont Principles: Ethical challenges, practices, and beliefs in the online data research community. In *Proceedings of the 19th ACM Conference on Computer-supported Cooperative Work & Social Computing* (pp. 941-953). DOI: 10.1145/2818048.2820078

Vollmer, S. H., & Howard, G. (2010). Statistical power, the Belmont Report, and the ethics of clinical trials. *Science and Engineering Ethics*, 16(4), 675–691. DOI: 10.1007/s11948-010-9244-0 PMID: 21063801

ADDITIONAL READING

Childress, J. F., Meslin, E. M., & Shapiro, H. T. (Eds.). (2005). *Belmont revisited: Ethical principles for research with human subjects*. Georgetown University Press.

Throne, R., & Hendon, M. (2024). Belmont 2.0: Ethical research expectations for PII in AI, ML, and data mining/scraping. In *Methodologies and ethics for social sciences research* (pp. 1-19). IGI Global.

Wester, K. L., Deaton, J. D., & Jayamohan, H. (2023). Ethics in research. In *Re-imagining research* (pp. 45–72). Routledge. DOI: 10.4324/9781003196273-3

Chapter 11
Belmont 2.0: Ethical Research Expectations for PII in AI, ML, and Data Mining/Scraping

Robin Throne
https://orcid.org/0000-0002-3015-9587
University of the Cumberlands, USA

Michalina Hendon
https://orcid.org/0009-0003-6088-2658
University of the Cumberlands, USA

ABSTRACT

With the growth of ubiquitous digital technologies, such as artificial intelligence and machine learning, specialized training and preparation are needed to best guide social science researchers for human research protections that involve protections for data with personally identifiable information. Some human research protection (HRP) officers have called for a Belmont 2.0 that offers more aligned guidance for HRP programs and institutional review boards (IRB) to address data ethics in this new era. This chapter presents an analysis of the shifting climate of HRP data ethics, Belmont Principles, and IRB and HRP implications for artificial intelligence, machine learning, data mining/scraping, and other ubiquitous technologies.

INTRODUCTION

In past work, the chapter authors, with others, have continued to examine the shifting climate of HRP and specific to PII data ethics within the era of emergent ubiquitous technologies and respect to the use of AI, ML, and DMS (Throne, 2022;

DOI: 10.4018/979-8-3693-3848-3.ch011

Throne et al., 2023). In 2022 and 2023, many regulatory bodies and other governmental groups globally issued new regulations and restrictions for PII, AI/ML, DMS, and personal data privacy for these continually evolving technologies (Throne, 2022). It remains essential in this time of momentous change for IRB and HRPP leadership to remain attentive to these emergent policy events in the context of the growing use of these ubiquitous digital technologies for HRP and PII protections.

Further, the Menlo Report[1] from the U.S. Department of Homeland Security (2012) in Science and Technology provided a set of guidelines and ethical principles for communication technology research (Throne et al., 2023). Built on the Belmont Report, the Menlo Report includes the three principles of the Belmont Report: Respect for Persons, Beneficence, and Justice. It includes a fourth principle of Respect for Law and Public Interest, and some authors have noted its relevance for biomedical computational research yet limited and dated for a contemporary data ethics framework (Hosseini et al., 2022). Many have since called for the scientific community and regulatory bodies to devote more attention and scrutiny to the ethical issues that surround big data (BD) research, including AI, ML, and DMS (Doerr & Meeder, 2022; Eto, 2022; Hosseini et al., 2022). More recently, Finn and Shilton (2023) noted the Menlo Report offered a case study in governmental ethics and the trade-offs sometimes necessary to establish ethical guidance amid rapid technological change. They explained, "The processes we observed at work during in the Menlo Report can be used as a framework not only for understanding how the development of ethics governance has proceeded in the past, but also for predicting tensions within, and perhaps improving, current and future projects that seek to bolster local ethics work with formalized governance" (Finn & Shilton, 2023, p. 335). Thus, ethical governance, such as revisiting the Belmont Report, may learn and gain understanding from these past attempts to establish contemporary ethical guidelines for this digital era.

Other researchers and legal scholars have noted the risks for PII in the use of these methods and, thus, the threats to data privacy (Wachter & Mittelstadt, 2019). Likewise, Ferretti et al. (2022) and Eto (2022) emphasized the challenges for research ethics committees in the review of BD research and the need for a consensus on systematic means to ensure ethical and responsible research involving AI, ML, DMS, and other ubiquitous technologies for recruitment, consent, risks and benefits, and data collection. Further, Eto (2022) specifically clarified the intention to define AI HSR[2] that may differ from the use of AI generally and noted the importance of consistency in definition and categorization for the use of these technologies in the context of human subjects research (HSR) can be especially helpful to the necessary policy monitoring and changes needed to regulations written in a prior context.

METHOD

As such, for this examination, the chapter authors conducted a literature review of the current scholarship surrounding the calls for Belmont 2.0 amid the contemporary climate of IRB and HRPP data ethics in the context of the Belmont Principles and IRB and HRP implications for AI, ML, DMS, and other ubiquitous technologies used for research. Literature review using a high-quality, rigorous selection of scholarly sources is an acceptable methodology to "allow researchers to enhance their understanding of prior work in their field, enabling them to identify gaps more easily in the body of literature and potential avenues for future research. More importantly, review articles may challenge established assumptions and norms of a given field or topic, recognize critical problems and factual errors, and stimulate future scientific conversations" (Kraus et al., 2022, p. 2578). In addition, a critical review can bring together these perspectives to thoroughly examine an existing challenge or problem within the discipline.

Thus, this chapter's objective is to present the results of analysis into the current scholarship surrounding the shifting climate of HRP and PII data ethics, Belmont Principles, and IRB and HRP implications for AI, ML, DMS, BD, and other ubiquitous technologies. Additional considerations are offered for the calls for Belmont 2.0 as a retrospective, considering today's use of ubiquitous technologies, such as generative AI, for the development of adequate HRP and PII protection policies in this evolving digital climate.

The Belmont Principles and Protections for Human Subject Personally Identifiable Information

Nagai et al. (2022) provided a regulatory history of the Belmont Report[3] and its three inherent principles—Respect for Persons, Beneficence, and Justice—and their specific requirements for informed consent, assessment of risks and benefits, and participant recruitment and selection. These principles are often interpreted as respecting autonomy (Respect for Persons), reducing harms (Beneficence), and promoting social justice (Justice) (Atenas et al., 2023). At the previous and 40th anniversary of the publication of the Belmont Report, several authors and researchers used these milestones to share their viewpoints on the need to revisit the Belmont principles, and the authors noted that opinions continue to differ on its ability to protect human subjects in research (Nagai et al., 2022).

For example, Brothers et al. (2019) noted the call for comment from 2011-2017 by the U.S. Office for Human Research Protections that resulted in a revised Common Rule, the federal regulations under 45 CFR 46[4] for human research protections, and in these opinions, called for a Belmont "reboot" at its 40th anniversary due to the change of contemporary context compared to the climate that existed when the Belmont Report was drafted. "More is at stake in a reboot of the *Belmont Report* than simply reforming the Common Rule. If we are to truly modernize the oversight and governance of research with human participants, we will need to expand the scope of policies that will receive attention in this broad effort" (Brothers et al., 2019, p. 171).

In prior work, the chapter authors stressed that the Belmont principle of Justice does hold implications for the 21st century if interpretations are considered within contemporary sociocultural contexts. As the Belmont authors stated in the fundamental ethical principle of Justice, *"Against this historical background, it can be seen how conceptions of justice are relevant to research involving human subjects."*[2] Likewise, the chapter authors previously reflected on these burdens and benefits in the context of contemporary ideologies that may challenge this long-standing HRP principle:

> *"Under the Belmont Principle of Justice, human protections professionals must examine race. We must consider disability, age, and gender. We must assess the susceptibility and sensitivity of the individual participant. We must consider bias and fairness. Equity as justice. Justice as inclusion. We must consider the vulnerable, the marginalized, the underrepresented. We must examine inclusion as well as exclusion...Belmont requires us to closely inspect informed consent so each individual can choose what shall or shall not happen to them. To each person, balanced and equitable risk and benefit. Choice as an individual decision empowers everyone, as everyone has the right to participate in research as they so choose rather than to be systematically excluded or included. Let us leave those injustices where they belong: in the pre-1974 history of human research protections" (Throne, 2023)*

Equally, Atenas et al. (2023) also stressed the need for considerations of Justice and inclusion for BD research, especially when examining the burden of harms carried historically by specific underserved and vulnerable groups, and offered various options beyond Belmont to do so:

> *"Thus, it is key to promote interdisciplinary dialogues and seek input from those who deal with sensitive topics, high-risk research situations, and/or vulnerable populations in the context of the epistemic and axiological shifts, where we are all potentially vulnerable, and all data are potentially sensitive" (Atenas et al., 2023, p. 22)*

Nevertheless, other researchers have noted that even the Justice principle and assessment of the burdens of research may be outmoded when considering current ubiquitous technologies (Costello et al., 2023). However, Belmont's ethical framework has served as an appropriate foundational ethical framework to guide new researchers (Atenas et al., 2023; Costello et al., 2023; Raman et al., 2023; Throne, 2023).

Further, Atenas et al. (2023) recommended that limited principles within an ethical framework, such as Belmont provides, may be ideal for undergraduate education.

Yet, Doorn (2022) highlighted the accessibility of researchers to access publicly available large datasets with identifiable information that has increased risks for the protection of PII:

> *"Nowadays more and more personal data of citizens are available. Many consumer electronics are equipped with sensors and internet technology, with which personal data can be collected, stored, and exchanged. Citizens partly share these data consciously, for example via social media, but governments and companies are also collecting a lot of data from citizens. There is an ever-increasing link between different data sources, sometimes blurring the line between public and private sources" (p. 2433)*

In this light, Atenas et al. (2023) highlighted that the issues surrounding research with ubiquitous technologies are further complicated by scholarship within for-profitdriven academic databases that are not free of algorithmic bias. Further, Costello et al. (2023) recommended that ethical review must extend beyond and after the research analyses and be considered again within the scholarly publication realm with explicit identifiers that lead back to the IRB or HRPP identification numbers as a form of traccability. "The identification number could be traced back to a committee/board, who could in certain circumstances validate whether the published study had indeed received approval as claimed. This would provide accountability for researchers" (Costello et al., 2023, p. 77).

With respect to ethical research with ubiquitous digital technologies and appropriate human subjects and their PII protections, Costello et al. (2023) also recommended improvements in ethics education for both new and veteran researchers, not as a mode of compliance but as an integrated aspect of the research design. Raman et al. (2023) echoed the need for ethical frameworks in the use of public secondary datasets beyond the regulations and principles followed by IRBs and HRPPs, especially considering the preparation of new investigators and the "topic-centric data repositories that are instantaneously and publicly accessible" for their use (p. 325). The authors also called for balancing the Belmont principles with other more applicable ethical frameworks for research with these types of public datasets as "students should evaluate if the proposed question potentially has elements that will not be ethical (formulate questions) and if the data are reasonable and appropriate for answering the question (collect/consider data). This is particularly relevant when secondary datasets are used" (Raman et al., 2023, p. 328).

Calls for Belmont 2.0

In 2018, many authors and researchers reflected on the 40th anniversary of the Belmont Report and its implications for HRP in a 21st-century digital world unforeseen in the decade Belmont was drafted. For example, Adashi et al. (2018) called for a reappraisal of Belmont at the anniversary. The authors suggested a Belmont addendum as the report never claimed "timelessness" and a need for scrutiny within the contemporary climate (p. 1347). "More recent reservations have centered on the necessity of "updates" impelled by evolving societal and scientific norms that could not have been foreseen by the Commission…Although the Belmont Report as such may not be a candidate for revision, serious consideration should be given to an addendum in recognition of the emerging issues of our time" (Adashi et al., 2018, p. 1347).

Further, authors, researchers, and HRP professionals continued to issue calls for a "Belmont 2.0" as a reappraised version of the enduring principles first offered in the Report that may not have anticipated the advanced digital age to come. For example, Raymond (2019) suggested, "Belmont 2.0 can ensure the clear benefits of big-data research are adequately balanced against poorly understood risks and harms. It should not require abuses on the scale of the Tuskegee syphilis study to prompt us to create guidelines fit for the digital world" (p. 277). Specifically, Raymond (2019) stressed that the "abrogation of the research community's ethical duties" continues under guidelines written well before the inception of the world wide web and BD research (p. 277). For example, the author noted:

> *"It is past time for a Belmont 2.0. We should not be asking those tasked with protecting human participants to single-handedly identify and contend with the implications of the digital revolution. Technological progress, including machine learning, data analytics and artificial intelligence, has altered the potential risks of research in ways that the authors of the first Belmont report could not have predicted" (Raymond, 2019, p. 277)*

In a *Nature* article published in a retrospective of the Belmont 40th anniversary, Raymond (2019) offered specific examples of risks to PII not addressed by current regulations or the Belmont principles. Research involving publicly aggregated data can threaten individual privacy and security as the data may be used to create new group classifications. Group data sets can make PII transparent through demographic classification and tracking and harms from mobile application data that reveal PII via time and place data collection (Raymond, 2019).

Likewise, Gross and Miller (2019) noted the impact of outmoded regulations in a digital age and specifically the limitations of the Health Insurance Portability and Accountability Act (HIPAA) to protect protected health information (PHI) and PII in an age of the Internet of Things (IoT) in support of other repeated calls for updated ethical guidelines in this context, such as Belmont 2.0. Gross and Miller (2019) emphasized the points raised by Raymond (2019). The authors added that the wholesale availability of PHI and PII has resulted from the "privacy-eroding power of machine learning in a digital world" despite the promises of healthcare technology for a better world" (p. 3). The authors recommended "privacy-preserving" technological solutions under a Belmont 2.0 could provide a balance between the need for ongoing research and data privacy protections and recommended:

"By making use of these privacy-preserving technologies the default within any Belmont 2.0" guidelines, we can reaffirm the primacy of health data privacy as a value, while supporting the principle of continuous learning to improve quality and efficiency of clinical care and equity of our healthcare systems. By rendering health data secure and its use transparent, and disproving the assumption that data must be shared for the utility of its analysis to be extracted, blockchain technology and affiliated privacy-preserving innovations could empower our answer to the call for ethical guidelines for the use of individuals' data for research" (Gross & Miller, 2019, p. 6)

Similarly, Stewart (2021) called out the challenges in BD research, specifically for consumer-based datasets, that Belmont's ethical definitions for human subject research fall short of the privacy challenges and informed consent implications for the use of these types of public data collection. For example, despite Belmont's ethical protections for vulnerable populations, including fair and equitable distribution of the burdens of research, incidents of harm have occurred by data brokers that disproportionally affected individuals within lower socioeconomic status or people of color (Stewart, 2021). Data brokers profile consumers by scouring social media, web browsing history, online and offline purchase history, credit card data, and government records such as driver's licenses, motor vehicle records, census data, birth certificates, marriage licenses, and voter registration information. Data brokers also use this information to categorize consumers into demographic groups based on susceptibility to certain advertising or likelihood to buy certain products, and current HRP policies do not protect PII in such instances as indirect identifiers can be aggregated for market research to assess the demographics of certain groups. IRBs, HRPPs, and social and health science researchers continue raising these concerns and other commercial purposes with calls for continued examination for policy change. Stewart (2021) explained,

"Once the participant has become a publicly accessible record of information within a dataset rather than physically partaking in a study, concerns regarding harm overall diminish. These secondary data sources are often considered to be far enough removed from the direct research participant to not pose any imminent risk to their wellbeing should their information be used in research" (Stewart, 2021, p. 6)

These informational risks inherent in AI, ML, and DMS research deserve PII risk protections for researchers and data sources. However, the principles of Belmont and even the Common Rule do not necessarily provide IRB and HRPP professionals with the oversight needed to do so (Stewart, 2021). Therefore, the author called for researchers to employ transparency and ethical standards, especially with data brokers, and to support enhanced understanding of these concerns by HRPP and IRB professionals, and said,

> *"Finally, supporting the further training of IRBs regarding the unique ethical concerns of both [consumer-based datasets] CBDs and contemporary big data sources broadly will reinforce the institutionalization of ethical standards regarding CBDs within the academy. Subjects such as eminent informational risks within public online data, the ethically questionable actions of third parties that contract with brokers, and the changing landscape regarding threats of data breaches and participant reidentification are three distinct topics researchers can support the widespread consideration toward within IRB reviews" (Stewart, 2021, p. 10)*

Most recently, Herington et al. (2023) noted the importance of Belmont for research with diverse populations, as Belmont requires individual assessment of consent, risk, and benefit for each individual research participant. As such, the focus on the individual within the existing ethical frameworks holds challenges for large demographically-based datasets. Like others before them, the authors called for a revisit of existing principles within the context of aggregated datasets, as "digital research often exposes different subjects to different autonomy, privacy, and reputational risks. These are not well mitigated by standard collection and consent procedures, and the risks themselves are often invisible to researchers" (Herington et al., 2023, p. 7). They concluded,

> *The explicit integration of justice in digital research recognizes that societies have a collective duty to help produce valuable knowledge under fair terms of cooperation. While more work needs to be done to fully explicate the institutional changes and individual practices that will realize justice in digital research, existing work in participatory action research, neuroethics, political philosophy, and indigenous research methods already provide an invaluable foundation for discharging this obligation (Herington et al., 2023, p. 8). Thus, "revisiting the principles of the Belmont Report for the 21st century will require a process that aligns with contemporary sensibilities" (Brothers et al., 2019, p. 168) and ongoing examination of HRP and PII policy implications.*

Human Research Protections Policy Implications for AI

As calls for action in ethical oversight of research and human protection of subjects, the evolving use of AI and ubiquitous technology continue to muddy the waters when tasked with providing ethical guidelines for this technology. Historically, academia and industry have noted concerns about HSR policies with the use of AI and the implications of the research, primarily within the constructs of dangers and risks to subjects and society (Salazar, 2023). Challenges arising from the policies governing AI and AI research, both upstream and downstream, have raised concerns in human research protection policy, policy formulation, and integration. Challenges highlight the necessity for researchers to observe a more unified and all-encompassing set of guidelines to accommodate the evolving requirements of AI research; however, the guidelines have complex considerations, as ethical research is a complex landscape that may still be published or bypass the peer-review process, with the assumption that the scientific method and research ethics were diligently followed in producing the study's outcomes. Without revising the Belmont principles, it can fall into what Rochel (2023) identifies as soft laws. Soft laws, as defined by Rochel (2023), found that ethical actors set ethical guidelines that represent this actor's conduct and can demonstrate good faith. While soft laws may lack immediate legal consequences and repercussions, they serve as a valuable framework for comprehending ethical norms. Understanding ethical and social norms can also serve as a model for addressing ethical issues when extending current policies to incorporate AI into ethical guidelines.

Algorithms, software programs, and networks are all created in a seminal form by an individual with ethical values, opinions, beliefs, biases, and perceptions that are used within logical decision-making when creating scripts that can enhance, restrict, or extend the learning model process. The question of AI implications is challenging as AI and the algorithms that run AI can be accurate and biased in parallel. Ethical guidelines should not be designed to block bad actors in the technology sector; instead, guidelines should illuminate the research's scope in reach and perspective, allowing the researcher to step outside their project to identify if the research holds an elevated risk or possible ethical consequences. As each country has initiated the process of acquiring policies surrounding AI use, the US is attempting to educate the American public with a framework to guide legislation.

Legislation and solidified laws are not currently being enacted nor discussed wholistically to keep pace with the advancement of technology. The United States government is promoting a framework surrounding "five principles that should guide the design, use, and deployment of automated systems to protect the American public in the age of artificial intelligence" (US Office of Science and Technology Policy, 2023b, para. 4). Promoting the Use of Trustworthy Artificial Intelligence [Executive

Order 13960], which set forth principles that govern the federal government's use of AI. As Blackman (2021) stated, "Other industries have already found ways to deal with complex ethical quandaries quickly, effectively, and in a way that can be easily replicated. Instead of reinventing this process, companies need to adopt and customize one of healthcare's greatest inventions: the Institutional Review Board, or IRB" (para. 2). Blackman's statement logically follows the question of the guidelines, decisions, and recommendations. Ethical guidelines and frameworks are only as good as the individuals using them to make the decisions. There is a need for not only guidelines but enhanced education to guide the researchers, organizations, and academia, as well as provide the IRB and HRPP with resources to glean the understanding of the research to understand the AI/cybersecurity research projects where transparency may not be as apparent within the review.

The number of case studies of research involving ubiquitous technologies is growing by the hour, and the challenge of pinpointing instances where AI algorithms and ML are utilized in human subject research is becoming increasingly complex to review by IRB and HRPP professionals without the understanding of the system or algorithms' potential for bias or future harm of human subjects. The IRB and HRPP are crucial in assisting researchers in navigating the complexities of HSR and integrating ethical standards for HRP, PII, data privacy, and other protections. Suppose there is a situation where an IRB or HRPP professional may not fully understand certain variables impacting AI bias, fairness, or transparency in HSR. In such cases, care must be taken to communicate with the researcher to discover the implications of the studies to ensure that beneficence, justice, respect for people, and nonmaleficence are applied for approval (Office for Human Research (OHRP), 2018). There is a need for open-source tools that can transparently and succinctly create a report that provides the potential bias discovered within the algorithm. An example of an open-source tool would be AI Fairness 360[5] (IBM Developer Staff, 2018).

The issue with open-source tools is that the IRB cannot run the algorithm. Can the process be holistic to assist the IRB reviewer in understanding that the AI can transparently run in the algorithm in a research population without noting the structure of the predictive algorithm? As stated within the discloser of the toolkit, the IBM Developer Staff (2018) noted, "It can be problematic, however, to base predictions, directly or indirectly, on attributes protected through social norms, organizational policies, or legal regulations" (para. 5). There is a growing need to justify an understanding of HRP policy and increased transparency so as to mitigate HSR privacy and confidentiality. Thus, increasing the possibility of bias and reidentification of the information may not address future harms.

The justification of large-scale models in deidentified data but with the capacity of the algorithmic or data sharing without careful consideration must also align with the Belmont principles to address the complexities now facing IRBs. There are issues with the toolkits; the vetting process is placed in a repository for reporting in the software designer's algorithm, which could entail an anti-trust issue. For example, there are concerns with using Natural Language Processing (NLP) by researchers and HSR reviewers. While AI and NLP reviews by the IRB and HRPP may be challenging, these groups may also use AI to streamline their process or NLP to write policies as well as discuss AI and ML for decision-making to scan applications before an official review by an IRB reviewer or board. Khurana et al. (2023) provided an example case of this for organizational use whereby NLP was used for ethical compliance with an "ACE Powered GDPR Robot Launched by RAVN Systems," which "used AI techniques to automatically analyze documents and other types of data in any business system which is subject to GDPR rules. It allows users to search, retrieve, flag, classify, and report on data, mediated to be super sensitive under GDPR quickly and easily" (p. 3728).

Although the tool of NLP may be used to provide reports and guidance, its use to substitute the need for human decision-making advancement may not yet be fully vetted, especially in the unique context of HSR-regulated protections. The contentious point in the discussion of the use of AI, NLP, ML, and DMS for organizational decision-making policies for protocol reviews is complex and dependent upon the system's user, such as the semantics of posed questions to analyze the complexities of the protocol. The calls for Belmont 2.0, including guidance and frameworks, may be justified in such cases to appropriately guide IRB and HRPP professionals for the use of ubiquitous technologies for HRP, coupled with accessible education on data transparency in research, to assist and enhance the researcher's ethical understanding of their research's reach, variables, and implications.

FUTURE RESEARCH RECOMMENDATIONS

Several researchers have called for ongoing examination of the current use of the Belmont principles for human research protections in a digital age. For example, Adashi et al. (2018), among many others, recommended that a reappraisal of the Belmont principles be conducted with more emphasis on transparency. Likewise, Stewart (2021) called for a collective of diverse scholars and researchers to consider the complexities of digital research, especially for consumer-based datasets, and the ethical frameworks required for BD research. Similarly, Herington et al. (2023) called for researchers to bring more transparency for purpose, method, and results and to determine appropriate stewardship of datasets and oversight of risks. Herington et al.

(2023) echoed the calls for further transparency. The authors encouraged researchers to "clearly report research ethics methodologies, including specifying methods of consent, review of risks to participants, reporting of adverse events, and expected costs and benefits of producing the knowledge" (p. 8).

Raymond (2019) suggested the involvement of the U.S. Congress to call on other international organizations to

> *"...authorize a national commission to write another Belmont Report to deal holistically with applications of data science that will only become more complicated as technology progresses. IRBs can help by capturing what big-data research has and has not been considered exempt, and publishing cases of how specific ethical challenges might best be addressed. This will not eliminate risks or rein in all bad practices. But it would be a pivotal step towards encouraging international harmonization of disparate approaches to difficult questions that face researchers around the world" (p. 277).*

Raymond (2019) concluded that the Belmont Report requires more than "tweaks" and deserves a complete and substantive reconsideration (p. 277). In addition, Brothers et al. (2019) stressed that once a revised Belmont based on contemporary interpretations is offered, continued testing is needed for policy and practice.

The chapter authors support these various recommendations for a collective approach to further examine the Belmont principles in the context of the growing use of ubiquitous digital technologies for research. In past work, the chapter authors concurred that a revisit of the Belmont Report may be warranted, specifically for the Justice principle in context with contemporary perspectives (Throne, 2023). It is essential for IRB and HRPP professionals to continue to integrate these considerations as they review research protocols involving PII data collection with AI, ML, DMS, and other technologies. Policy examination and subsequent changes are needed to align with this evolving climate of digital forms of research. As they have noted in past work, IRB and HRPP leadership must continue to be attentive to this fluctuating digital landscape whereby the contranym of sanction applies. At the same time, IRBs, HRPPs, and regulatory organizations grapple with the refinement of policy and procedures amid new guidance and best practices for PII used within research that involves these technologies, including AI, ML, DMS, BD, and IoT. The social good and advances in scientific knowledge from these technologies have been widely recognized across the scholarship. However, the risks to HRP and PII remain and may outweigh the benefits of research when not considered.

CONCLUSION

While this analysis of the current scholarship for data ethics in the use of AI, ML, DMS, and other ubiquitous technologies in research may be limited due to the sourcing methods used, an inductive thematic analysis can be effective in devising summary inferences. Many have stressed the contemporary challenges faced by IRBs and HRPPs in the protection of PII in an ever-evolving digital era. Others have called for improved transparency from researchers regarding methods and risks to PII using these ubiquitous technologies for research. IRB and HRPP professionals must continue to grapple with questions of PII protections surrounding the use of these methods for research. Specialized training and policy changes are needed to best guide social science researchers for HRP and protections for human subjects data and PII. Whether or not a Belmont reboot or 2.0 is conducted, IRB and HRPP professionals, in concert with other research ethics leadership, continue to monitor the debate, guidance, and new thought that arises as these diverse groups continue to grapple with the need for standards and boundaries for HRP, PII protection, and data privacy within HSR, leaders must continue to consider best practices and the establishment of standards for these protections in digital spaces not overtly addressed within the Common Rule and Belmont Principles in the contemporary digital era.

This work was previously published in *Methodologies and Ethics for Social Sciences Research*, edited by Aytekin Demircioğlu, pp. 1-19, copyright 2024 by IGI Global

REFERENCES

Adashi, E. Y., Walters, L. B., & Menikoff, J. A. (2018). The Belmont Report at 40: Reckoning with time. *American Journal of Public Health, 108*(10), 1345– 1348. DOI: 10.2105/AJPH.2018.304580

Atenas, J., Havemann, L., & Timmermann, C. (2023). Reframing data ethics in research methods education: A pathway to critical data literacy. *International Journal of Educational Technology in Higher Education*, 20(1), 1–27. DOI: 10.1186/s41239-023-00380-y PMID: 36846483

Blackman, R. (2021). If your company uses AI, it needs an institutional review board. *Harvard Business Review*.https://hbr.org/2021/04/if-your-company-usesai-it-needs-an-institutional-review-board

Brothers, K. B., Rivera, S. M., Cadigan, R. J., Sharp, R. R., & Goldenberg, A. J. (2019). A Belmont reboot: Building a normative foundation for human research in the 21st century. *The Journal of Law, Medicine & Ethics*, 47(1), 165–172. DOI: 10.1177/1073110519840497 PMID: 30994072

Costello, E., Brunton, J., Bolger, R., Soverino, T., & Juillerac, C. (2023). Massive omission of consent (MOOC): Ethical research in educational big data studies. *Online Learning: the Official Journal of the Online Learning Consortium*, 27(2). Advance online publication. DOI: 10.24059/olj.v27i2.3759

Developer Staff, I. B. M. (2018, November 14). AI fairness 360. *IBM Developer*. https://www.ibm.com/opensource/projects/ai-fairness-360/

Doerr, M., & Meeder, S. (2022). Big health data research and group harm: The scope of IRB review. *Ethics & Human Research*, 44(4), 34–38. DOI: 10.1002/eahr.500130 PMID: 35802789

Doorn, N. (2022). Wastewater research and surveillance: An ethical exploration. *Environmental Science. Water Research & Technology*, 8(11), 2431–2438. DOI: 10.1039/D2EW00127F PMID: 36353217

Eto, T. (2022). *Conducting an effective IRB review of artificial intelligence human subjects research (AI HSR)*. Tech InHSR. https://techinhsr.com/wp-content/ uploads/2022/08/AI-HSR-WHITE-PAPER-TechInHSR-08.2022-1.pdf

Ferretti, A., Ienca, M., Velarde, M. R., Hurst, S., & Vayena, E. (2022). The challenges of big data for research ethics committees: A qualitative Swiss study. *Journal of Empirical Research on Human Research Ethics; JERHRE*, 17(1-2), 129–143. DOI: 10.1177/15562646211053538 PMID: 34779661

Finn, M., & Shilton, K. (2023). Ethics governance development: The case of the Menlo Report. *Social Studies of Science*, 53(3), 315–340. DOI: 10.1177/03063127231151708 PMID: 36802925

Goksel, N., & Bozkurt, A. (2019). Artificial intelligence in education: Current insights and future perspectives. In *Handbook of research on learning in the Age of Transhumanism* (pp. 224–236). IGI Global., DOI: 10.4018/978-1-5225-8431-5.ch014

Gross, M. S., & Miller, R. C.Jr. (2019). Preserving privacy & promoting learning: Technology to empower Belmont 2.0. *SSRN*. https://papers.ssrn.com/sol3/papers.cfm?abstract_id=3390979

Herington, J., Connelly, K., & Illes, J. (2023). Ethical imperatives for working with diverse populations in digital research. *Journal of Medical Internet Research*, 25, e47884. DOI: 10.2196/47884 PMID: 37721792

Hosseini, M., Wieczorek, M., & Gordijn, B. (2022). Ethical issues in social science research employing big data. *Science and Engineering Ethics*, 28(3), 29. DOI: 10.1007/s11948-022-00380-7 PMID: 35705883

Hurley, E., & Pierce, H. H. (2019, July 30). The Belmont Report at 40 – A moment of admiration. *Ampersand*. https://blog.primr.org/belmont-report-40/

Khurana, D., Koli, A., Khatter, K., & Singh, S. (2023). Natural language processing: State of the art, current trends and challenges. *Multimedia Tools and Applications*, 82(3), 3713–3744. DOI: 10.1007/s11042-022-13428-4 PMID: 35855771

Kraus, S., Breier, M., Lim, W. M., Dabić, M., Kumar, S., Kanbach, D., Mukherjee, D., Corvello, V., Piñeiro-Chousa, J., Liguori, E., Palacios-Marqués, D., Schiavone, F., Ferraris, A., Fernandes, C., & Ferreira, J. J. (2022). Literature reviews as independent studies: Guidelines for academic practice. *Review of Managerial Science*, 16(8), 2577–2595. DOI: 10.1007/s11846-022-00588-8

McCutchan, S. (2022). *Research data governance for global health*. Duke Global Health Institute.

Nagai, H., Nakazawa, E., & Akabayashi, A. (2022). The creation of the Belmont Report and its effect on ethical principles: A historical study. *Monash Bioethics Review*, 40(2), 157–170. DOI: 10.1007/s40592-022-00165-5 PMID: 36357708

Office for Human Research (OHRP). (2018, January 15). *Read the Belmont Report [Text]*. OHRP. https://www.hhs.gov/ohrp/regulations-and-policy/belmont-report/read-the-belmont-report/index.html

Petermann, M., Tempini, N., Kherroubi Garcia, I., Whitaker, K., & Strait, A. (2022). *Looking before we leap: Expanding ethical review processes for AI and data science research.* Ada Lovelace Institute.

Raman, R., Utts, J., Cohen, A. I., & Hayat, M. J. (2023). Integrating ethics into the guidelines for assessment and instruction in statistics education (GAISE). *The American Statistician*, 77(3), 1–8. DOI: 10.1080/00031305.2022.2156612

Raymond, N. (2019). Reboot ethical review for the age of big data. *Nature*, 568(7752), 277. DOI: 10.1038/d41586-019-01164-z PMID: 30988466

Rochel, J. (2023). Learning from the ethics of AI - A research proposal on soft law and ethics of AI. *Tilburg Law Review*, 27(1). Advance online publication. DOI: 10.5334/tilr.297

Salazar, T. (2023). A new frontier: Human subject research ethics in an artificial intelligence world. *The Health Lawyer*, 35(4), 23–52.

Stewart, R. (2021). Big data and Belmont: On the ethics and research implications of consumer-based datasets. *Big Data & Society*, 8(2). Advance online publication. DOI: 10.1177/20539517211048183

Throne, R. (2022). Adverse trends in data ethics: The AI Bill of Rights and human subjects protections. Information Policy & Ethics eJournal. doi:DOI: 10.2139/ssrn.4279922

Throne, R. (2023, August 28). Belmont Justice and fair, equitable, and inclusive research recruitment. Ampersand. https://blog.primr.org/belmont-justice-and-fairequitable-and-inclusive-research-recruitment/

Throne, R., Hendon, M., & Kozinski, J. (2023). Graduate student investigator: Best practices for human research protections within online graduate research. In Walters, K. (Ed.), *Dynamic curriculum development and design strategies for effective online learning in higher education.* IGI Global., DOI: 10.4018/978-1-6684-8646-7.ch001

U.S. Department of Homeland Security (DHS), Science and Technology, Cyber Security Division (CSD). (2012). *The Menlo Report: Ethical principles guiding information and communication technology research (No. 249;* p. 17). DHS. https://www.dhs.gov/sites/.../CSD-MenloPrinciplesCORE-20120803_1.pdf

Wachter, S., & Mittelstadt, B. (2019). A right to reasonable inferences: Re-thinking data protection law in the age of big data and AI. *Columbia Business Law Review*, 2019(2), 494–620.

ADDITIONAL READING

Brown, N. (2023). Research ethics in a changing social sciences landscape. *Research Ethics*, 19(2), 157–165. DOI: 10.1177/17470161221141011

Marsolek, W., Barrick, K., Kubas, A., McBurney, J., & Mohr, A. H. (2022). Making data right: Embedding ethics and data management in data science instruction. In Mani, N. S., & Cawley, M. A. (Eds.), *Handbook of research on academic libraries as partners in data science ecosystems* (pp. 189–208). IGI Global., DOI: 10.4018/978-1-7998-9702-6.ch010

Martinez, J. (2023). Ethics and equity: Lessons learned from an institutional review board (IRB) chair of color. In Thomas, U. (Ed.), *Cases on servant leadership and equity* (pp. 112–131). IGI Global.

Salah, A. A., Canca, C., & Erman, B. (2022). Ethical and legal concerns on data science for large scale human mobility. In *Data science for migration and mobility*. Oxford University Press., DOI: 10.5871/bacad/9780197267103.003.0002

Schwester, R. W. (2019). Research ethics. In *Scholarly ethics and publishing: Breakthroughs in research and practice* (pp. 633–645). IGI Global.

Stegenga, S. M., Steltenpohl, C. N., Lustick, H., Meyer, M. S., Renbarger, R., Standiford Reyes, L., & Lee, L. E. (2023). Qualitative research at the crossroads of open science and big data: Ethical considerations. *Social and Personality Psychology Compass*. Advance online publication. 10.1111pc3.12912

Throne, R. (2022). Free the data but not the human subjects identifiers. DOI: 10.2139/ssrn.4304179

Vasiliu-Feltes, I., & Thomason, J. (Eds.). (2021). *Applied ethics in a digital world*. IGI Global., DOI: 10.4018/978-1-7998-8467-5

KEY TERMS AND DEFINITIONS

Data Mining/Scraping: While data mining and data scraping are often used synonymously across disciplines, the distinction between mining versus scraping typically refers to data mining as the gathering of large datasets for analysis, while data scraping involves the process of gathering large datasets. Due to the overlap in meaning, for the purposes of this chapter, the terms are concomitant.

Human Research Protections Program (HRPP): A human research protections program (HRPP) is typically a collaboration between all individuals and groups involved in human subjects research including IRBs and IRB staff, principal investigators, institutional leaders, faculty research supervisors, student researchers, and others among the research community. The HRPP typically offers training, education, and other guidance to ensure research with human subjects and human subjects data is ethical, responsible, and scientifically sound (Throne et al., 2023).

Institutional Review Board (IRB): The Institutional Review Board (IRB) is an administrative board established within an organization or institution to ensure the protection of the rights and welfare of human research subjects recruited to participate in research activities conducted under the organization or institutional affiliation (Throne et al., 2023).

Natural Language Processing: "Natural language processing (NLP) refers to "a subfield of computer science, information engineering, and artificial intelligence concerned with the interactions between computers and human (natural) language" (Goksel & Bozkurt, 2019, p. 236).

Ubiquitous Digital Technologies: Ubiquitous digital technologies refer to all everyday objects embedding intelligent microprocessors able to create architectures that underpin ubiquitous computing, such as networking, artificial intelligence, and wireless computing. Ubiquitous technologies employ sensitive human-computer interaction without spatial or temporal constraints (Throne et al., 2023).

ENDNOTES

1. U.S. Department of Homeland Security (DHS), Science and Technology, Cyber Security Division (CSD). (2012). The Menlo Report: Ethical principles guiding information and communication technology research (No. 249; p. 17). https://www.dhs.gov/sites/default/files/publications/CSDMenloPrinciplesCORE-20120803_1.pdf

2. Tamiko Eto's White Paper from August 2022 that offered an AI decision tree for human subjects research (HSR) as adapted from the Office for Human Research Protection (OHRP) post-2018 IRB review decision trees and intended to further the discussion of human subjects protections within AI for IRB capacity building. See https://techinhsr.com/wp-content/uploads/2022/08/AI-HSR-WHITE-PAPER-TechInHSR-08.2022-1.pdf

3. US National Commission for the Protection of Human Subjects of Biomedical and Behavioral Research. (1978). *The Belmont Report: Ethical principles and guidelines for the protection of human subjects of research* (Vol. 1): Department

of Health, Education, and Welfare, National Commission for the Protection of Human Subjects of Biomedical and Behavioral Research.

4 Protection of Human Subjects 45 CFR §46 (2018). https://www.ecfr.gov/on/2018-07-19/title-45/subtitle-A/subchapter-A/part-46

5 See https://www.ibm.com/opensource/open/projects/ai-fairness-360/

Chapter 12
Graduate Student Investigator:
Best Practices for Human Research Protections Within Online Graduate Research

Robin Throne
https://orcid.org/0000-0002-3015-9587
University of the Cumberlands, USA

Michalina Hendon
https://orcid.org/0009-0003-6088-2658
University of the Cumberlands, USA

James Kozinski
Western Governors University, USA

ABSTRACT

This chapter presents the best practices used by institutional review boards (IRBs) and human research protections programs (HRPPs) to prepare online graduate student investigators for human research protections specific to research within online graduate degree programs or where research supervisors are not proximal to graduate student investigators and their research protocols. In recent years, advances in artificial intelligence (AI), machine learning (ML), and other data mining/scraping forms have adversely impacted individual privacy and the unintended sharing of personally identifiable information (PII). With this growth of ubiquitous digital technologies, such as AI, ML, and data mining/scraping, used across online graduate degree programs, specialized training and preparation are needed to best prepare graduate student researchers for human research protections involving data with PII.

DOI: 10.4018/979-8-3693-3848-3.ch012

INTRODUCTION

A graduate student's submission of their research to the institutional review board (IRB) or some form of a research ethics committee is often a rite of passage remembered long after the student has completed the graduate degree. The purpose of the IRB is to ensure the safety of human subjects involved in research, privacy, and confidentiality for human subjects identifiers, fairness and equity in research recruitment, and to ensure risks are minimized for all research involving human subjects and/or their data, and to make certain no physical or psychological harm comes to the research participants (Lewis & Throne, 2021). In short, graduate student researchers must be best prepared to follow the rules and regulations to ensure ethical and responsible graduate research is conducted. However, some graduate student researchers blamed their research supervisors or the institution for lacking confidence and preparedness (Ciampa & Wolfe, 2019). Other graduate student researchers shared concerns over the constraints on their research due to the IRB and called for reimagining ethical research preparation in graduate education (Slovin & Semenec, 2019).

While submission of the graduate research protocols to the IRB or other ethical review processes may often be seen as a critical milestone for graduate student researchers, some past scholarly researchers have noted the impediments that may be encountered in the submission of the research to the IRB (Lewis & Throne, 2021; Lynch & Kuntz, 2019; Slovin & Semenec, 2019). Some institutions maintain "normalized research" expectations that may lead to unintended obstacles for graduate student researchers designing research protocols that may deter from what may often be perceived as this normalized course of inquiry (Lynch & Kuntz, 2019, p. 12). For example, Lynch and Kurtz (2019) reported on past research that noted this positioning of the IRB or other academic review as the gatekeeper for traditional expectations had been found to lead graduate student researchers to employ research protocols to ensure such reviews are "without interrogation," thereby deviating from their initial inquiry intentions (p. 12). At the same time, others have referred to this normalized or traditional IRB journey as a "post-positivist" view to maintain a status quo in the expectations for graduate research (Slovin & Semenec, 2019, p. 15).

Yet, still, others have noted that some faculty researchers perceive the IRB as an unnecessary process and conduct research out of bounds even though they understand the IRB and their aims are consistent with a desire to advance knowledge while protecting human subjects from harm or risk (Reisig et al., 2022). Unfortunately, these intentional IRB boundary violations may be passed on within the mentoring process of graduate student researchers when the online graduate program does not provide an otherwise sound and adequate foundation for ethical preparations to ensure responsible research. In this context, responsible and ethical research requires the training of graduate student researchers to comply with the policies and expectations of the IRB and/or human research protections program (HRPP) or other ethical research review processes.

Concurrently, other past researchers have reported on graduate students' frustration over their unpreparedness when submitting their research protocols to the IRB (Ciampa & Wolfe, 2019). In a study of a Doctor of Education program, Ciampa and Wolfe (2019) noted that graduate students reported receiving no preparation for the intricacies of submission of their research to the IRB or an understanding of the expectations for data collection and analysis in their submissions. Graduate students shared these feelings of unpreparedness from designing their research to completing the IRB application, and thereby, risks in the submission of the IRB application were identified as they had little confidence in whether their research protocols would be accepted.

When graduate student researchers' preparation for ethical and responsible research and human subjects protections is not addressed intentionally within the graduate program, the problems reported in the past scholarship for these students' experience may persist. However, with the emergent use of artificial intelligence (AI), machine learning (ML), and other ubiquitous technologies in graduate research, this preparation becomes even more essential to ensure the practice of ethical and responsible graduate research across disciplines. When well synthesized within the design of the graduate program, a research component that includes preparation and understanding of the ethical requirements of IRB research can best prepare graduate students to become post-graduate independent researchers. The purpose of this chapter is to present the latest trends and best practices for the design of online graduate degree programs that include graduate research and the best strategies to prepare student investigators for human research protections. These considerations may foster both research supervisors and graduate student researchers to design better human research protections within graduate research and reconsider the submission of the researcher protocols to the IRB, when required, as an obstacle or barrier to research administration.

BACKGROUND

The purpose of the IRB, HRPP, and/or other research ethics review processes in graduate research is to protect research participants, participant data, and to support the preparation of graduate student investigators for ethical and responsible research with human subjects and data with human subject identifiers. Many past researchers have reported on the experiences of stress and frustration from the perceived expectations and pressures surrounding graduate research, including the perceived burdens of the online graduate program's research ethics review process. For example, Eliasson and DeHart (2022) called for the need to consider student researcher trauma in addition to human subjects' protections for research participants as a component of methodological training. If review procedures considered student investigator trauma as part of the IRB submission experience, the authors noted that the trauma experienced by researchers would be placed more at the forefront as "a key consideration among risks imposed by the research process" (p. 8).

The ever-evolving nature of digital data collection methods has further complicated the challenges for ethical research review within online graduate programs. As technology is used in all facets of individuals' lives, graduate students' ethical research should be cognizant that ethical research is not exclusionary to disciplines. As student researchers embark on their academic writing journey and develop their methodologies for conducting research, students often employ various forms of data collection. Methods include online surveys or interviews, recruitment via social networking platforms, web scraping, and semantic analysis, with the component of understanding human behavior in some facets. Behavioral data sets can be collected with the use of different mediums (IoT, social media, facial recognition, intelligent homes) amassed through methods akin to data classification, data mining, and pattern recognition and then patterned through AI and machine learning in which the output can be analyzed through many multi-disciplined theoretical lenses (psychology, economics, sociology) (Bhatia-Lin et al., 2019; Saura et al., 2021a; Saura et al., 2021b). A common theme set within the research on the privacy of data is the question of who is considered the owner of user-generated content (UGC) or user-generated data (UGD) (Saura et al., 2021b). The question can be raised depending upon the collection method to whom behavioral data belongs. The data ownership, be it UGC or UGD, is sometimes elusive and needs ethical oversight and conceptual understanding by both graduate student researchers and their research supervisors.

For example, the Educational Goods Framework (Brighouse et al., 2018) provides a lens for how higher education institutions prioritize and consider values in preparing graduate students to engage in research ethically and responsibly. Such a framework may allow us to explore the knowledge, skills, attitudes, and dispositions institutions want graduate students to develop as they navigate their research roles. The framework may also allow us to explore what institutions value – in conjunction with or in contrast to the values associated with federal IRB policy – as graduate students develop into researchers.

Educational goods, then, can be used as an evaluative framework for future research to discover how an institution's IRB policy impacts the educational goods of an institution's graduate student researchers, to discover what values about research are prioritized by an institution in the implementation of IRB, and to discover what values about research are not prioritized by an institution. Doing so may expand our knowledge about how IRB policy impacts graduate student researchers and potentially broaden the theoretical framework of educational goods (Hseih & Shannon, 2005). Primarily, this and other models' distributive value of adequacy may be important to focus on within the educational goods framework to ensure adequate preparation for new investigators. In an era where graduate student researchers, as well as their research supervisors, may view the IRB or other ethics review as "bureaucratic," it is also an era where ethics review is an essential institutional task to protect individual data privacy and reduce risks and harms to research participants (Brown, 2023, p. 161).

This trends analysis included two assumptions. First, it was assumed that a transdisciplinary perspective was appropriate as, despite differences among doctoral programs, IRB processes remain similar amid other programmatic distinctions. Second, it was assumed that despite doctoral programmatic distinctions, graduate research supervisors are less programmatically reliant as they view human research protections, which allowed for transdisciplinary trends to be considered.

The Belmont Report and Developing Ethical Researchers

The preparation of graduate student researchers by the HRPP, IRB review, and/or other ethical review process involves quintessential steps consistently across fields and disciplines to ensure human participants and their data are protected within the research process. In addition to the Common Rule[1] that mandates the U.S. federal regulations for human research protections, these steps are particularly guided by *The Belmont Report*[2] to ensure three ethical principles for research with human participants are upheld: Respect for Persons, Beneficence, and Justice. The *Belmont Report* was written by the National Commission for the Protection of Human Subjects of Biomedical and Behavioral Research (1979) following the National Research Act of 1974, which was charged to identify the basic ethical principles necessary to underlie the conduct of human subjects research (HSR) and to develop guidelines to ensure ethical and responsible research is conducted in accordance with the principles.

Of course, not all graduate student research is research subject to IRB or another ethical review. Per federal regulations, this research must meet the federal definition of research and involve HSR or data with PII (Pater et al., 2022). Graduate program leadership must first distinguish whether the graduate research project involves actual HSR or the research expectations defined by federal regulations, such as PII, before incorporating the ethical aspects of graduate research training. These distinctions require careful consideration, especially within a contemporary landscape whereby "computational capacity has greatly increased in the last 20 years, but new capabilities in fields like AI and machine learning are changing the nature of research and leaving significant gaps in both ethical norms and oversight" (Pater et al., 2022, p. 1). These computational nuances must be closely considered in the design of the graduate research component to ensure appropriate ethical oversight and investigator training has been included within the graduate program to best prepare ethical and responsible graduate student researchers. These decisions can be complicated by the allowances for ubiquitous digital technologies within the graduate research component. They often must be considered from a current perspective as these methods may introduce increased harm/risk for PII. As Huh-Yoo and Rader (2020) stressed:

> *Being digital meant increasing possibilities of confidentiality breach, unintended collection of sensitive information, and unauthorized data reuse. Concurrently, interviewees found it difficult to pinpoint the direct harm of those risks. The ambiguous, messy, and situated contexts of digital research data did not fit neatly into current human subjects' research protection protocols. (p. 1)*

Thus, graduate program leadership must carefully consider how risk assessment will be evaluated when digital technologies are allowed within the graduate research component (Huh-Yoo & Rader, 2020). The following section, Ubiquitous Technologies and Data Ethics provides more on this.

Preparation for online graduate students to submit their research protocols to the IRB or other ethical review processes is often considered as solely a process-oriented, documental preparation rather than securing the conceptual foundations needed to ensure ethical and responsible research. Reisig et al. (2022) called for educational departments and institutions to make more concerted efforts to ensure sound ethical practices and training are provided to faculty and students, especially surrounding data management techniques and data sources to store and secure data archives properly. The authors noted the prior research that has shown:

> *It is critical that researchers are made aware of such services and that these data management methods meet the needs of faculty for their full potential to be realized (Whitmire et al., 2015). Efforts to this end encourage an organizational culture rooted in academic integrity and proper research conduct. (pp. 10-11)*
>
> *Even when IRB or HRPP policies are established to ensure directives exist for human subject protection, adequate researcher ethical preparation is necessary for the risk of psychological and/or physical harm to be elevated. (Reisig et al., 2022)*

Specifically, the Belmont principle of *Justice* can prepare graduate student researchers to understand fairness, equity, and inclusion in human subjects recruitment and data collection. As Friesen et al. (2022) noted, it is essential for IRBs to ensure investigators understand that the principles that protect human subjects from harm may also exclude certain populations from research. Assurances for diversity require intentional "context-specific analysis," especially when involving vulnerable populations within the research protocols (Friesen et al., 2022, p. 10). When graduate student researchers understand research is a social good and research participation is a benefit, risks may be mediated in this context, whereby research participation is also a benefit to participants, not only a risk. However, in the use of ubiquitous digital technologies for research purposes, this context must also be considered for gaps in ethical oversight in the use of data collection methods that did not exist when The Belmont Report and Common Rule were established (Friesen et al., 2021). Therefore, the balance between risks and benefits for vulnerable populations and other research participants must be further considered in the wake of the contemporary climate with these emerging technologies for research inclusion and participant awareness of PII gathered (Brown, 2023; Friesen et al., 2021; Friesen et al., 2022; Throne, 2022).

Further, Hite et al. (2022) called for intentionality in responsible conduct of research (RCR) efforts for research involving graduate student investigators. Ethical dilemmas should be posed to new investigators within an educational setting to elevate their ethical and responsible judgment as to what constitutes RCR in specific situations such as predatory authorship, IRB violations, and the degree to which ethical challenges can pose harm and risk to research participants (Hite et al., 2022). The authors suggested that these learning activities may best occur within the training setting, such as "infused throughout the university curriculum and supported at the programmatic, departmental, and college levels," to better prepare graduate student researchers prior to facing them within the field (Hite et al., 2022). For example, Stevens and Caskey (2022) suggested this can be accomplished through writing an IRB proposal as an authentic, mentored, and scaffolded assignment to introduce ethical research within graduate coursework. Similarly, Brown (2023) encouraged justice and other ethical principles to be embedded within the curriculum, assuring that ethical reviews are necessary for all involved in the research process or those who oversee the work of new investigators who may be novices to what constitutes ethical research practices. The authors concluded:

> *Instead of considering ethics as a matter of litigation, libel, complaints, and tribunals, they must encourage and enable research ethics committees to become more supportive, developmental, and dynamic, as otherwise, the ethics approvals processes will fail to keep in step with the changing landscape of the social sciences research. (p. 163)*

Ubiquitous Technologies and Data Ethics

Data ownership and across all facets of individuals' lives, graduate students' ethical research should be cognizant that ethical research is not exclusionary to disciplines. As student researchers embark on their academic writing journey and develop their methodologies for conducting research, students often employ various forms of data collection. Methods include online surveys or interviews, recruitment via social networking platforms, web scraping, and semantic analysis, with the component of understanding human behavior in some facets. Behavioral data sets can be collected with the use of different mediums (IoT, social media, facial recognition, intelligent homes) amassed through methods akin to data classification, data mining, and pattern recognition and then patterned through AI and machine learning in which the output can be analyzed through many multi-disciplined theoretical lenses (psychology, economics, sociology) (Bhatia-Lin et al., 2019; Saura et al., 2021a; Saura et al., 2021b). A common theme set within the research on the privacy of data is the question of who is considered the owner of user-generated content (UGC) or user-generated data (UGD) (Saura et al., 2021b). The question can be raised depending upon the collection method to whom behavioral data belongs. Data ownership, whether UGC or UGD, is sometimes elusive and needs ethical oversight and conceptual understanding.

As many attempts to comprehend how AI, ML, data mining/scraping, and other ubiquitous digital technologies are changing a graduate student's everyday life, research using these data collection methods is also being conducted at a rapid pace. Graduate students are reaching to understand, learn, and contribute to the literature in the void of what technology will provide in the coming years in ubiquitous digital technologies, such as artificial intelligence, machine learning, and data mining. Ubiquitous technologies are no longer considered just an Information Technology issue in academic research; the question of how ethical computing in research is disseminated without the researcher having a clear understanding of the technology or how the data is accumulated or gleaned from datasets (Drolet et al., 2022). Researchers are grappling with the ethical effects of research and the complexity of utilization, repercussions, and the application process from an ethical perspective.

The Melo Report from the U.S. Department of Homeland Security in Science and Technology provided a set of guidelines and ethical principles for communication Technology Research; the proposed framework was developed by both organizational leaders, lawyers, and scholars across the internet and communication technology (ICT) field (U.S. Department of Homeland Security, Science, and Technology, Cyber Security Division [CSD], 2012). The Melo Report has built the ethical framework for ICT research on the Belmont Principals, including respect for human rights and privacy, the informed consent process, confidentiality and anonymity, minimization of harm and risks, and transparency and accountability. The Report was supported by both Association for Computing Machinery (ACM) and the Institute of Electrical and Electronics Engineers (IEEE) Computer Society; the Office for Human Research Protections (OHRP) did not explicitly state its support or opposition to the Menlo Report. The Menlo Report was focused on cybersecurity research ethics and did not directly relate to HSR, which falls under the purview of the OHRP (Fiske &

Hauser, 2014). Although the ICT ethical framework has been frequently cited, it has yet to be considered current since 2012 (Finn & Shilton, 2023). Instead, the Report and framework can be found as Archived Content in the Cyber Security Division (CSD) of the U.S. Department of Homeland Security. To ensure that technology is used appropriately for research across various fields, including ICT research, it is crucial to have a review and a call to action that provides clear guidance for the current research landscape.

As institutions are not required to apply regulatory and federal compliance policies to research that is not federally funded or identified as human subject research, as such, there can be many research protocols that need to be submitted to IRB for protocol review (Hine, 2021). Investigators should not rely on an ethical board. However, the researcher should first conceptually explore the study to understand the ethical needs from the participant recruitment stage to reporting research outcomes. Ethical issues such as identification of the difference between public and private data are issues that Williams (2023) found to be in question from a student perspective. Williams's conceptual article compares public data in newspaper archives as public data to what society may now view within public social media sites. Postulating that if data were available in a private but accessible social media group, does the data remain private? Can thc connection of public data in the public domain virtually differ from the relinquishment of the private element when the user provides the information with the understanding of joining and posting information within a social media group? The additional question of informed consent must also be addressed as to whether the social media user is aware of how personally identifiable information (PII) is to be gathered and analyzed. Simple practices like social media scraping for sentiment analysis can throttle the ether between public and private data; as Williams (2023) found, delineating between public and private data depends upon the student's university's ethical policies to define that gray area. As technology ethics are expanding in complexity and understanding, the loosely demarcated nature of what constitutes participant privacy expectations must be clearly defined.

Reshaping Online Graduate Research with the Use of Technology Ethics

Can a comprehensive IRB review and ethical oversight be fully realized if the student or the reviewer needs to have intimate knowledge of the research outcomes? One of the significant concerns in research is the training provided to graduate students, emphasizing their focus and development within their core discipline with an understanding of ethical guidelines when utilizing technology in research. Scientists are utilizing technology in most facets of their research, thinking about how technology is being integrated into personal and social research as well as the consequences of computing, application, theory, and implementation may have an impact beyond the creation of the research itself (Head, 2020). As ethics should not be applied explicitly to one discipline but all disciplines, the question of the student's exploration of the data with ethical insight is questioned during the ethics review. As graduate student researchers develop their research strategies and methodologies, they need to prioritize the privacy protection of the study's participants with consideration of digital methods. To achieve this, researchers should consider asking themselves context-specific digital privacy protection questions to ensure the due diligence of participant privacy. Examples of such questions can be simple and straightforward. More questions than answers persist:

- Can the retrieval of the data process be explained thoroughly, as well as the ethical reasoning for using the data?
- In a constantly connected and accessible virtual society, can graduate student researchers collect data ethically and responsibly from participants in familiar, explainable language as to what PII is to be collected?

The utilization of ubiquitous digital technologies in research has brought about an elevated awareness of ethical considerations across various elements of research stemming from recruitment to the narration of research results. Most of the standard policies and procedures that researchers implement begin with using clearly defined, informed consent documents written in understandable language to convey transparency of research processes. Shulman et al. (2022) reported issues with communicating instructions and educating users to protect their data. Like an informed consent document used as a method to educate the user/research participant on the privacy of their data and how it will be used, Shulman et al. noted that when it comes to privacy policies, "users may skip or disregard the messages the notifications and warnings convey" (p. 5). However, it is the responsibility of the ethical graduate student investigator to explore how the methods that will be used to collect, analyze, disseminate, and store research data to provide the research participant with both education and respect in keeping the data collected private to the extent of the agreement written in the informed consent. As researchers use ubiquitous technology in research, governance depends on guidance. It should adhere to the guiding principles of an ethical review, which should entail the components of an understanding of the privacy protection methods the researcher will use to safeguard the data when and if the data will be used again, and then move beyond the ethical review guidelines and to educate users with their research through participant empowerment (Ben-Ari & Enosh, 2013). Until guidance and data protection legislation are enacted akin to the GDPR in the United States, the question of the misuse and privacy of data exists in the back of a research participant's mind.

As noted, an example of such guidance stemmed from The Melo Report, hosted and supported by the U.S. Department of Homeland Security in Science and Technology to provide guidelines and ethical principles for communication Technology Research. The proposed framework was developed by organizational leaders, lawyers, and scholars in the Information and Communication Technology (ICT) field (DHS, 2012). The Melo Report built the ethical framework for ICT research upon the Belmont Principals, including respect for human rights and privacy, the informed consent process, confidentiality and anonymity, minimization of harm and risks, and transparency and accountability. The Report was supported by both the Association for Computing Machinery (ACM) and the Institute of Electrical and Electronics Engineers (IEEE) Computer Society. In a reflection on why the Menlo report failed, Finn and Shilton (2023) stated:

> *In the case of the Menlo Report, the work of the report was shaped by forward- and backward-looking goals. The report authors had future researchers in mind, hoping to enable new forms of research through data sharing. But they were also concerned with repairing a record of past controversial research practices and resolving an uncertainty disrupting their field: whether to treat their research data as human subjects data. (pp. 2-3)*

It may have been for the repair of past research and the uncertainty of disruption that the Office for Human Research Protections (OHRP) did not explicitly state its support or opposition to the Menlo Report, or as other researchers suggest, the Menlo Report was focused on cybersecurity research ethics and did not directly relate to human subject research that falls under the purview of the OHRP (Fiske & Hauser, 2014). Although the ICT ethical framework has been frequently cited, it has yet to be considered current since 2012 (Finn & Shilton, 2023). Instead, the Report and framework can be found as archived content in the U.S. DHS government website's Cyber Security Division (CSD). Guidelines and frameworks are needed to guide students' progress, and just as technology is dynamic, ethical policies cannot remain static.

Finally, the General Data Protection Regulation (GDPR) enacted in 2018 by the European Union provides a uniform framework that offers consumers greater control over their PII and other data, fosters organizational accountability, and applies strict penalties for noncompliance (Fox et al., 2022; ICO, 2023). GDPR focuses on protecting consumer data privacy and is concerned with the organizational culmination of users' private data; the research purpose or context defines research provisions. Although GDPR is read as a policy with a penalty for organizational collection of private citizen data without the user's knowledge, researchers are polarized on the application of GDPR to research, stating policies are too broad or too strict surrounding the use of secondary data and global data sharing (Soini, 2020). However, the seven fundamental principles of the research under the GDPR include "lawfulness, fairness, and transparency; purpose limitation; data minimization; accuracy; storage limitation; integrity and confidentiality; and accountability," which are tantamount to the application of the Common Rule policies (ICO, 2023, para. 2). Therefore, the broad application in data collection needs further review and revised to follow a structure with more ethical oversight, accountability and consequences for researchers and organizations that do ethical harm.

Online Graduate Student Ethical Reflection

Suggestions for a path to encourage ethical research practice using ubiquitous technology begin with the online graduate student educational preparation. Training for HSR must cover technology and the ethics of using technology understandably and incorporate reflexivity to ensure ethics training is iterative across the graduate research process. Graduate students may desire to challenge the status quo of ethical review boards, or as Reisig et al. (2022) noted, even prompted by the research supervisor to do so and invite subject matter experts to the table to define and defend the ethics of the process with the use of technology. Students should merge disciplines and learn from each other in ethical computing practices. The defense of having an epistemic challenge in research and needing to understand the ethical ramifications with or without technology is no longer valid. With technological advancements, research in the computing field or integrated technology now requires a greater emphasis on ethical reflection than in previous decades. As ethical norms and guidelines tend to be broadly defined, they exist in academic research as community standards. However, educational institutions promote and educate in responsible research practices, whereby ethics is a shared responsibility (Head, 2020). With guidance from their graduate research supervisor and/or dissertation/thesis committee, the graduate student researcher must consider the methodology to ascertain the ethical risk versus the benefits of the research outcome and whether the IRB performs an ethical review of the study (Fiske & Hauser, 2014).

Thus, in situations where contextual factors and ethical dilemmas are present, it is vital for the graduate research advisor and/or dissertation/thesis committee members responsible for overseeing graduate students to acknowledge and address any potential technology ethics that the student may overlook. However, if the committee members lack ethical insight, there is a risk that issues related to researcher misconduct may arise later. For example, Throne (2022) highlighted Hosseini et al., (2022) recommendations for IRBs and HRPPs to establish research ethics and data use frameworks to minimize inherent data biases embedded within datasets and the sustained social reproduction in the use of such data. Therefore, graduate student researchers must be trained to understand these intrinsic biases from the concurrent use of archival data and not solely rely on others to identify these risks and biases.

Further, research supervisors should proactively seek and acquire knowledge about ethical considerations in using technology, ensuring they can provide informed guidance and oversight to students conducting research in this domain.

By remaining current with the ever-evolving ethical practices and technological advancements, research supervisors and/or dissertation/thesis committee members can fulfill their roles effectively and contribute to maintaining ethical standards in research. As such, any graduate research ethics program should encompass the entire research process, starting from the conception of research and extending to the publication of findings or attainment of a degree. Specifically, the online graduate research preparation program should instill in newly established researchers the methods to compose research with ethical considerations, fostering a culture of ethical practice in the context and understanding of the non-proximal oversight by the graduate research supervisor and/or the IRB and ethics review of the research protocols. This includes promoting ethical awareness, education on ethical principles, and encouraging reflective practices. When a program can sustainably provide an ethical framework for graduate student research to build their future research agenda, the chances of ethical research, whether within or outside academic research, may prove ethical practices are sustained and prioritized throughout a post-graduate researcher's career.

Recommendations and Future Research

The chapter authors call for two key questions to be considered for future research:

1. Do graduate programs best prepare graduate students as post-graduate researchers, ensuring that their preparation enables them to flourish as ethical and responsible independent researchers in both academic and professional contexts?
2. How important is it to incorporate specialized training related to ubiquitous digital technologies into online graduate programs so graduate student researchers can thrive as post-graduate ethical and responsible independent researchers?

Other researchers have also called for consideration of these trends in the need for continued investigation into best practices and preparation for graduate student investigators to ensure they leave graduate institutions as independent ethical, and responsible researchers. Specifically, Reisig et al. (2022) called for continued examination of their IRB violations scale to improve self-reporting capabilities as additional protection from HSR harm and risk. Likewise, Hite et al. (2022) called for future research to explore how training and education for graduate student researchers can address RCR and mitigate unethical issues within a higher education setting. For digital data collection, Huh-Yoo and Rader (2020) stressed the need to examine further the likelihood of participant data risk/harm and the uncertain characteristics involved in ethical oversight of digital research, which remains largely unknown and under ongoing debate.

CONCLUSION

It is clear from past scholarships that academic institutions should promote and educate in responsible research practices and make ethics a shared responsibility across graduate programs and the institution's research community, which includes IRBs, HRPPs, and any other ethical review body. As such, any graduate research ethics program should instill in new investigators the ethics, principles, and methods appropriate to ensure ethical, responsible research is conducted within an ethical research culture fostered in partnership with the IRB, HRPP, or other ethics review. The online graduate program must consider whether the research component meets the definition of research subject to IRB or other ethical oversight and whether digital methods, such as AI, ML, and data mining/scraping, are allowed before which ethical research learning outcomes are to be included within the graduate curriculum. Thereafter, the best practices will consist of authentic research projects mentored by a graduate research supervisor whereby ethical principles are embedded within the learning activities. These steps help to promote ethical and responsible research for graduate student investigators who will be best prepared for a contemporary evolving research climate and encourage ongoing reflective practices to sustain ethical research practices throughout a post-graduate researcher's career. Future research is needed to continue to examine trends and best practices for online graduate research programs to ensure graduate student researchers flourish as post-graduate ethical and responsible independent researchers.

NOTE

This work was previously published in *Dynamic Curriculum Development and Design Strategies for Effective Online Learning in Higher Education,* edited by Kelley Walters, pp. 1-15, copyright 2023 by IGI Global.

REFERENCES

Bailey, M., Dittrich, D., Kenneally, E., & Maughan, D. (2012). The Menlo Report. *IEEE Security and Privacy*, 10(2), 71–75. DOI: 10.1109/MSP.2012.52

Ben-Ari, A., & Enosh, G. (2013). Power relations and reciprocity: Dialectics of knowledge construction. *Qualitative Health Research, 23*(3), 422–429. DOI: 10.1177/1049732312470030

Brighouse, H., Ladd, H. F., Loeb, S., & Swift, A. (2018). *Educational goods: Values, evidence, and decision-making*. University of Chicago Press. DOI: 10.7208/chicago/9780226514208.001.0001

Brown, N. (2023). Research ethics in a changing social sciences landscape. *Research Ethics*, 19(2), 157–165. DOI: 10.1177/17470161221141011

Ciampa, K., & Wolfe, Z. (2019). Preparing for dissertation writing: Doctoral education students' perceptions. *Studies in Graduate and Postdoctoral Education, 10*(2), 86-108. DOI: 10.1108/SGPE-03-2019-0039

Finn, M., & Shilton, K. (2023). Ethics governance development: The case of the Menlo Report. *Social Studies of Science*, 53(3), 315–340. Advance online publication. DOI: 10.1177/03063127231151708 PMID: 36802925

Fiske, S. T., & Hauser, R. M. (2014). Protecting human research participants in the age of big data. *Proceedings of the National Academy of Sciences of the United States of America*, 111(38), 13675–13676. DOI: 10.1073/pnas.1414626111 PMID: 25157175

Fox, G., Lynn, T., & Rosati, P. (2022). Enhancing consumer perceptions of privacy and trust: A GDPR label perspective. *Information Technology & People, 35*(8), 181–204. DOI: 10.1108/ITP-09-2021-0706

Friesen, P., Gelinas, L., Kirby, A., Strauss, D. H., & Bierer, B. E. (2022). IRBs and the protection-inclusion dilemma: Finding a balance. *The American Journal of Bioethics*, 1–14. PMID: 35482887

Head, G. (2020). Ethics in educational research: Review boards, ethical issues, and researcher development. *European Educational Research Journal, 19*(1), 72–83. DOI: 10.1177/1474904118796315

Hite, R. L., Shin, S., & Lesley, M. (2022). Reflecting on responsible conduct of research: A self study of a research-oriented university community. *Journal of Academic Ethics*, 20(3), 399–419. DOI: 10.1007/s10805-021-09418-0 PMID: 34104126

Hsieh, H. F., & Shannon, S. E. (2005). Three approaches to qualitative content analysis. *Qualitative Health Research*, 15(9), 1277–1288. DOI: 10.1177/1049732305276687 PMID: 16204405

Huh-Yoo, J., & Rader, E. (2020). It's the Wild, Wild West: Lessons learned from IRB members' risk perceptions toward digital research data. *Proceedings of the ACM on Human-Computer Interaction, 4*(CSCW1), (pp. 1-22). ACM. DOI: 10.1145/3392868

Information Commissioner's Office (ICO). (2023, May 19). *A guide to the data protection principles.* A Guide to the Data Protection Principles; ICO. https://ico.org.uk/for-organisations/uk-gdpr-guidanceand-resources/data-protection-principles/a-guide-to-the-data-protection-principles/

Lewis, C., & Throne, R. (2021). Autoethnography and other self-inquiry methods for practice-based doctoral research. In *Practice-based and practice-led research for dissertation development* (pp. 87-107). IGI Global.

Lynch, S., & Kuntz, A. (2019). A critical autoethnography of a doctoral students' research journey: Learning to take risks in the academy. *Curriculum Studies in Health and Physical Education*, 10(2), 156–171. DOI: 10.1080/25742981.2019.1588762

National Commission for the Protection of Human Subjects of Biomedical and Behavioral Research. (1979). *The Belmont Report: Ethical principles and guidelines for the protection of human subjects of research.* U.S. Department of Health and Human Services. https://www.hhs.gov/ohrp/regulations-andpolicy/belmont-report/read-the-belmont-report/index.html

Pater, J., Fiesler, C., & Zimmer, M. (2022). No humans here: Ethical speculation on public data, unintended consequences, and the limits of institutional review. *Proceedings of the ACM on Human-Computer Interaction, 6*(GROUP), (pp. 1-13). ACM. DOI: 10.1145/3492857

Reisig, M. D., Flippin, M., & Holtfreter, K. (2022). Toward the development of a perceived IRB violation scale. *Accountability in Research, 29*(5), 309–323. DOI: 10.1080/08989621.2021.1920408

Saura, J. R., Palacios-Marqués, D., & Ribeiro-Soriano, D. (2021). Using data mining techniques to explore security issues in smart living environments in Twitter. *Computer Communications*, 179, 285–295. DOI: 10.1016/j.comcom.2021.08.021

Saura, J. R., Ribeiro-Soriano, D., & Palacios-Marqués, D. (2022). Assessing behavioral data science privacy issues in government artificial intelligence deployment. *Government Information Quarterly*, 39(4), 101679. DOI: 10.1016/j.giq.2022.101679

Shulman, Y., Kitkowska, A., & Meyer, J. (2022, July 5). *Informing users: Effects of notification properties and user characteristics on sharing attitudes*. ArXiv.Org. DOI: 10.1080/10447318.2022.2086592

Soini, S. (2020). The GDPR, secondary research purposes and global data sharing—One-wheel too many. *European Journal of Human Genetics, 28*(6), 694. doi:DOI: 10.103841431-020-0608-x

Stevens, D. D., & Caskey, M. M. (2022). Building a foundation for a successful doctoral student journey: A scholarship of teaching and learning investigation. Innovative Higher Education, 1–23. PMID:36467630 14 Graduate Student Investigator.

Throne, R. (2022). Adverse trends in data ethics: The AI Bill of Rights and human subjects protections. Information Policy & Ethics eJournal. doi:DOI: 10.2139/ssrn.4279922

U.S. Department of Homeland Security (DHS), Science and Technology, Cyber Security Division (CSD). (2012). The Menlo Report: Ethical principles guiding information and communication technology research (No. 249; p. 17). DHS. https://www.dhs.gov/sites/.../CSD-MenloPrinciplesCORE-20120803_1.pdf

Williams, R. (2023). Think piece: Ethics for the virtual researcher. *Practice*, 5(1), 41–47. DOI: 10.1080 /25783858.2023.2179893

ADDITIONAL READING

Bates, B. J., & Birch, B. (2013). Online research and obtaining human subjects/IRB approvals. In *Online instruments, data collection, and electronic measurements: Organizational advancements* (pp. 40–55). IGI Global. DOI: 10.4018/978-1-4666-2172-5.ch003

Cresswell-Yeager, T. J., & Bandlow, R. J. (2020). Transformation of the dissertation: From an end-ofprogram destination to a program-embedded process. In *Creating a framework for dissertation preparation: Emerging research and opportunities* (pp. 41-66). IGI Global.

Durdella, N. R. (2022). Conducting research with human participants: An IRB guide for students and faculty. SAGE.

Nucera, S., Tartarisco, G., Epasto, A., Smeriglio, D., Mazzeo, A., Pioggia, G., & Anastasi, A. (2018). Ubiquitous, wearable, mobile: Paradigm shifts in e-learning and diffusion of knowledge. In *Handbook of research on mobile devices and smart gadgets in K-12 education* (pp. 286–307). IGI Global., DOI: 10.4018/978-1-5225-2706-0.ch018

Resnik, D. B. (2018). *The ethics of research with human subjects: Protecting people, advancing science, promoting trust* (Vol. 74). Springer., DOI: 10.1007/978-3-319-68756-8

Taylor, P. M. (2022). Being a doctoral student: The dissertation stage. In *Research anthology on doctoral student professional development* (pp. 64-68). IGI Global.

Toncich, D. (2022). *Key factors in postgraduate research supervision: A guide for supervisors*. Routledge. DOI: 10.1201/9781003338741

KEY TERMS AND DEFINITIONS

Human Research Protections Program (HRPP): A human research protections program (HRPP) is typically a collaboration between all individuals and groups involved in human subjects research including IRBs and IRB staff, principal investigators, institutional leaders, faculty research supervisors, student researchers, and others among the research community. The HRPP typically offers training, education, and other guidance to ensure research with human subjects and human subjects data is ethical, responsible, and scientifically sound.

Institutional Review Board (IRB): The Institutional Review Board (IRB) is an administrative board established within an organization or institution to ensure the protection of the rights and welfare of human research subjects recruited to participate in research activities conducted under the organization or institutional affiliation.

Ubiquitous Digital Technologies: Ubiquitous digital technologies refers to all everyday objects embedding intelligent microprocessors able to create architectures that underpin ubiquitous computing, such as networking, artificial intelligence, and wireless computing. Ubiquitous technologies employ sensitive human-computer interaction without spatial or temporal constraints (Nucera et al., 2018).

ENDNOTES

1 The Common Rule [45 CFR 46] sets forth the basic policy requirements for the composition and function of the IRB, criteria for IRB approval, informed consent requirements, and definitions, which has been adopted by multiple federal agencies. In addition to the basic policy, additional protections for vulnerable populations are regulated.

2 The full text of *The Belmont Report* and details related to the history of the Report is available from the U.S. Health and Human Services Office for Human Research Protections at https://www.hhs.gov/ohrp/regulations-and-policy/belmont-report/read-the-belmont-report/index.html

Chapter 13
Role of Marketing Ethics in Sustainable Marketing:
Responsible and Ethical Strategies and Initiatives

Pratap Chandra Mandal
https://orcid.org/0000-0002-0300-4002
Indian Institute of Management, Shillong, India

ABSTRACT

It is imperative for companies to adopt ethical policies for sustainable marketing. The objective of the study is to analyze the marketing ethics policies adopted by companies and their roles in creating a sustainable organization. The methodology adopted is a conceptual analysis of the corporate marketing ethics policies. Various controversies and rules and regulations in ethical marketing practices are discussed. In the digital age, consumer privacy is under threat. The policies may help in developing a sustainable company. Initiatives adopted by companies towards remaining ethical and adopting sustainable marketing are highlighted. Academicians may analyze the existing policies and suggest relevant and effective marketing ethics policies. Practicing managers may evaluate the existing marketing ethics policies in their organizations and adopt policies that are effective. These strategies allow companies to adopt ethical policies and sustainable marketing research objectives and for the ethical and responsible protection of human subjects and their data.

DOI: 10.4018/979-8-3693-3848-3.ch013

INTRODUCTION

Companies must be sustainable in their actions (De Oliveira Neto et al., 2018). Companies should adopt sustainable marketing principles. Sustainable marketing may be defined as “a process of planning, implementing and controlling the development, price-formation, and distribution of a product in a way that guarantees adherence to the following three criteria: (1) satisfying consumer needs; (2) guaranteeing the achievement of the organization’s goals; (3) the whole process being in harmony with the ecosystem” (Fuller, 1999). Companies can adapt sustainable marketing principles when they have excellent ethics in place. This is true for marketing ethics. Companies are required to pursue ethical marketing practices. Ethical marketing involves the application of ethics into the process of marketing. Ethics in marketing has the potential to have a long-term benefit for the entire society. It examines whether the company is painting a truthful picture of the product or service. It can be said that societal and ethical marketing are overlapping concepts. Harming society would be considered by most to be an unethical practice. By using ethical practices, everyone can reap the benefits (Rawat et al., 2015). Customers and society at large will be affected adversely in the long run if companies continue pursuing unethical marketing practices (Aboelenien & Naaman, 2023). The reputation of companies is at stake when they adopt such practices. The survival of such companies may be jeopardized (Leonidou et al., 2020). Companies must adopt ethical marketing practices. Consequently, companies should focus on their customers and other stakeholders and sustainable marketing goals for long-term consumer and business welfare. Companies can achieve these through ethical marketing practices (Chong & Patwa, 2023).

Companies may face moral dilemmas when adopting sustainable marketing and ethical practices (Reddy et al., 2023). Practicing managers sometimes are not clear about the most appropriate thing to do. Companies may not have proper ethical practices (Valls Martínez et al., 2021). However, at the same time, managers are well aware that customers prefer those companies that they perceive as adopting sustainable marketing principles and ethical marketing practices (Peterson et al., 2021). So, companies must adopt sustainable marketing and ethical marketing practices.

The above discussions indicate that those companies that adopt sustainable marketing and ethical marketing principles will succeed in the long run. At present, customers and society at large tend to favor such companies. So, companies need to focus on sustainable marketing and ethical practices. Although the topic is essential and requires discussions, few studies focus on analysis. The study aims to address this research gap. The study's objective is to analyze the various aspects of sustainable marketing principles adopted by companies and the roles of marketing ethics, as well as to understand the links between corporate marketing ethics and the various aspects of sustainable marketing.

The methodology adopted is a conceptual analysis of the various aspects of sustainable marketing principles adopted by companies and businesses and the roles of marketing ethics in such marketing activities. Research studies on the topic and company practices are analyzed from various academic journals. Primary data was not collected, and empirical analysis was not conducted.

The novelty and the contributions lie in the fact that a conceptual analysis of the various aspects of marketing ethics and the roles of marketing ethics in sustainable marketing are discussed. Various corporate marketing ethics policies adopted by companies are highlighted. Rules and regulations related to ethical marketing practices are discussed. Organizations should develop a sustainable company to succeed and thrive in the long run. Various initiatives companies adopt towards ethical and sustainable marketing practices are discussed. Academicians may analyze the various marketing ethics policies adopted by companies and suggest relevant and effective policies. Practicing managers may analyze the marketing ethics policies existing in their companies and suggest improvements in such policies or the adoption of effective policies. All these will allow companies to be ethical in all their marketing actions and adopt sustainable marketing.

CORPORATE MARKETING ETHICS POLICIES

Conscientious marketers face moral dilemmas when they adopt sustainable marketing and marketing ethics policies. They often cannot ascertain the appropriateness of many of their activities (Sattar et al., 2023). Again, many managers may lack acceptable moral sensitivity. So, companies must have proper policies in place. Companies must develop corporate marketing ethics policies. The policies define the broad guidelines that all employees in the organization are expected to follow. These policies should cover distributor relations, advertising standards, customer services, pricing, product development, and general ethical standards (Mohamed, 2013).

Companies may develop and implement policies and guidelines. However, even the finest guidelines cannot resolve all the problematic ethical situations and issues related to sustainable marketing that marketers face. Marketers are in a dilemma between choosing actions that generate immediate sales and those that benefit society. The marketing behavior of companies may be described as immoral or even amoral if they choose actions that aim to produce immediate sales. Again, if they do not act, they might be viewed as ineffective as marketing managers. They may be unhappy because of the constant moral tension. From the discussions, it is evident that managers need to develop a set of principles that will help them understand

and realize the moral importance of the situation. Based on this understanding, they must decide the courses of action that will ensure the welfare of their customers and society at large.

Free Market and Legal System

Companies and marketing managers require specific principles that will guide them on ethics and social responsibility issues. Companies may adopt the philosophy of the free market and the legal system to decide and solve the issues. The principle states that companies and marketing managers are not responsible for moral judgments. The market and the legal system determine what managers can do. However, the major issue with this principle is that since the market and the legal system decide everything, companies may be irresponsible even while remaining within legal domains (Mokha, 2018).

Responsibilities of Individual Companies

Critics and experts believe that individual companies and managers are responsible for their actions rather than the systems. This philosophy suggests that companies and managers should have a social conscience and should act based on such conscience. Corporate discussions should be taken based on the high standards of ethics and morality exhibited by companies and managers. Corporate decisions should be based on social conscience and not on *what the system allows* (Elhajjar, 2023).

Companies and managers must understand the philosophy of socially responsible and ethical behavior. The societal marketing concept suggests that each manager must look beyond what is legal and what is allowed. They should develop standards based on personal integrity, corporate conscience, and long-term consumer welfare (Tripathi & Sharma, 2023).

Issues related to ethics and social responsibility should be discussed in an open, proactive, and forthright manner. The discussions help companies and businesses to develop, build, and maintain strong customer relationships based on honesty and trust (Iglesias et al., 2020). For example, SC Johnson is a company that makes familiar home-product brands such as Pledge, Shout, Windex, Saran Wrap, and Ziploc. The philosophy of SC Johnson lies in the belief that “Integrity is part of our DNA. It has been our family way since 1886” (Johnson, 2007). This philosophy allows SC Johnson to have a long tradition of doing and supporting what is right, even at the expense of sales. For example, Saran Wrap is one of the best-known and most prominent brands of SC Johnson. The brand is known for doing what is right at any cost (Godfrey & Jacobus, 2022).

Business Standards and Practices

International marketers face unique challenges regarding the issue of ethics while dealing with environmentalism. Different business standards and practices are prevalent in different countries. Business standards and practices vary widely from one country to another. For example, bribes and kickbacks are illegal, and various treaties against bribery and corruption have been signed and ratified by more than 60 countries (Lu et al., 2019). However, standard business practices exist in many countries. The World Bank estimates that bribes totaling more than $1 trillion per year are paid out worldwide (Feil, 2021). Research also reveals that firms in some countries like Russia, China, Mexico, and Indonesia pay more bribes than in other countries. Corruption is relatively more common in countries like Yemen, Kenya, and Sierra Leone. Companies from countries like Australia, Denmark, Japan, and Finland are the least corrupt (Gauthier et al., 2021). Companies must evaluate whether they need to lower their ethical standards to compete effectively in countries with lower standards. Companies should be committed to a common set of shared standards worldwide.

Individual and professional associations develop and suggest codes of ethics. Again, many companies are proactive and adopt their codes. For example, the American Marketing Association, an international association of marketing managers and scholars, developed a code of ethics that marketers are expected to adopt (Weatherly, 2021). The code of ethics includes various ethical norms. These include:

- *Do no harm*: Marketers should consciously avoid harmful actions or emissions by embodying high ethical standards. They should also adhere to all applicable laws and regulations in customers' choices (Audi, 2022).
- *Foster trust in the marketing system*: Marketers should strive for good faith and fair dealing. They should ensure that the efficacy of the exchange process is maintained. They should also avoid deception in product design, pricing, communication, and delivery or distribution (Nill & Laczniak, 2022).
- *Embrace ethical values:* Companies should build relationships and enhance consumer confidence in marketing integrity. They should abide by the core values of honesty, responsibility, fairness, respect, transparency, and citizenship (Hagenbuch & Mgrdichian, 2020).

Programs Developed by Companies

Companies develop programs emphasizing the importance of remaining ethical in all business actions. Managers are encouraged to understand ethical issues and then find proper responses while remaining within the realm of ethics. Companies organize ethics workshops and seminars and create ethics committees (Barrett, 2018). Many major companies in the United States appoint high-level ethics officers. The responsibilities of the officers include championing ethical issues and helping to resolve ethical problems and concerns facing employees. Companies also develop their codes of ethics (Saygılı et al., 2022).

Companies develop programs and codes of ethics so that everyone in the organization displays ethical behavior. However, written codes of ethics and ethics programs may not always ensure ethical behavior. Employees are required to exhibit total corporate commitment. Companies can ensure ethics and social responsibility when such aspects are components of the overall corporate culture (Amaral et al., 2023). For instance, the Google Code of Conduct concludes: "It is impossible to spell out every possible ethical scenario we might face. Instead, we rely on one another's good judgment to uphold a high standard of integrity for ourselves and our company. Remember … don't be evil. If you see something that isn't right, speak up!" (Zhong et al., 2023).

Development of Ethical Marketing Communications

Companies face ethical dilemmas when they do marketing communications. Companies require breaking through the clutter of messaging by consumers. Advertisers believe that they have to push the boundaries of what could be considered ethical and sometimes even legal (Shahbazi et al., 2024). The United States is concerned about marketing communications and tries to ensure that marketing communications are ethical. A substantial body of U.S. laws and regulations governs advertising (Chong & Patwa, 2023). The laws and regulations require that advertisers not make false claims, use false demonstrations, or create advertisements with the capacity to deceive, even when no one is deceived. For example, a floor wax advertiser cannot claim that the product protects six months unless it does so under typical conditions. Similarly, a baker of diet bread cannot claim that its product has fewer calories simply because its slices are thinner (Chong & Patwa, 2023). It is difficult to establish the differences between deception and *puffery*. Puffery is simple exaggeration that is not meant to be believed and is permitted by law. However, reputed brands are involved in unethical advertisements and in promoting false claims about usage, product ingredients, and benefits (Kagendo-Kiarie, 2022). Such brands include Fair and Lovely of Unilever, Coronil of Patanjali, Dairy Milk of Cadbury, and Maggi of Nestle (Singh & Mishra, 2019).

Controversies in Marketing Communications

Many companies and brands get entangled in controversies when they promote their offerings through marketing communications. Many brands are subject to controversies about unethical advertisements, promoting false claims about usage, ingredients, and benefits. Companies and brands are subject to controversies when they do bait-and-switch advertising marketing communications in business markets, and take unfair advantage of vulnerable groups, and target kids and children in online marketing.

Bait-and-Switch Advertising

Sellers are accused of doing bait-and-switch advertising to attract buyers under false pretenses. For example, a seller may advertise a mobile phone for $149. Consumers try to purchase the advertised phone, but the seller cannot refuse to sell it for $149. The seller cannot downplay its features, show a faulty handset, or promise unreasonable delivery rates to switch the buyer to a more expensive phone (Gangjee, 2024).

Marketing Communications in Business Markets

Companies that sell their offerings in business markets must abide by several regulations. For example, it is illegal for salespeople to misinform customers or mislead them about the advantages of buying a product. It is illegal to offer bribes to purchasing agents or others who are in a position to influence B2B sales. Their statements must match advertising claims. It is expected that their statements will match advertising claims. They should not obtain or use competitors' technical or trade secrets through bribery or industrial espionage. They also cannot disparage competitors or their products by suggesting things that are not true (Brennan, 2021).

Unfair Advantage of Vulnerable Groups

Companies and businesses generate public controversy when they do marketing communications that unfairly advantage vulnerable or disadvantaged groups (Hamby et al., 2024). Vulnerable groups may include kids, children, and senior citizens. Disadvantaged groups may include individuals with low-income or ethnic minority residents. Companies also generate controversy when they try to promote potentially harmful offerings. For example, the cereal industry has been criticized for marketing efforts directed toward kids and children through the years. Kids and children have a shallow level of resistance. Critics worry that high-powered appeals presented by animated characters will overwhelm children's defenses. Such marketing communications will lead kids and children to want sugared cereals or breakfast with a poor balance (Eagle & Dahl, 2018). Toy marketers have been similarly criticized. Again, experts believe that a critical area of concern for many consumer-protection advocates is the millions of kids who are online (Kennedy et al., 2019).

Targeting Kids and Children in Online Marketing

The usage of cell phones, tablets, software apps, and social networking sites has exploded in the last few years. An important concern is protecting unknowing or unsuspecting children in an increasingly complex technological world. The market today consists of 8-to-12-year-old tweens who are highly mobile and are happy to share locations through apps and communicate with others by phone. Experts characterize such kids as *SoLoMo* (Social Local Mobile). However, only one in five parents use basic content-control features on smartphones, tablets, and game consoles (Rezníčková & Zaušková, 2020). All these challenges indicate that it is tough to establish ethical and legal boundaries in marketing to kids and children who are online and offline.

Many kids and children watch television. So, television may be an effective medium to reach children. Consequently, marketers target kids and children at younger ages with product tie-ins for many offerings. Such offerings include *Disney Princess* characters, pajamas, retro G.I. Joe toys and action figures, *Dora the Explorer* backpacks, and *Toy Story* playsets (Preston, 2016).

Criticisms and Marketing Ethics in Targeting Kids and Children

Companies are increasingly putting an increasing emphasis on marketing pushes involving kids and children. Teachers and parents are divided in their opinions regarding such promotions. Some favor groups such as *Campaign for a Commercial-Free Childhood* (Pecora, 2022). The members of such groups believe that kids and children are incredibly susceptible to advertising. Also, the endorsements of schools for such products make kids and children believe that the product is suitable for them. This holds for almost all product categories. Nevertheless, many schools and daycare centers that operate on tight budgets favor the free resources and promotional materials that marketers offer. Such materials may include Care Bear worksheets, Pizza Hut reading programs, and Nickelodeon magazines (Morton & Treviño, 2021). Marketers do not always attract criticism when they try to target kids and children, minorities, or other particular segments of society. For example, Colgate Junior toothpaste, which Colgate-Palmolive markets, contains special features designed to encourage kids and children to brush longer and more often (Mandal, 2019).

Thus, the issue is not about who is targeted but how and for what purpose. Socially responsible marketing calls for targeting that serves the company's interests and those who are targeted.

Rules and Regulations for Ethical Marketing Practices

Critics wish that companies and businesses should encourage responsible and ethical marketing. Critics are worried about wrong messages being communicated to children and kids through various company promotions. The Children's Advertising Review Unit is the self-regulatory agency of the advertising industry. It has published guidelines that companies need to abide by while doing promotions to kids and children (Koester, 2003). Some critics feel that companies should find other ways to communicate with kids and children. Again, many critics and experts are of the opinion that there should be a complete ban on advertising to kids and children (Dumitrescu et al., 2016).

The Advertising Standards Authority (ASA) in the United Kingdom takes strict actions against unethical marketing communications. It bans some advertisements outright for being socially irresponsible (Hardy, 2018). Drop Dead, a British clothing line, ran an advertisement that showed a shockingly thin model. The advertisement prompted ASA to condemn the image of the brand because the advertisement communicated a dangerously inappropriate style for teenage girls. ASA issued a statement that complained that the "hip, rib, and collar bones of the models were highly visible." It also noted that she had visible (and unnatural) "hollows in her thighs" (Taylor et al., 2018). Many companies in the United Kingdom have started using socially responsible marketing communications. Companies have begun to respond to such pressures. For example, the fashion brand Topshop removed an advertisement from its website after advocates complained that the featured model was dangerously gaunt. This might negatively influence young shoppers (Mandal, 2021).

In the United States, the Council of Fashion Designers of America (CFDA) is an organization that takes strict actions against advertisement agencies that violate rules and regulations in hiring underage models. However, violations continue to take place. At present, designers and modeling agencies are required to verify the identification of models and to confirm that the models are at least 16 years old (Johann, 2015). It also encourages industry members to recognize early signs of eating disorders, to call for healthy snacks backstage, and to ban the use of models under 18 years of age at fittings or at photo shows held after midnight. There is widespread skepticism that the industry can regulate itself. Nevertheless, the National Eating Disorders Associations praise the CFDA guidelines (Gehlhar, 2021).

The responses from designers and modeling agencies are slow. However, changes are taking place. Designers such as Tommy Hilfiger and Tory Burch complain that models who apply with them appear young and thin. Burch says, "I still use some emaciated girls coming in. It is still a problem" (Mogaji & Mogaji, 2021). Critics are concerned about the ultra-thin style promoted by fashion advertising for young girls. Critics and experts keep a close watch on such aspects. However, the issue is much more in-depth and challenging than just checking IDs and offering up a snack.

The UK Advertising Standards Agency is serious about marketing communications. The agency states that using touching-up photos in advertisements is unethical (Hörnle et al., 2019). The agency banned two advertisements by L'Oréal because it felt they were misleading. One advertisement featured Julia Roberts promoting Teint Miracle, a new Lancôme skin product. The company claimed that the product provides *luminosity to the skin*. The other advertisement featured the model, Christy Turlington. The advertisement promoted a cosmetic concealer called Eraser. The product promises to hide wrinkles and skin dislocation. The company tried to convince the agency that the advertisements were not misleading and did not exaggerate the effectiveness of its product. However, the UK Advertising Standards were unconvinced (Hörnle et al., 2019). Whether consumers want realistic images and measured promises or prefer and want to see exaggerated claims and unrealistic images of beauty is debatable.

The company's trade promotion activities are closely monitored. The Robinson-Patman Act states that it is illegal for companies to favor specific customers over others in trade promotions. The act requires that companies follow similar promotional allowances and services to all customers on proportionately equal terms (Yonezawa et al., 2020).

Different rules and regulations are in place to protect customers who are called on at home or buy products at a location that is not the permanent place of business for the sellers. These rules are enforced because such customers are more vulnerable than other customers. The rules protect such customers compared to customers who physically visit company stores to make a purchase. Again, individuals may be vulnerable to high-pressure selling techniques. The Federal Trade Commission (FTC) has adopted a three-day cooling-off rule for vulnerable individuals. The rule states that an individual who buys a product at a temporary location like a hotel, workplace, dormitory, or home can take remedial action. Individuals can return anything that costs more than $25 within 72 hours of buying if they are unsatisfied. Companies are bound to refund without asking any questions (Sánchez et al., 2018).

Personal selling takes place in the business-to-business domain. In such instances, it is illegal for any salesperson to offer bribes to purchasing agents or anyone who can potentially influence a sale. Salespeople cannot apply bribery or industrial espionage to obtain or use competitors' technical or trade secrets (Uğurlu, 2023).

Marketing Research and Public Policy Issues

Marketing research conducted by companies benefits both the sponsoring company and its consumers. In the digital age, it is easier and more convenient for companies to conduct marketing research and to generate customer insights (Gibson et al., 2022). Marketing research allows companies to gain insights into customer requirements and preferences. However, companies may misuse marketing research. It is easier for companies to misuse, misinterpret, and misrepresent the marketing research findings in the digital age. Companies should gain insights to deliver satisfying products and services and develop strong customer relationships. However, misusing market research and its findings can harm or annoy customers. There may be various public policy issues related to marketing research conducted by companies. Such public policy issues may also result in ethical issues. Three central public policy and ethical issues in marketing research related to intrusions on consumer privacy, consumer data security and protection, and the misuse of research findings (Gibson et al., 2022).

Intrusions on Consumer Privacy

Consumers are optimistic about marketing research and believe that marketing research serves a valuable purpose. They enjoy responding to customer surveys and providing opinions to researchers. However, some customers strongly resent or even mistrust marketing research. They worry that companies will collect customer information, store it in databases, and analyze it. The databases may be full of personal information about customers. Customers also fear researchers might use sophisticated techniques to probe and analyze their deepest feelings. Customers fear that researchers may track their internet and mobile device usage. Researchers may observe and analyze the behavior of consumers when they shop and then apply this knowledge to manipulate consumers' buying behavior (Saritas et al., 2021). Recent surveys indicate that more than 90 percent of Americans feel that they have lost control over the collection and use by companies of their data and information they share on social media sites (Saritas et al., 2021).

A classic example of misuse of consumer data is Target. Target used the buying histories of consumers to understand whether their female customers were pregnant or not. Target also tried to accurately estimate the child's gender and the due date of delivery (Mandal, 2021). Target provides every customer with a Guest ID number tied to his or her name, credit card, or email address. Target also tracks customers' purchase details and demographic details from other sources. Target studied the buying histories of women who had previously signed up for baby registries. Target developed a pregnancy prediction score for each customer based on the information collected. This score was based on the purchasing patterns of female customers across 25 product categories (Hill, 2012). The company then used the score to predict which females might be pregnant. Based on the prediction, Target started sending personalized books of coupons for baby-related items to expectant parents. The coupons were keyed to their pregnancy stages (Duhigg, 2012).

The strategy worked initially for Target because it had a good marketing sense. Based on the prediction, Target could identify the parents-to-be, and based on the subsequent offers, Target tried to build relationships with such customers and tried to convert them into loyal customers as their families developed (Hill, 2012). However, Target had to face the backlash from angry customers. An angry man accused his high school-aged daughter of receiving Target coupons for cribs, strollers, and maternity clothes. "Are you trying to encourage her to get pregnant?" he demanded. The Target store manager apologized. However, when he called again to apologize after a few days, he learned that Target's marketers had known about the young woman's pregnancy before her father did (Hill, 2012). Many other customers also complained that Target knew about their pregnancies even before they revealed their pregnancies to even family and close friends. Such customers also started wondering what other information Target might be keeping track of. They were also worried that based on the data collected and the prediction done, Target might be developing their profiles (Duhigg, 2012). As one reporter concluded: "The store's bull's-eye logo may now send a shiver … down the closely-watched spines of some (Target shoppers)" (Hill, 2012). Critics and customers themselves are worried about consumer privacy and the security of the information collected by companies from them. Marketers should be careful not to cross over the privacy line while collecting, storing, and mining customer information. They should understand that there is a fine line of demarcation between ethical and unethical.

Intrusions on Consumer Privacy and Technoethics

Marketers find it difficult to draw a line regarding marketing research and privacy related to technical data (Mandal, 2021). For example, it is difficult to answer whether it is good or bad when some retailers use mannequins with cameras hidden in one eye to record customer demographics and shopping behavior. Marketers argue that such recording is done to understand customers better and serve them better (Najafi et al., 2021). Similarly, companies monitor posts made by consumers on Facebook, Twitter, Instagram, YouTube, or other social media to be more responsive. Companies also track individuals' mobile phone usage. Such tracking may be done to issue location-based information, offers, and advertisements (Bhandari & Bansal, 2018).

SAP provides Consumer Insight 365 service, which helps mobile service providers "extract data about subscribers (and their) mobile-centric lifestyles" (Plattner & Leukert, 2015). It keeps track of more than 300 mobile web surfing, text messaging, phone calls, and other mobile events per day for each of its 20 to 25 million mobile subscribers across multiple carriers (Christl et al., 2017). Marketers can generate customer insights based on the data. According to one analyst, by combining mobile data with other information, the service can tell businesses, "Whether shoppers are checking out competitor prices on their phones or just emailing friends. It can tell them the age ranges and genders of people who visited a store between 10 a.m. and noon and link location and demographic data with shoppers' web browsing histories. Retailers might use the information to arrange store displays to appeal to certain customer segments at different times of the day or to help determine where to open new locations" (Kaye, 2015). Such information allows marketers to understand customers and their requirements and preferences better. However, such information might be *a little close for comfort* for those customers who respect privacy (Plattner & Leukert, 2015).

Market researchers should ensure consumer privacy. Increasing concerns about consumer privacy is a significant issue for the marketing research industry. Companies are required to collect, store, and analyze sensitive customer data to generate customer insights. However, it becomes difficult for companies always to maintain consumer trust when it comes to the collection and analysis of data in marketing research. Consumers desire personalization. Personalization requires the collection and analysis of huge amounts of customer data. However, customers are worried about privacy. Consumers should make a trade-off between personalization and privacy (Gouthier et al., 2022). Consumers want to receive relevant and personalized offers that meet their needs. However, they worry about or resent that companies may track them too closely. Companies should be cautious when they cross the line when gathering, using, and analyzing customer data (Mandal, 2019).

Responsibilities of Governments and Businesses for Consumer Privacy

Companies and businesses should try to ensure consumer privacy. Otherwise, failure to do so may result in angry and frustrated customers and increased government intervention (Libaque-Sáenz et al., 2021). Consequently, the marketing research industry considers several options for responding to issues related to intrusion and privacy. For example, the Marketing Research Association has undertaken initiatives like *Your Opinion Counts* and *Respondent Bill of Rights* to educate consumers about the benefits of marketing research. The initiatives also enlighten consumers to distinguish marketing research from telephone selling and database building (Mandal, 2019).

Enlightened companies consider consumer privacy to be a severe issue. Such companies try to protect consumers' privacy (Quach et al., 2022). Companies like Facebook, American Express, Apple, IBM, Microsoft, and even the U.S. government have appointed a chief privacy officer (CPO). The primary responsibility of the CPO is to safeguard and ensure consumers' privacy. The CPO of IBM claims that her job requires *multidisciplinary things and attitude*. She needs to coordinate with all the departments in the company, from technology, legal, and accounting to marketing and communications. She needs to ensure that all the departments work together towards the common goal of safeguarding customer privacy (Mandal, 2019).

Consumers do not hesitate to share information if they are convinced that marketing research will provide value in the exchange of information. For example, Amazon customers are not worried when Amazon develops and maintains a database of previous purchases. The customers are assured that Amazon analyzes those purchases to provide future product recommendations and to serve them better. The process also saves time, energy, money, and resources and provides value (Lin, 2022). Similarly, users of Bizrate gladly complete surveys rating online seller sites because they can view the overall ratings of others when making purchase decisions (Duan & Zhang, 2021). Researchers should ask for only the information they require, take responsibility for the customer information collected, use it responsibly to provide customer value, and avoid sharing information with third parties without customers' explicit consent.

Measures to Protect Consumer Privacy and Ensure Technoethics

Companies and various organizations in the marketing research industry adopt initiatives to protect consumer privacy and ensure technoethics (Mandal, 2021). TRUSTe is an organization that monitors the privacy practices of websites. TRUSTe conducted a survey which revealed that more than 90 percent of respondents view online as a *really* and *somewhat* important issue. More than 75 percent of the respondents agreed with the statement, "The internet is not well regulated, and naïve users can easily be taken advantage of." Respondents also expressed their desire to not be tracked when they are online. More than 66 percent of the respondents said they do not want companies to tailor advertisements to their interests (Slepchuk et al., 2022). It is not surprising that a lesser number of individuals are willing to reveal personal information on websites.

The marketing research industry adopts several measures to respond to such issues. Several associations have come together and worked in collaboration to protect consumer interests. The European Federation of Associations of Market Research Organizations and the European Society for Opinion and Marketing Research (ESOMAR) have recently jointly suggested a range of amendments to the European Commission's Directive 95/46/EC. The directive deals with data protection (Lachapelle, 2022). The research industry also has considered adopting broad standards. These standards are based on the ESOMAR and International Chamber of Commerce worldwide code of ethical practices (the ICC/ESOMAR International Code on Market and Social Research). The code of ethics outlined here includes researchers' responsibilities to respondents and the general public. For example, the code requires that researchers reveal their names, addresses, and other necessary details to respondents. Respondents should be able to contact the researchers if they feel necessary. It also bans companies and research agencies from representing activities such as database compilation or sales and promotional pitches as research (Madan et al., 2022).

Consumer Data Security and Protection

Companies develop and maintain large customer databases. Such large databases can pose serious data security challenges for companies. Consumers tend to believe that the data they share with companies is in safe custody. They believe that companies will adopt sufficient measures to protect the personal data they collect. However, in many cases, data breaches happen. Recently, many reputed companies and brands have experienced significant consumer data breaches. Sometimes, such data breaches have affected millions of consumers (Negri & Ieva, 2022).

For example, Facebook reported that two separate major data breaches happened. The data breaches affected its users and exposed the personal information of 87 million and 50 million users, respectively. The information related to the personalities of users, social networks, and users' engagement on the platform (Brown & Peterson, 2022). Major retailers such as Target, Macy's, and Saks Fifth Avenue have reported data security breaches. The breaches compromised millions of customers' payment card information (Nadeem et al., 2021). A massive data breach was reported by Equifax, a credit reporting agency. The breach exposed the sensitive data of more than 146 million customers. The information contained the customers' passport numbers, Social Security Numbers, and driver's license information (Choi, 2021). Yahoo! also announced a significant data breach a few years back. The data breach affected the personal information in more than 3 billion Yahoo! accounts (Daswani & Elbayadi, 2021).

Likewise, Marriott International reported that hackers breached its Starwood Reservation system. Hackers stole the personal information of more than 500 million customers of Marriott International (Shukla et al., 2022). Marriott-owned Starwood hotels comprise Sheraton, Westin, W Hotels, St. Regis, Aloft, and others. Hotels are a rich source of information about guests. The information may contain combinations of names, mailing addresses, phone numbers, email addresses, credit card numbers, Starwood Preferred Guest (SPG) account information, dates of birth, gender arrival and departure information, reservation dates, and communication preferences. The information may also contain some guests' travel histories and passport numbers. The data breach involved customers who booked reservations for Starwood hotel brands from early 2014 to September 2018. Marriott was prompt and took immediate steps when the data breach was discovered. However, it was too late, and the damage had already been done to Marriott's reputation. One data privacy advocate said, "They can say they take security seriously, but they don't if you can be hacked over four years without noticing" (Paraskevas, 2020).

Failures to secure data may harm consumers. Moreover, data breaches may affect and damage the reputation and images of companies and brands. Firstly, data breaches and data theft can be costly. For example, Equifax reported recovery costs of $400 million from the data breach incident. Marriott had to experience lawsuits and recovery costs. The data breach for Marriott was massive and long. Marriott faced hefty European fines under the EU's General Data Protection Regulation (GDPR) (Barrett, 2019). Such incidents of data breaches and data thefts will require government intervention. Governments should enforce rules and regulations so that such incidents do not happen. Also, the massive data security lapses erode consumer trust. Customer-brand relationships that are built and developed over a long period may be destroyed. Thus, companies must take the security of consumer data seriously.

Misuse of Research Findings

Market researchers can employ research studies as a powerful persuasion tool. Companies often use the results of research studies to substantiate their claims in advertising and promotion. However, marketers may misuse, misinterpret, and misrepresent research findings. Research results may be used as vehicles for pitching sponsors' products. In many cases, research surveys and studies are designed to support the claims marketers want to make (Mandal, 2021). For example, a Black Flag survey once asked, "A roach disk … poisons a roach slowly. The dying roach returns to the nest and is eaten by other roaches after it dies. In turn, these roaches become poisoned and die. How effective would this type of product be in killing roaches?" As expected, 79 percent of the respondents said the process would be effective (Mandal, 2021).

Another example is the UK phone and broadband service provider TalkTalk, which claims that customers can save €160 per year (Chapman et al., 2018). Through their television advertisements, they urged individuals to join the community of customers *who are already saving an average of over £140 (€160) a year* (Yang et al., 2013). Competitors of TalkTalk reacted strongly to such claims. They accused that such claims were misleading exaggerations because they were based on the average savings of TalkTalk customers and not on savings that new target customers could achieve. TalkTalk argued that question marks surrounded the on-screen text of their claims. So, the claim was more of a question than a definitive statement. Nevertheless, the marketing tactics of TalkTalk were severely criticized by the UK Advertising Standards Authority. The authorities accused TalkTalk of misleading the public with information that was not true (Chapman et al., 2018).

Some marketers and advertisers openly rig their research designs or blatantly misrepresent the research findings. Most abuses tend to be more subtle stretches. In many cases, disputes arise regarding the validity, reliability, interpretation, and use of research findings. Again, any research result can be interpreted how the researchers want to interpret it. So, research results are dependent largely on the biases and viewpoints of the researchers.

Ensuring Ethics for Marketing Research

Critics and experts believe that ethics should be ensured in marketing research. Several associations recognize that marketing research can be abused, misinterpreted, and misused. Associations like the American Marketing Association, the Marketing Research Association, and the Council of American Survey Research Organizations (CASRO) are of the opinion that codes of ethics should be developed and enforced for marketing research (Hopkins et al., 2021). Such organizations have developed codes of research ethics and standards of conduct. For example, the CASRO Code of Standards and Ethics for Survey Research outlines the responsibilities of researchers to respondents. Such responsibilities include confidentiality, privacy, and avoidance of harassment. It also outlines significant responsibilities in reporting results to clients and the public (Council of American Survey Research Organizations, 2001).

Rules and regulations are enforced to prevent unethical behavior in marketing research. However, rules and regulations may not be sufficient to prevent unethical behavior. Companies and market researchers should accept responsibility for policing the conduct and for the reporting of their marketing research. They should work towards protecting the best interests of consumers and their own.

The Sustainable Company

Companies must fulfill the wants and demands of customers properly. Those companies which can satisfy and delight their customers thrive and prosper in the long run. Those companies that are unable to fulfill customer requirements, which cause harm intentionally or unintentionally to their customers, to others in society, or to future generations, will decline and cease to exist in the long run. One critic observes, "Sustainability is an emerging business megatrend, like electrification and mass production, that will profoundly affect companies' competitiveness and survival." Another critic observes, "Increasingly, companies and leaders will be assessed not only on immediate results but also on … the ultimate effects their actions have on societal well-being. This trend has been coming in small ways for years but now is surging. So pick up your recycled fair-trade coffee, and get ready" (Löser, 2015).

Companies become sustainable when they generate value for their customers in the long run. Such values should be created through socially, environmentally, and ethically responsible actions. Sustainable marketing requires companies and businesses to think not only about customers' present requirements but also about the impact of their actions on future generations. It means that companies should be concerned about customers in the future. This will ensure the survival and success of the business, shareholders, employees, and society at large. Companies should pursue the philosophy of shared value and the triple bottom line of people, planet, and profit. Companies should focus on sustainable marketing to engage with customers and build profitable relationships. Such relationships should be developed by creating value for customers to capture value from customers in return – now and in the future.

Initiatives by Companies

Companies aim to address and deal with issues related to ethics and social responsibility proactively, openly, and forthrightly. This helps companies to build, develop, and maintain strong customer relationships based on honesty and trust. SC Johnson believes in doing what is right, even at the expense of sales. Saran Wrap was previously manufactured for more than 50 years from polyvinylidene chloride (PVDC), an ingredient responsible for two major differentiating features of the product – impenetrable odor barrier qualities and superior microwavability. Saran Wrap was able to differentiate itself from Glad and Reynolds' competing wraps because of PVDC. Glad and Reynolds did not contain PVDC. However, in the early 2000s, regulators, environmentalists, and consumers voiced concerns about materials containing chlorine, specifically polyvinyl chloride (PVC) (Johnson, 2015). SC Johnson has its own Greenlist analysis. The analysis helps SC Johnson rate product ingredients based on their environmental and human health impact. The analysis confirmed the hazards caused by PVCs. SC Johnson promptly pledged to eliminate PVCs from its products and packaging (Godfrey & Jacobus, 2022).

SC Johnson became proactive and eliminated PVDCs in 2004, even though PVDCs were not banned till then. The company developed a polyethylene version of Saran Wrap, free from PVDCs (Godfrey & Jacobus, 2022). The market share of Saran Wrap dropped from 18 percent in 2004 to only 11 percent at present. Over the years, such decisions have hurt the sales. However, such initiatives have helped SC Johnson earn and keep customers' trust. The CEO of SC Johnson comments, "I don't regret the decision. Despite the cost, it was the right thing to do, and ... I sleep better at night because of it. We gained a surer sense of who we are as a company and what we want SC Johnson to represent" (Johnson, 2015).

Google has developed its code of ethics, the Google Code of Conduct. The conduct emphasizes the basic philosophy of Google: "Don't be evil" (Oravec, 2018). The core message of the code of conduct is simple: Google employees (known as *Googlers*) need to earn the faith and trust of users. For this, employees must exhibit the highest possible standards of ethical business conduct. The Google Code of Conduct emphasizes "providing our users unbiased access to information, focusing on their needs, and giving them the best products and services that we can. However, it is also about doing the right thing more generally – following the law, acting honorably, and treating each other with respect" (Pendergraft, 2019).

All Googlers, from board members to the newest employee, are expected to take personal responsibility for practicing the code's spirit and letter and encouraging other Googlers to do the same. Employees should report violations to their managers, human resources representatives, or an Ethics and Compliance hotline. The code states: "If you have a question or ever think that one of your fellow Googlers or the company as a whole may be falling short of our commitment, don't be silent. We want – and need – to hear from you" (Mamaysky, 2021).

Calls for Future Research

Researchers may collect primary data and conduct empirical analysis to understand the various ethical issues related to sustainable marketing. They may analyze the existing corporate marketing ethics policies and develop and suggest policies, codes of marketing ethics, and relevant and appropriate policies. They may analyze markets other than that of the United States and develop relevant codes and policies for other markets. All these will allow companies and businesses to remain ethical in their marketing actions and adopt sustainable marketing principles in the long run.

DISCUSSION AND CONCLUSION

Companies and businesses require developing policies and guidelines to help their managers deal with questions about marketing ethics and sustainable marketing. However, the best policies and guidelines cannot ensure that all the ethical decisions employees must make are resolved. Nevertheless, some principles are enforced that will guide the marketers. One principle states that the free market and the legal system should decide such issues. A second principle states that individual companies and managers are responsible for their actions. Consequently, each firm and marketing manager must work out a socially responsible and ethical philosophy. The sustainable marketing concept requires that managers should not only focus on what is legal and allowable. They should develop standards based on personal integrity, corporate conscience, and long-term consumer welfare. All these are discussed in the study. The discussions have both theoretical and managerial implications. Academicians may study and analyze the various aspects of marketing ethics and their roles in sustainable marketing. They may analyze the various situations and suggest and formulate codes of conduct and ethics that companies and businesses may apply. Academicians may also analyze the effectiveness of such codes of conduct and ethics in specific situations. Companies and businesses should operate within the realms of such codes and principles to remain ethical and earn the trust of their customers and society's trust.

Further, companies and businesses must remain ethical in all their activities and actions. The discussions will emphasize the importance of ethical behavior for company managers and employees. Companies that customers perceive as ethical will thrive and succeed in the long run. The discussions will sensitize managers to develop codes of conduct and ethics for their organizations. Managers may evaluate and decide the best activities and principles for their organizations and implement those codes of conduct and ethics. All these will help companies to develop trust and belief in the minds of their customers and society at large and succeed in the long run.

Finally, good ethics are a cornerstone for companies and businesses regarding sustainable marketing. Companies face moral dilemmas when dealing with ethical issues and sustainable marketing. They require the development of corporate marketing ethics policies, which are broad guidelines everyone in the organization must follow. Some approaches include the free market and legal system, responsibilities of individual companies, appropriate business standards and practices, and programs developed by companies. Based on such approaches, managers may develop sustainable companies. Companies like SC Johnson and Google, among others, adopt various initiatives concerning marketing ethics and sustainable marketing. The study focused on conducting a conceptual analysis of various ethical issues, sustainable marketing, and the formulation of corporate marketing ethics policies. Primary data is not collected, and empirical analysis is not performed. The study also focused mainly on analyzing the markets in the United States.

REFERENCES

Aboelenien, A., & Naaman, K. (2023). Ethical considerations in marketing: Academics, practitioners, and other marketplace actors. In *Research Handbook on Accounting and Ethics* (pp. 328–342). Edward Elgar Publishing. DOI: 10.4337/9781800881020.00033

Amaral da Luz, K. C., & Musial, N. T. K. (2023). Can organizational culture encourage corporate social responsibility and improve environmental performance in [B] companies? *Journal of Management and Sustainability*, 13(1), 215. DOI: 10.5539/jms.v13n1p215

Audi, R. (2022). The marketing of human images is a challenge to ethical leadership. In *Leadership and business ethics* (pp. 297–311). Springer Netherlands. DOI: 10.1007/978-94-024-2111-8_18

Barrett, C. (2018). *Everyday ethics for practicing planners*. Routledge. DOI: 10.4324/9781351179461

Barrett, C. (2019). Are the EU GDPR and the California CCPA becoming the de facto global standards for data privacy and protection? *Scitech Lawyer*, 15(3), 24–29.

Bhandari, R. S., & Bansal, A. (2018). A comparison of psychological and socio-economic factors affecting individual user's social media behaviour. *FIIB Business Review*, 7(2), 126–134. DOI: 10.1177/2319714518789759

Brennan, R. (2021). Business-to-business (industrial) marketing ethics. In *The Sage Handbook of marketing ethics*, 238-247. SAGE

Brown, G., & Peterson, R. S. (2022). The Subordinated Board: Facebook. In *Disaster in the Boardroom* (pp. 89-101). Palgrave Macmillan, Cham.

Chapman, J., Chinnaswamy, A., & Garcia-Perez, A. (2018, January). The severity of cyber attacks on education and research institutions: a function of their security posture. In *Proceedings of ICCWS 2018 13th International Conference on Cyber Warfare and Security. Academic Conferences and Publishing Limited* (pp. 111-9).

Choi, Y. B. (2021). Organizational cyber data breach analysis of Facebook, Equifax, and Uber cases. [IJCRE]. *International Journal of Cyber Research and Education*, 3(1), 58–64. DOI: 10.4018/IJCRE.2021010106

Chong, W. K., & Patwa, N. (2023). The value of integrity: Empowering SMEs with ethical marketing communication. *Sustainability*. 15(15), 11673. DOI: 10.3390/su151511673

Christl, W., Kopp, K., & Riechert, P. U. (2017). Corporate surveillance in everyday life. *Cracked Labs*, 6, 2017–10.

Council of American Survey Research Organizations (2001). CASRO code of standards and ethics for survey research.

Daswani, N., & Elbayadi, M. (2021). The Yahoo breaches of 2013 and 2014. In *Big Breaches* (pp. 155–169). Apress. DOI: 10.1007/978-1-4842-6655-7_7

De Oliveira Neto, G. C., Pinto, L. F. R., Amorim, M. P. C., Giannetti, B. F., & de Almeida, C. M. V. B. (2018). A framework of actions for strong sustainability. *Journal of Cleaner Production*, 196, 1629–1643. DOI: 10.1016/j.jclepro.2018.06.067

Duan, W., & Zhang, J. (2021). The comparative Performance of online referral channels in e-commerce. *Journal of Management Information Systems*, 38(3), 828–854. DOI: 10.1080/07421222.2021.1962598

Duhigg, C. (2012, February 19). Psst, You in Aisle 5. *New York Times,* MM30.

Dumitrescu, C., Shaw Hughner, R., & Shultz, C. J.II. (2016). Policy and marketing changes to help curb childhood obesity: Government ban vs. industry self-regulation. *International Journal of Consumer Studies*, 40(5), 519–526. DOI: 10.1111/ijcs.12296

Eagle, L., & Dahl, S. (2018). Product placement in old and new media: Examining the evidence for concern. *Journal of Business Ethics*, 147(3), 605–618. DOI: 10.1007/s10551-015-2955-z

Elhajjar, S. (2023). The current and future state of the marketing management profession. *Journal of Marketing Theory and Practice*, 1–18.

Feil, H. (2021). The cancer of corruption and World Bank project performance: Is there a connection? *Development Policy Review*, 39(3), 381–397. DOI: 10.1111/dpr.12503

Fuller, D. A. (1999). *Sustainable marketing: Managerial-ecological issues.* Sage publications. DOI: 10.4135/9781452220611

Gangjee, D. S. (2024). Law (s) against unfair competition: The legitimacy of the UK approach. *Journal of Intellectual Property Law and Practice*, 19(2), 90–100. DOI: 10.1093/jiplp/jpad105

Gauthier, B., Goyette, J., & Kouamé, W. A. (2021). Why do firms pay bribes? Evidence on the demand and supply sides of corruption in developing countries. *Journal of Economic Behavior & Organization*, 190, 463–479. DOI: 10.1016/j.jebo.2021.06.041

Gehlhar, M. (2021). *The fashion designer survival guide: start and run your own fashion business.* Simon and Schuster.

Gibson, S., Hsu, M. K., & Zhou, X. (2022). Convenience stores in the digital age: A focus on the customer experience and revisit intentions. *Journal of Retailing and Consumer Services*, 68, 103014. DOI: 10.1016/j.jretconser.2022.103014

Godfrey, P. C., & Jacobus, L. E. (2022). *Ethics and business: An integrated approach for business and personal success.* Wiley.

Gouthier, M. H., Nennstiel, C., Kern, N., & Wendel, L. (2022). The more the better? Data disclosure between the conflicting priorities of privacy concerns, information sensitivity and personalization in e-commerce. *Journal of Business Research*, 148, 174–189. DOI: 10.1016/j.jbusres.2022.04.034

Hagenbuch, D. J., & Mgrdichian, L. M. (2020). Mindful marketing: A strategy-based, branded Approach for encouraging ethical marketing. *Marketing Education Review*, 30(1), 15–28. DOI: 10.1080/10528008.2019.1686993

Hamby, A., Orazi, D., & Moreau, P. (2024). Whose story is this? Source reveal as a communication tactic to increase consumers' advocacy for social causes. *Journal of Business Research*, 170, 114309. DOI: 10.1016/j.jbusres.2023.114309

Hardy, J. (2018). Advertising regulation. In *The advertising handbook* (pp. 123–140). Routledge. DOI: 10.4324/9781315558646-9

Hill, K. (2012, February 16). How Target figured out a teen girl was pregnant before her father did. *Forbes.*

Hopkins, C. D., Ferrell, O. C., Ferrell, L., & Hopkins, K. H. (2021). Changing perceptions of marketing ethics and social responsibility in principles of marketing. *Journal of Marketing Education*, 43(2), 244–259. DOI: 10.1177/0273475321995553

Hörnle, J., Schmidt-Kessen, M., Littler, A., & Padumadasa, E. (2019). Regulating online advertising for gambling–once the genie is out of the bottle.... *Information & Communications Technology Law*, 28(3), 311–334. DOI: 10.1080/13600834.2019.1664001

Iglesias, O., Markovic, S., Bagherzadeh, M., & Singh, J. J. (2020). Co-creation: A key link between corporate social responsibility, customer trust, and customer loyalty. *Journal of Business Ethics*, 163(1), 151–166. DOI: 10.1007/s10551-018-4015-y

Johann, M. (2015). Marketing Communications in the Network Organizations. *Communication and Information Technology in Society: Volume 2 Information and Communication Technologies (ICT) in Management, 2*, 35.

Johnson, F. (2015). SC Johnson's CEO on doing the right thing, even when it hurts business. *Harvard Business Review*, •••, 33–36.

Johnson, S. (2007). SC Johnson builds business at the base of the pyramid. *Global Business and Organizational Excellence*, 26(6), 6–17. DOI: 10.1002/joe.20170

Kagendo-Kiarie, R. (2022). Ethical marketing: Concepts and reasons. In *Products for conscious consumers: Developing, marketing and selling ethical products* (pp. 15–33). Emerald Publishing Limited. DOI: 10.1108/978-1-80262-837-120221002

Kaye, K. (2015, October 26). The $24 Billion Data Business That Telcos Don't Want to Talk About. *Advertising Age*, 12–14.

Kennedy, A. M., Jones, K., & Williams, J. (2019). Children as vulnerable consumers in online environments. *The Journal of Consumer Affairs*, 53(4), 1478–1506. DOI: 10.1111/joca.12253

Koester, J. V. (2003). Legal briefing: Advertising to children in the USA. *Young Consumers*, 4(1), 65–71. DOI: 10.1108/17473610310813735

Lachapelle, A. (2022). Data protection enforcement in the era of the Directive on Whistleblowers: Towards a collective approach? In *Research Handbook on EU Data Protection Law* (pp. 600–619). Edward Elgar Publishing. DOI: 10.4337/9781800371682.00032

Leonidou, L. C., Aykol, B., & Eteokleous, P. (2020). Internal and external drivers of an ethical international marketing strategy: Implications on reputational advantage and performance. *The SAGE Handbook of marketing ethics*, 20.

Libaque-Sáenz, C. F., Wong, S. F., Chang, Y., & Bravo, E. R. (2021). The effect of fair information practices and data collection methods on privacy-related behaviors: A study of mobile apps. *Information & Management*, 58(1), 103284. DOI: 10.1016/j.im.2020.103284

Lin, X. (2022, April). Big data analysis of personalized recommendation in e-commerce. In *2022 International Conference on Social Sciences and Humanities and Arts (SSHA 2022)* (pp. 768-771). Atlantis Press. DOI: 10.2991/assehr.k.220401.147

Löser, F. (2015). *Strategic information systems management for environmental sustainability: enhancing firm competitiveness with Green IS* (Vol. 6). Universitätsverlag der TU Berlin.

Lu, J., Ren, L., Qiao, J., Yao, S., Strielkowski, W., & Streimikis, J. (2019). Corporate social responsibility and corruption: Implications for the sustainable energy sector. *Sustainability*, 11(15), 4128. DOI: 10.3390/su11154128

Madan, A., Rosca, L. D., Dumitru, I., & Canda, A. (2022). The value of art in persuasive marketing communication and its sustainable effect on the country of origin. *Sustainability (Basel)*, 14(3), 1228. DOI: 10.3390/su14031228

Mamaysky, I. (2021). Understanding ethics and compliance: A practitioner's guide to effective corporate compliance programs. *Loy. U. Chi. J. Reg. Compl.*, 6, 58.

Mandal, P. C. (2019). Ethical issues and concerns in collection of marketing information and marketing intelligence: Ethical issue in collection of information. [IJBIR]. *International Journal of Business Intelligence Research*, 10(2), 16–28. DOI: 10.4018/IJBIR.2019070102

Mandal, P. C. (2021). Socially-responsible marketing communications: Concerns, strategies, and initiatives. [JMME]. *Journal of Media Management and Entrepreneurship*, 3(1), 1–17. DOI: 10.4018/JMME.290301

Mogaji, E., & Mogaji, E. (2021). Brand Extension. *Brand Management: An Introduction through Storytelling*, 181-205.

Mohamed, K. (2013). Evaluation of societal marketing (sustainable marketing) in maintaining ethics in marketing goods or services to the customers. *International Journal of Management Research and Reviews*, 3(11), 3703.

Mokha, A. K. (2018). Social responsibility and ethics in marketing. *IITM Journal of Management and IT*, 9(2), 50–56.

Morton, F., & Treviño, T. (2021). Targeting kids in the digital age: The ethics of online marketing towards children. In *Humanistic management in Latin America* (pp. 50–71). Routledge. DOI: 10.4324/9780429351174-4

Nadeem, W., Tan, T. M., Tajvidi, M., & Hajli, N. (2021). How do experiences enhance brand relationship performance and value co-creation in social commerce? The role of consumer engagement and self brand-connection. *Technological Forecasting and Social Change*, 171, 120952. DOI: 10.1016/j.techfore.2021.120952

Najafi, K., Heidarzadeh Hanzaee, K., & Khounsiavash, M. (2021). Exploration of the role of visual merchandising (mannequin) on women's shopping behavior through the moderating role of type of store and customers' intention: An experimental research approach. *Journal of Business and Management*, 13(1), 247–272.

Negri, F., & Ieva, M. (2022). The interplay between privacy failure, recovery and crisis communication management: An integrative review and research agenda. *Sinergie Italian Journal of Management*, 40(1), 77–101. DOI: 10.7433/s117.2022.05

Nill, A., & Laczniak, G. (2022). Direct-to-consumer genetic testing and its marketing: Emergent ethical and public policy implications. *Journal of Business Ethics*, 175(4), 669–688. DOI: 10.1007/s10551-020-04632-z

Oravec, J. A. (2018). "Don't be evil" and beyond for high tech organizations: Ethical statements and mottos (and responsibility). In *Handbook of research on civic engagement and social change in contemporary society* (pp. 220–237). IGI Global. DOI: 10.4018/978-1-5225-4197-4.ch013

Paraskevas, A. (2020). Human trafficking and modern slavery in Europe's hotels. In *Slavery and liberation in hotels, restaurants and bars* (pp. 70–86). Routledge. DOI: 10.4324/9780367855383-6

Pecora, N. (2022). Media policies for children: Issues and histories in the US. In *The Routledge International Handbook of children, adolescents, and media* (pp. 425–431). Routledge. DOI: 10.4324/9781003118824-54

Pendergraft, G. (2019). Controversy over gender differences and free speech at Google. *Sage (Atlanta, Ga.)*. Advance online publication. DOI: 10.4135/9781526488664

Peterson, M., Minton, E. A., Liu, R. L., & Bartholomew, D. E. (2021). Sustainable marketing and consumer support for sustainable businesses. *Sustainable Production and Consumption*, 27, 157–168. DOI: 10.1016/j.spc.2020.10.018

Plattner, H., & Leukert, B. (2015). Quality time at work. In *The In-memory revolution* (pp. 222–242). Springer. DOI: 10.1007/978-3-319-16673-5_10

Preston, C. (2016). Pre-school children and marketing communications. *International Journal of Consumer Studies*, 40(5), 618–623. DOI: 10.1111/ijcs.12307

Quach, S., Thaichon, P., Martin, K. D., Weaven, S., & Palmatier, R. W. (2022). Digital technologies: Tensions in privacy and data. *Journal of the Academy of Marketing Science*, 50(6), 1–25. DOI: 10.1007/s11747-022-00845-y PMID: 35281634

Rawat, S. R., Bhatia, K., Hegde, M., Bhat, N., & Tewari, S. (2015). The importance of ethical marketing practices. *Journal of Business Management and Economics*, 3(2), 23–30.

Reddy, P., Chandu, V., Ryali, N., Gangula, A., Mishra, D., & Mishra, S. (2023). Ethical Business Marketing: The Practices to Compete Successfully in Business. In *Data-Driven Approaches for Effective Managerial Decision Making* (pp. 259-276). IGI Global.

Rezníčková, M., & Zaušková, A. (2020, July). Use of Social Media in the Presentation of Eco-Innovations of Slovak Businesses. In *7th European Conference on Social Media ECSM 2020* (p. 232).

Sánchez Abril, P., Oliva Blázquez, F., & Martínez Evora, J. (2018). The Right of Withdrawal in Consumer Contracts: A comparative analysis of American and European law. *InDret*, 3, 18–13.

Saritas, O., Bakhtin, P., Kuzminov, I., & Khabirova, E. (2021). Big data augmentated business trend identification: The case of mobile commerce. *Scientometrics*, 126(2), 1553–1579. DOI: 10.1007/s11192-020-03807-9 PMID: 33424052

Sattar, M. M., Barr, J. C., & Sfodera, F. (2023). Cause-Related Marketing. In *Cultural Marketing and Metaverse for Consumer Engagement* (pp. 71–95). IGI Global. DOI: 10.4018/978-1-6684-8312-1.ch006

Saygılı, E., Ozturkoglu, Y., & Ozturkoglu, O. (2022). Corporate responsibilities and values in codes of ethics: An exploratory study of the Global Fortune companies. *European Journal of International Management*, 18(1), 76–101. DOI: 10.1504/EJIM.2022.123752

Shahbazi Takabi, A., Amiri, H., & Karami, R. (2024). The Effect of Ethical Marketing in Auditing on Auditors' Job Performance. *International Journal of Finance & Managerial Accounting*, 9(35), 43–64.

Shukla, S., George, J. P., Tiwari, K., & Kureethara, J. V. (2022). Data Security. In *Data Ethics and Challenges* (pp. 41–59). Springer. DOI: 10.1007/978-981-19-0752-4_3

Singh, G., & Mishra, A. (2019). Ethics and Celebrity Advertising: Cases in the Indian Advertising Industry. In *Ethical Branding and Marketing* (pp. 173-184). Routledge.

Slepchuk, A. N., Milne, G. R., & Swani, K. (2022). Overcoming privacy concerns in consumers' use of health information technologies: A justice framework. *Journal of Business Research*, 141, 782–793. DOI: 10.1016/j.jbusres.2021.11.073

Taylor, C. R., Cho, Y. N., Anthony, C. M., & Smith, D. B. (2018). Photoshopping of models in advertising: A review of the literature and future research agenda. *Journal of Global Fashion Marketing*, 9(4), 379–398. DOI: 10.1080/20932685.2018.1511380

Tripathi, S., & Sharma, D. (2023). Green Marketing: Sustainable Strategies for Success. In *Climate Change Management and Social Innovations for Sustainable Global Organization* (pp. 76-92). IGI Global.

Uğurlu, K. (2023). Personal Selling in Accommodation Businesses. In *International Handbook of Skill, Education, Learning, and Research Development in Tourism and Hospitality* (pp. 1-18). Springer Nature Singapore.

Valls Martínez, M. D. C., Martín-Cervantes, P. A., & Peña Rodríguez, S. (2021). Ethical banking and poverty alleviation banking: The two sides of the same solidary coin. *Sustainability*, 13(21), 11977. DOI: 10.3390/su132111977

Weatherly, N. L. (2021). The ethics of organizational behavior management. *Journal of Organizational Behavior Management*, 41(3), 197–214. DOI: 10.1080/01608061.2021.1890664

Yang, A., Lee, D., Hwang, J., & Shin, J. (2013). The influence of regulations on the efficiency of telecommunications operators: A meta-frontier analysis. *Telecommunications Policy*, 37(11), 1071–1082. DOI: 10.1016/j.telpol.2013.02.004

Yonezawa, K., Gómez, M. I., & Richards, T. J. (2020). The Robinson–Patman act and vertical relationships. *American Journal of Agricultural Economics*, 102(1), 329–352. DOI: 10.1093/ajae/aaz049

Zhong, S., Zhao, X., & Song, J. (2023). MNEs' Ambidexterity Strategies and Moral Conflicts: The Case of Google in China. *Journal of Business Ethics*, 186(4), 781–796. DOI: 10.1007/s10551-023-05437-6

ADDITIONAL READING

Bulley, C. A., Braimah, M., & Blankson, F. E. (2018). Ethics, neuromarketing and marketing research with children. [IJCRMM]. *International Journal of Customer Relationship Marketing and Management*, 9(2), 79–95. DOI: 10.4018/IJCRMM.2018040105

Murphy, P. E., Laczniak, G. R., & Harris, F. (2016). *Ethics in marketing: International cases and perspectives*. Taylor & Francis. DOI: 10.4324/9781315626642

Palmer, J. E., Morgan, J. W., Hinojosa, S., Olomi, J. M., Ayala, L., & Rosay, A. B. (2022). Anti-racist and intersectional approaches in social science and community-based research. In *Social justice research methods for doctoral research* (pp. 222–243). IGI Global.

Chapter 14
Ethics of Health Research Data Ownership and Global Regulations

Nisha Daniel
https://orcid.org/0009-0006-6484-0940
University of Edinburgh, UK

ABSTRACT

In the era of Artificial Intelligence and Big Data, is data really the new oil? Within health research data governance, this chapter presents a compelling discussion of healthcare data ethics and the need for data safety and privacy. Data ownership debates public or private ownership and details an ethical society's global divergent perspectives and philosophical foundations. The building of a consistent data governance model may be beneficial to all and may address the issues of data bias, achieve data justice, and articulate individual autonomy and individual data rights.

INTRODUCTION

"Data is the new oil" (Humbly, 2006).

This quote essentially initiates the question of whether data is personal or public goods. Between these ways of epistemic judgments, there can be anticipated benefit or harm, calling forth an ethical stand right at the juncture. It begs yet another question: Is data really a commodity? There are as many threats to identity, exploitation, and violation of human rights arising from the use of data as the benefit it brings to oneself. By weighing the advantages and disadvantages of data use, it is possible

DOI: 10.4018/979-8-3693-3848-3.ch014

to provide high-quality services, allocate public resources efficiently, and achieve favorable results that benefit the respective community.[1]

Big data, machine learning, artificial intelligence, and other innovative and ubiquitous technologies are creating opportunities and challenges for different social practices at a rapid pace. Many questions remain ongoing such as, who benefits from these opportunities and advantages, and who pays the price? The ethical consequences of how we handle information are the focus of the field of data ethics. Certain definitions go deeper, while others concentrate more broadly on preventing harm during the data lifecycle (collection, sharing, and use). To identify morally sound solutions and best practices, Floridi and Taddeo's (2016) definition of data ethics considers not only the data itself but also the algorithms that evaluate it and the customs surrounding its use.

BACKGROUND

Recognizing data as a public good can lead to numerous societal benefits, including increased innovation, economic growth, social impact, transparency, and collaboration while necessitating careful attention to privacy and ethical concerns. What is the relevance of the "new oil" today? How do we quantify the air we breathe to determine individual boundaries when anyone can claim that the air they breathe in is theirs? Similarly, how can we differentiate between personal health information derived from our bodies and how can it be tuned for public purposes? These are key questions that require answers in a digital society. Further, healthcare patients have an inherent right to govern their personal information. There is transparency in using a private model of ownership that is patient-centric, as it empowers individuals to determine the utilization, purpose, and accessibility of their data. However, challenges arise regarding individual ownership. Patients might not have the knowledge to make wise decisions about how to use their data, and managing data flow within intricate healthcare networks can be challenging. Furthermore, overly stringent individual controls may make it more difficult to advance public health research and other initiatives.

However, institutional and public ownership puts the good of society as a whole first. This model simplifies data access for research and development, which could hasten the advancement of healthcare. It can also help with outbreak management and public health surveillance. Issues with public ownership, such as persistent concerns about accountability and transparency, further undermine public trust. Nevertheless, there is a need for crucial security measures due to the possibility of data breaches or unauthorized access by third parties. The objectives of this chapter include:

1. Establish an ethical data-position so as to be shared in the public interest.
2. Demonstrate how data poses the risk of being shared or used without consent.
3. Identify and list the public attitudes towards data-sharing.

History of Data Ethics

The field of data ethics has evolved in tandem with technological advancements, societal shifts, and changing perspectives on privacy and individual rights. A brief classification of its historical development follows.

Early Era (Pre-20th Century): Philosophical Foundations

While the term "data ethics" wasn't coined until the 20th century, the philosophical principles underlying it have their roots in classical and modern philosophy. Thinkers like Immanuel Kant, John Stuart Mill, and Jeremy Bentham laid the groundwork for discussions about individual rights, privacy, and the ethical implications of actions.

The Rise of Computers and Data Storage (Mid-20th Century): Privacy Concerns

As computers became more powerful and capable of storing vast amounts of data, concerns about privacy and surveillance began to emerge. Governments and corporations started collecting and storing personal information, leading to discussions about the ethical implications of data collection and use.

The Information Age and the Internet (Late 20th Century)

The internet revolutionized the way we collect, store, and share data. The explosion of data and the proliferation of online services and social media platforms led to an unprecedented increase in the amount of personal data being collected and analyzed. At the same time, high-profile data breaches and scandals exposed the vulnerabilities of data systems and raised public awareness about the importance of data protection.

The Era of Big Data and Analytics (21st Century)

The rise of big data and advanced analytics techniques presented new ethical challenges. The ability to analyze vast datasets has led to concerns about discrimination, surveillance, and the misuse of data. Governments and international organizations began to develop regulations and guidelines to address data privacy and security concerns. Examples include the General Data Protection Regulation (GDPR) in the European Union and the California Consumer Privacy Act (CCPA) in the United States.

Contemporary Developments: Ethical AI

The development of Artificial Intelligence (AI) has introduced new ethical questions related to data bias, transparency, and accountability. Organizations are increasingly adopting data governance frameworks and privacy-by-design principles to ensure that data is collected, stored, and used ethically. Technologies like blockchain and edge computing are raising new ethical considerations related to data ownership, control, and privacy. As technology continues to evolve, the field of data ethics will undoubtedly face new challenges and opportunities. It is essential for individuals, organizations, and governments to stay informed about the latest developments and to work together to develop ethical frameworks and practices for the responsible use of data.

Sources of Health and Social Care Data

There are many different sources of health and social care data, as detailed in Table 1.

Table 1. Types of data

Data Type	Description
Clinical data	Produced in hospitals and stored in hospital information systems, picture archiving and communication systems, laboratory databases, etc.
Primary care data	Produced in primary care (e.g. GP practices in the UK) and stored in local systems.
Biomedical research data	Produced by academia and pharmaceutical companies and stored in databases and libraries.
Health business data	Produced by healthcare providers, insurance companies and other organizations, and is typically stored in local information and transaction systems.
Patient-generated data	This includes data produced by wearable devices (e.g. activity trackers that capture data including heart rate, steps and calories) and web-based apps and platforms (e.g. self-management platforms)
Social work data	Includes electronic systems (software) to facilitate day-to-day work of social workers, online resources for professionals and people using social work services, assistive technology for people using services, social media and social networking interfaces, informatics, personal identifiable data, information management, hardware, online learning, and AI/ML for processing large amounts of data about the population to predict their needs.

Data Relevance

Health Data Ownership is the tight relationship between individual rights and societal benefits. Public health and government oversight require complete information, unlike businesses, which only require data for specific purposes. Governments may find it difficult to purchase all this data, and when they cannot, some businesses may decline to sell them in order to stay out of legal trouble, negative publicity, or rivalry. The services and insights that the data helps generate are what really make it valuable, but private ownership impedes this growth.

Companies can bundle their data with analysis services to create monopolies. If the government publishes this data publicly, anyone can use it or create competing analysis tools, thereby breaking these monopolies. Therefore, the core ideas that we need to consider are as follows:

- Public health and governmental comprehensive data ownership.
- Private ownership of data limits usefulness.
- Public availability of data fosters competition and innovation (Rodwin, 2009).

Health Research tests the public attitudes toward the organization conducting it and the public trust it holds. The ethical obligation of any health research is to allow for public engagements that will take research towards socially valuable goals and reach their inherent value. Data, as a public good, enables access and benefit-sharing, whereas data, as a commercial gain, makes it opaque and increases its stature.

EXPOSITIONS AND CLARIFICATIONS OF DATA ETHICS

The field of data ethics is relatively new but, like individual privacy and autonomy, is historically rooted in established philosophical theories. These theories provide frameworks for understanding and addressing the ethical implications of data collection, storage, analysis, and use.

Utilitarianism
Core principle: The greatest good for the greatest number.
Application to data ethics: This theory suggests that data should be used in a way that maximizes overall benefit to society, even if it means sacrificing the interests of a few. For instance, using data to develop life-saving medical treatments could justify some privacy infringements.
Deontology

Core principle: Adherence to moral rules and duties, regardless of the consequences.
Application to data ethics: Deontological ethics emphasizes respecting individual rights and autonomy. This means that data should be collected and used only with explicit consent and that privacy should be protected as a fundamental right.
Virtue Ethics
Core principle: Developing good character traits and habits.
Application to data ethics: Virtue ethics focuses on the moral character of individuals involved in data handling. It encourages organizations and individuals to cultivate virtues like honesty, integrity, and responsibility in their data practices.
Social Contract Theory
Core principle: Individuals agree to give up some freedom in exchange for the benefits of living in a society governed by rules.
Application to data ethics: This theory suggests that data subjects have a right to expect that their data will be used in ways that are consistent with the social contract. For example, organizations may be expected to protect data privacy in exchange for the ability to collect and use data for legitimate purposes.
Feminist Ethics
Core principle: Challenges traditional power structures and emphasizes the importance of caring, empathy, and relationships.
Application to data ethics: Feminist ethics draws attention to the ways in which data practices can reinforce gender biases and inequalities. It encourages organizations to consider the impact of their data practices on marginalized groups.

These theories provide a foundation for ethical decision-making in the context of data. While they may sometimes conflict, they can be used to identify and address the ethical challenges posed by data collection, storage, analysis, and use.

Objections and Responses

Minorities, disadvantaged groups, and Indigenous people are the targets of many research projects, including genetic studies. These communities' lack of participation—which includes improper consent, lack of consultation, and improper interpretation or misuse of their data—is too frequently the cause of this.

However, breakthroughs in genetic analysis and less expensive genome sequencing have led to a surge in projects for studying people, plants, animals, and environments. Powerful data analysis tools also help researchers gain more insight into these projects. Combining datasets and making them reusable appears to be a solution to obtain the most results from this research.

The open science movement promotes data sharing because it feels that doing so will increase research and benefit society. This viewpoint centers on the data on individuals and their environments that the scientific community is interested in. But it's becoming more and more obvious that not everyone has benefited equally from these goals (Claw et al., 2018). Yet, aggregating datasets to facilitate their reuse is a logical path toward deriving more value from the initial investment (Johnson et al., 2020). This section presents a variety of possible objections to these cases.

There are 'Personal Data' or There can be 'Commodification' of These Data

The Cambridge Analytica data breach (2018) has highlighted the commercial and politicized misuse of data that further instigates considerations such as special protection for health data and the harms of its misuse. There's agreement that health data deserves special protection due to the potential for significant harm if misused. Regulations like the General Data Protection Regulation (GDPR, 2016) in the European Union classify health data as a "special category" for this reason. Misuse of data can violate basic rights such as privacy and result in discrimination, which affects both people and entire communities. Even non-sensitive data has the potential to worsen already-existing disparities and cause harm through stigmatization.

While individual control over data is important, there are arguments for ethical obligations to share data for positive outcomes in healthcare. Policy initiatives like "Data Saves Lives"[2] highlight the potential benefits of data sharing. Scholars such as Prainsack et al. (2022) argued that focusing solely on individual control might overshadow bigger questions. We need to consider whether data collection is always justified and who benefits from its use.

These pose a key question: Can prioritizing other principles (like public health benefits) sometimes outweigh individual control over health data? It leaves the answer open for further discussion, highlighting the need to weigh these competing interests carefully. Allowing patients or healthcare providers to own their medical data sounds good at first, but this might diminish overall health research. Putting the data under private ownership may make it difficult and expensive to gather information about the entire population, which is crucial for public health studies, especially when hospitals or insurers hold the data. Furthermore, firms can sell anonymized data with little restrictions due to existing legal frameworks, which raises privacy concerns irrespective of ownership.

For clinical data, data ownership refers to a more complex idea than just who owns the data. Although some scholars contend that because a patient's property belongs to them, they should have exclusive rights over it; this argument falls short of explaining the reality. Clinical data is frequently co-constructed, incorporating contributions from multiple sources. Acknowledging the patient's stake in their data while refraining from asserting full ownership is a better strategy. Rather, the objective ought to be to reunite patients with the research process that utilizes that information.

Genetic Data is Unique to One Person

Globally, there are about 370 million Indigenous people. Genetic research, however, frequently disregards their rights and interests. These communities seldom benefit from such research, even though they may be exposed to risks. This is particularly true now that genetic data is publicly accessible thanks to the growth of open science. Native American communities are cautious about disclosing genetic information due to bad experiences with previous research projects. They need to be more involved in genetic research in order to guarantee justice. When sharing genetic data, Indigenous people have three main concerns: trust, responsibility, and a fair approach (Hudson et al., 2020).

According to Andrews (2002), Genetic sequence patents are problematic because they grant exclusive control over the fundamental building blocks required for scientific inquiry. This may result in circumstances that impede significant discoveries. This is demonstrated by the example of Athena Neuroscience patenting a gene associated with Alzheimer's disease. In a similar vein, free research on a gene sequence connected to breast cancer is prohibited by Myriad Genetics' patent. The fact that these companies must grant researchers permission to use the particular gene sequences they have patented ultimately hinders the search for cures. Ballantyne and Schaefer (2018) suggested it is like *a joint account model, where the individual genetic data is of significant interest to* the family members in case of genetic disorders. This is when the unique genetic data must be shared.

The Component of 'Identifiable Personal Information' is at Risk

Scientific Data issued the "FAIR Guiding Principles for Scientific Data Management and Stewardship"(Wilkinson et al., 2016). The goal of the writers' guidelines was to increase the findability, accessibility, interoperability, and reuse of digital assets. Following these guidelines, we can derive at a method to protect sensitive data from misuse.

Data Loss can be Minimized/Eliminated by Legal Protection, but how can we Erode the Stigma Created by it?

Insufficient or poor-quality data with false conclusions can lead to the stigmatization of certain characteristics of particular communities. This may lead to discriminatory treatment, inappropriate targeting, or even a reduction in the allocation of resources to that region. This must be prevented by following ethical research methods in an efficient manner (Bharadwaj et al., 2017).

What Counts as a Harm?

A liberal perspective offers the most plausible answer. Liberals subscribe to John Stuart Mill's theories, which hold that people should be allowed to live their lives as they see fit, with the only restriction being that their decisions and behaviors should not injure other people. This sums up the idea that while we should respect each person's rights, we should also set clear boundaries for those rights to avoid doing harm (Ashcroft, 2006).

Overcoming Objections and Challenges

Creating an Environment of Accountability for All Stakeholders of the Data utilized/concerned

Principle: Power balance

Health data cooperatives (Hafen et al., 2014) function as a health data repository designed to provide access to health data for the benefit of society. In this cooperative structure, members have the ability to contribute, store, and oversee their health information, asserting their ownership and control over data-sharing decisions, such as with research groups, businesses, and government entities. This arrangement allows individuals and communities to potentially gain financial benefits or other advantages by granting access to their data to health-related institutions, including research bodies and government services. Whether in the Personal or public interest, how benefit-sharing is made successful among socio-cultural and ethnographic groups when the institution sharing the data is trustworthy, transparent, and accountable. Otherwise, opaque commercial data power may be set as a rule.

Figure 1. Participatory data stewardship model

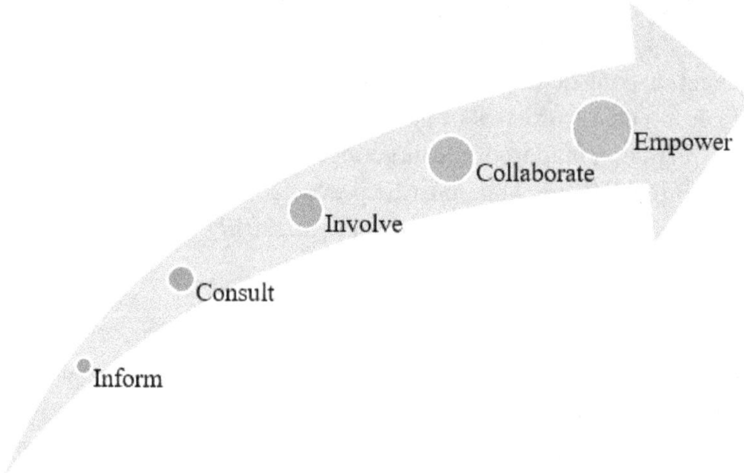

Figure 1 illustrates that a participatory Data Stewardship model of the Lovelace Institute (2021), which must form the foundation for safe, transparent data collection that will enable data sharing to the public benefit and be a solution to attenuated public trust. Further, Carter et al. (2015) offer a solution via an informal consultation: "Social License, which ought to be satisfied for balancing power struggles in the data ecosystem." Carter and colleagues described this social license as: 'the expectations of society regarding the conduct and activities of corporations that go beyond the requirements of formal regulation."

The concept of social license in health research data has been explored in various contexts. Burt (2024) and Stephenson et al. (2022) both emphasize the need for public support and consensus in the use of health data, with Burt's study identifying specific uses within social license. Somerville (2019) and Cullen (2017) extended this concept to other health-related areas, such as alcohol licensing and driver licensing, highlighting the importance of community involvement and the impact of these decisions on social and cultural factors. These studies collectively underscore the significance of public support and community engagement in health data use and related decision-making. Finally, inclusive and participatory data practices like mini-publics and focus groups for controversy-free data engagement must be employed.

Raising Awareness and Teaching the Public the Value of "Sharing is (in Fact) Caring"

The concept of a "data commons" refers to a shared digital space where data are openly available and accessible to a community for various purposes, such as research, innovation, and social good. This typically involves pooling diverse datasets, often from different sources or contributors, under open-access principles. This shared resource allows for collaboration, analysis, and development of insights or solutions that might not be possible with fragmented or siloed data. The data commons model promotes the transparency, collaboration, and democratization of data, enabling broader participation and potentially fostering innovation across various domains (Grossman, 2023).

Data Governance Strategies Based on Solidarity can Aid in the Responsible Use of Big Data

Prainsack and Buyx (2012) defined solidarity as the enacted pledges to accept costs to aid others with whom a person sees relevant similarities. Likewise, solidarity drives individuals towards collective responsibility, which, in turn, overcomes notions of personalization or commodification. In contrast, justice requires us to make the claims of those left out, excluded, or disadvantaged in and around the use of data visible. Interplay of data solidarity with data justice becomes imperative as datafication shapes and gradually transform societies. Issues of justice around data-driven practices have received more and more attention in recent years, like WHO's (2022) guidelines on artificial intelligence and data ethics. Justice tends to center primarily on conceptions of fairness and sometimes beyond. So far, data solidarity is largely overlooked, although it is often indispensable (Braun & Hummel, 2022).

The concept of solidarity is key to including those left out by systemic injustices in discussions on fairness. Solidarity refers to the shared practices between individuals or groups; they feel connected because of similar views and goals. These goals can range from sharing risks and benefits to forming social bonds or recognizing claims that have been previously excluded. Essentially, solidarity strengthens the social fabric through shared experiences. For example, during a pandemic, governments have recommended that people use contact tracing apps out of solidarity with their community. Solidarity, such as using a contact tracing app, is not a guaranteed path to data justice. As in the case of the Black Lives Matter protests following George Floyd's murder in the USA, there were similar concerns that tracing data could be used by the state or the police to target protestors. Despite these potential downsides, solidarity remains a crucial element in achieving justice. It provides an initial spark that can bring people together and pave the way for fairer outcomes.

Even when marginalized groups become visible and their concerns are reflected in the data, injustice can creep in later stages. These groups might remain underrepresented or inadequately considered in decision-making processes and outcomes. This can occur both overtly and through underlying structures. A real-world example is the disproportionate impact of COVID-19 on Black and Indigenous communities in North America. Data clearly show this disparity, yet these communities often face further marginalization when allocating scarce resources, such as vaccines.

Commentators have pointed out that vaccine allocation plans often overlook socioeconomic factors, which are crucial determinants of health outcomes. Addressing these health inequities should be factored into vaccine rollout strategies. This case underscores how even data-supported evidence of injustice might not be sufficient to create change unless it is coupled with action based on solidarity. The examples illustrate how shared practices, fueled by a sense of solidarity, can have a catalytic and corrective effect. By actively listening to and addressing the concerns of marginalized groups, we can work towards more equitable data practices. Solidarity can play a crucial role in identifying injuries. Often, individuals do not realize that they have been wronged until they compare their experiences with those of others. However, determining whether injustice has occurred, whether personally or witnessed, can be a complex task. Although it does not define definitive criteria for assessing injustice, it suggests that personal experiences are valuable because they bring potential injustices to light. Sharing these stories can help identify those who need solidarity and why, which is essential for data justice efforts.

This challenge is even greater in a "black box society" in which data-driven and automated injustices occur behind the scenes, impacting both those who create and those affected by injustice. This emphasizes the importance of data justice practices that adhere to established standards of fairness. Ideally, these practices should be translated into formal rules, regulations, and structures that guide and influence real-world actions. While feelings of solidarity may spark the formalization process, it is crucial to move beyond emotions and convert them into consistent actions. In essence, this means taking concrete steps based on solidarity to achieve lasting change. In addition, data justice is intertwined with solidarity in the sense of making the concerns and interests of others one's own and committing to including them in social endeavors mediated by data. For data governance, which invokes justice as one of its central principles, the reflections on data solidarity proposed a call for concrete experiences of marginalization and experiences of injustice in data-intensive contexts as a conceptual place. Data solidarity requires us to continually check the procedures and arrangements of justice to ensure that all claims for recognition and respect have been heard and considered.

Further Ethical Discussions over Health Data

Health research has presented a complex ethical landscape. Researchers must grapple with potential conflicts between individual participant autonomy and the perceived public good of research. Biases can creep based on factors such as gender or participant capacity, and the trustworthiness of institutions conducting research, especially those involving decentralized data collection, is crucial. Additionally, ethical debates surround how long health data should be stored and who ultimately owns it.

Autonomy of Participants

Autonomy means the ability to form independent beliefs and desires, is crucial for individuals to control how their data are used (Laurie et al., 2021). Informed consent is key here as it requires balancing individual interests with the broader public good. However, ensuring that people have the capacity to truly give consent, especially in healthcare research, is essential and therefore raises the question: how can we best guarantee meaningful autonomy in data usage?

Informed Consent and Autonomy

Principle: Respect for Persons[3]

Informed consent was not required. Relational autonomy, which considers the real-world context of a patient's choices, raises a bar for ethical research. Ongoing consent is crucial because of changing circumstances and a patient's capacity to understand the research. However, several challenges remain. Open-ended research might require broader or deferred consent, which could limit patient autonomy. In addition, factors such as power imbalances between researchers and patients can make it difficult to ensure that decisions are voluntary. The Nuremberg Code emphasizes this, highlighting the need to avoid coercion and carefully examine the context in which consent is given.

De-Identification in Big Data Doesn't Equate to Complete Anonymization

Principle: Group Privacy

The growth of anonymized datasets for health research has presented a double-edged sword. While anonymization protects individual privacy, it also raises concerns about re-identification and potential misuse of data, as seen in the Cambridge Analytica scandal (Laurie et al., 2021). This is further complicated by the concept of group privacy, which Nissenbaum's theory suggests can be compromised when data are transferred between contexts, such as from health research to political campaigns. These issues are particularly relevant in projects funded by both public and private entities, where volunteer group privacy and preferences may be exposed in unintended ways.

Ethical Dilemmas Between Autonomy and Beneficence

Principle: Beneficence[4]

The ethical principles of research include beneficence (benefits outweigh risks), non-maleficence (minimizing harm), and respect for autonomy (participant choice). A key debate centers on individual versus collective interests. While individual rights are important, there can be situations where limitations may be justified, such as when research benefits others, avoids harm to others, or promotes justice. The question then becomes: When is it ethically acceptable to prioritize these other principles over individual autonomy? Is there a universal rule, or does it depend on a specific context? This leaves this question unanswered, highlighting the complexity of navigating these ethical principles in research.

Autonomy Is Impossible Without Transparency

Principle: Proportionality (Laurie et al., 2021)

Proportionality in health research regulation can, at its broadest level, be understood as an attempt to balance two considerations that sometimes compete: the protection of individuals affected by research – especially, but not limited to, human subjects – and the promotion of socially valuable research. Informed consent, a cornerstone of ethical research, hinges on transparency, autonomy, and voluntariness being interconnected. The ICH guidelines[5] emphasize this by requiring researchers to disclose details, such as potential individual benefits, compensation for harm, participant identity disclosure practices, study duration, and expected level of involvement. Without clear and complete information, volunteers would not be able to make autonomous choices about participating in the research, jeopardizing the foundation of voluntariness.

Bias in Data Collection and Analysis

The dangers of bias in health and social care algorithms are caused by the under-representation of different ethnicities throughout the data collection process. This bias could have negative consequences. Several questions arise regarding how to address this issue. How can we collect data that minimizes ethnic and gender gaps, thereby reducing bias in the final algorithms? How can we ensure that datasets include different languages to avoid discrimination against those with language barriers?

Datafication Often Leads to Data Injustice

Principle: Specially considered protection

Datafication in Health Research often leads to data injustice. Informed consent is especially important for vulnerable groups that may not be able to provide it (Mayer-Schönberger & Cukier, 2013). Data and algorithms can involve bias, albeit unintentionally, at every stage in their creation and use. This can have harmful consequences. For example, science can perpetuate racist ideas even though race is a social construct. Some authors have argued that racial demographics in healthcare create problems as well and that we should use other methods to understand health differences between groups. Finally, innovative ubiquitous technologies can replicate existing inequalities even though they may seem objective or progressive. Instead of looking for one-size-fits-all solutions, the best approach may be to design technology with a focus on benefiting the overall society.

Disaggregated Data Collection Agitates Inequity and Further Bias

Principle: Insist on Justice[6]

While data collection centers offer possibilities for better decision-making, there are concerns about access and bias. Vulnerable populations might have been excluded, similar to the limitations of Crawford's 2013 experiment. This lack of inclusion can lead to biased data, as demonstrated by Obermeyer et al. (2019). Their findings suggest that biased data can worsen existing disparities in healthcare by diverting resources away from marginalized groups who may be systematically excluded from data collection processes. Likewise, Taylor's (2017) capabilities approach emphasizes the importance of ensuring that everyone has the necessary skills and resources to participate meaningfully in these processes (see Figure 2). This aligns with Grisso and Applebaum's definition of capacity as possessing the abilities required to make informed decisions (Palmer & Harmell, 2016).

Figure 2. Capabilities approach of Talyor (2017)

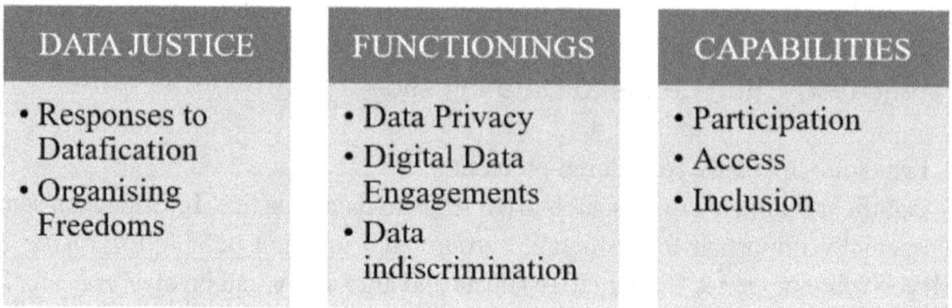

Data Neutrality

Principle: Gender Parity[7]

Bias present in data collection can be drawn from Caroline Criado Perez's book "Invisible Women." Data often reflects the unequal realities of society. The book argues that historically, women have been excluded from data collection, leading to a significant gender data gap. This gap creates a bias towards male experiences in many areas. It criticizes the "male-unless-otherwise-indicated approach", which treats male experiences as the norm. This approach ignores the unique perspectives and needs of women, especially those of color, disabilities, and working-class backgrounds. This also highlights a crucial issue in data collection: women's invisibility. This is due to two factors: 1) a lack of data collection on women altogether and 2) existing data often not separated or parsed by gender (sex). Several solutions have been suggested to address this gender data gap. These include risk assessments by proactively identifying potential biases in data collection and analysis through risk assessments. "Red Teaming" for data bias is by simulating adversarial scenarios to expose potential biases in data that might be otherwise overlooked or disregarded. Data audits and algorithmic fairness checks must be done regularly to evaluate data collection practices and algorithms to identify and rectify any gender biases that may be present.

GLOBAL HEALTH DATA REGULATIONS

The rapid advancements in technology and the increasing reliance on digital health solutions have necessitated the development of robust regulations to protect individuals' health data privacy and security. Different authorities have adopted varying approaches to regulating health data, reflecting their unique cultural, legal, and political contexts. There are also some key considerations when comparing regulations across jurisdictions. Some regulations may apply to all personal data, while others may focus specifically on health data. The rights granted to individuals may vary, including the right to access, rectify, erase, and object to the processing of their data. Data controller and processor obligations impose some requirements on organizations that collect, process, and store health data. The penalties for non-compliance with regulations may vary across jurisdictions. It is important to note that the regulatory landscape for health data is constantly evolving, and organizations must stay informed about the latest developments in their operating jurisdictions.

European Union (EU)

- *General Data Protection Regulation (GDPR):* The GDPR is a comprehensive data protection law that applies to all personal data processed within the EU, including health data. It imposes strict requirements on data controllers and processors, such as obtaining explicit consent, ensuring data security, and providing individuals with specific rights regarding their data. Being the legal framework for personal data protection in the EU, it governs data protection, acccss, and (re)use, which are fundamental for Al development and use. GDPR violations can lead to significant fines and reputational damage.
- *eHealth Directive:* This directive sets out common principles for the interoperability of electronic health records across EU member states. It aims to facilitate the sharing of health data between healthcare providers while ensuring patient privacy and security.
- *EU AI Act:* The overarching regulatory framework for Al systems in the EU. It aims to ensure ethical Al development while protecting health, safety, and fundamental rights. Non-compliance could result in legal issues and market access restrictions.
- *MDR*: Medical Device Regulation ensures medical devices perform consistently and comply with safety and performance requirements. Non-compliance can prevent market access and pose risks to patient safety.
- *Data Governance Act*: Increases data availability and facilitates data sharing. It specifies conditions for re-use of certain protected data held by public sector bodies, including data protected by intellectual property rights. It also provides processes and structures to facilitate voluntary data sharing.
- *Interoperable Europe Act:* Specifies the framework for cross-border interoperability of public services, including health systems. Ensures interoperability standards for public health systems across EU member states. Crucial for seamless data exchange and integration of Al tools across different healthcare systems.

- *NIS2 Directive*: Aimed at enhancing cybersecurity across critical sectors, including healthcare, by imposing stricter security requirements and incident reporting obligations on organizations. Its importance for Al health medical device tools lies in its focus on protecting sensitive health data and ensuring the resilience of healthcare systems against cyber threats.
- *IVDR:* The In Vitro Diagnostic Regulation establishes specific requirements for Al-based diagnostic tools. It defines risk classes and conformity assessment procedures. Non- compliance can prevent market access and pose risks to patient safety.
- *Data Act*: Fosters fair access to and use of data generated by products or services. It specifies requirements for accessibility and transfer of generated data, which is crucial for Al development and use in healthcare.
- *Open Data Directive*: Promotes the use of open data and stimulates innovation in products and services. Sets rules for re-use of public sector information, which can be valuable for Al development in healthcare. Provides access to potentially valuable public sector data for Al development and training.
- *GPSR:* General Product Safety Regulation is a broad framework for the safety of products placed on the EU market. It covers Al technology embedded in other products not covered by sector-specific regulations.

United States of America

- *Health Insurance Portability and Accountability Act (HIPAA)*[8]: While not specifically focused on data privacy, HIPAA establishes national standards for the protection of certain personally identifiable health information (PHI). It requires covered entities, such as healthcare providers and health insurers, to implement safeguards to protect PHI from unauthorized access, use, or disclosure.
- *State-Level Regulations*: Several states have enacted their own data privacy laws, which may provide additional protections for health data. For example, California's Consumer Privacy Act (CCPA) and Nevada's Revised Statutes Chapter 603A both apply to health data.

China

- *Cybersecurity Law*: This law imposes obligations on organizations that process personal data, including health data. It requires entities to obtain consent, implement security measures, and notify authorities of data breaches.
- *Personal Information Protection Law*: This law, which came into effect in 2021, provides more specific protections for personal data, including health

data. It strengthens individuals' rights and imposes stricter requirements on data controllers and processors.

Canada

The Personal Information Protection and Electronic Documents Act, 2000 (PIPEDA) provides a framework for the protection of personal information, including health data.

Australia

The Privacy Act 1988 sets out principles for the handling of personal information, including health data.

India

The Information Technology Act 2000, along with various guidelines and regulations, provides a legal framework for data protection, including health data.

Digital Personal Data Protection Act (DPDP Act) is a law that regulates the processing of digital personal data in India. Parliament passed it in August 2023 and came into effect on September 1, 2023. The DPDP Act aims to balance the rights of individuals to protect their personal data with the need to process that data for lawful purposes. It establishes data subject rights, i.e., the rights and duties of individuals whose data is processed, known as "data principals," and data fiduciary obligations or the obligations of entities that process data, known as "data fiduciaries." A maximum penalty of INR 250 crore (approx. $31 million) is imposed for breaches according to the Act. There is also the appointment of a consent manager. Parental consent is required to process the data of minors under 18 years of age. The DPDP Act also includes exemptions and exceptions, such as, processing for legal rights and claims, processing by courts or tribunals, processing for the prevention, detection, investigation, and prosecution of offenses, processing for research, archiving, and statistical purposes and processing for debt recovery purposes.

Data Ownership Across Jurisdictions

Ownership of individual-level health data, data sharing, and data governance by Mittelstadt et al. discuss the limitations of both private and public ownership models for health data (Piasecki & Cheah, 2022). IMS Health, the largest medical information organization, operated in more than 100 countries and earned more than $2 billion in selling medical data in 2006. Emanuel et al. explored the ethical challenges related to data access, sharing, and use (as cited in Open Data Institute, 2022).

Academic Health Research and Open Access

The scientific community is rapidly embracing open access, in which research data are widely shared. This offers exciting possibilities: faster scientific progress, getting more out of data, and ensuring that the research is trustworthy. However, this also creates new challenges, both practical and ethical, especially in research involving human participants.

Open access to research data has a significant impact on the core principles of research ethics: respecting participants, doing well (beneficence), and ensuring fairness (justice). This suggests that these principles need to be reinterpreted in light of open access to ensure that the rights and well-being of the research participants are still protected (Ross et al., 2018).

Although open access promotes the sharing of genomic data and related information, it clashes with the principles of indigenous data sovereignty. It emphasizes the control of Indigenous communities over their own data. This movement focused on protecting the collective interests of Indigenous peoples and ensuring that they benefited from their data.

Their vision is “data for governance” and “governance of data,” meaning that indigenous communities should have the authority to decide how their data are used and governed. This is in contrast to completely open access, which might not consider the specific needs and concerns of Indigenous populations (Johnson et al., 2020).

In addition, a faculty member or institutional researcher does not have any legal right to review the data developed by a colleague. The concept of work-for-hire may apply, whereby employers own the rights to employee creations, including research data, which may be legally clear but not necessarily ethically clear. Thus, challenges lie in ethics, not in law. There is no agreed-upon standard for ethical behavior in research, making it difficult to determine the penalties for misconduct. Accessing the data is another sticky point. Legal cases highlight the complexities of who can access research information, and for what reasons. Ethical considerations and institutional procedures play a crucial role, but currently, researchers do not have the legal right to see colleagues' data. The author suggests that institutions should develop clear data ownership policies and provide guidance for doing so (Fishbein, 1991).

Clinical Trials

Clinical trials offer complex examples of access to data. Drug companies have conducted these trials to prove that their medications are safe and effective, gaining approval from the FDA. However, they often oppose competitors to use their data to develop generic versions of the same drugs. The 1994 international treaty limited generic manufacturers' access to this data, potentially delaying the availability of affordable generic drugs. While rewarding companies for research investment is understandable, keeping data secret can hide potential health risks. Some pharmaceutical firms have manipulated data by publishing only partial results and skewing the overall picture. To address this, medical journal editors agreed in 2005 to publish only studies based on registered trials. Later, in 2007, the US mandated clinical trial registration. Despite these advancements, some argue for full public access to all the research data. This ongoing tension between commercial interests and public health raises the question of whether all research data, not only clinical trial data, should be made completely open (Commissioner, 2024).

Digital Health Research

The integration of AI into health care has introduced further complexities. Algorithmic bias can perpetuate existing inequalities in healthcare if the training data reflect historical bias. Ensuring the availability of diverse datasets for AI training and implementing ethical frameworks to monitor algorithmic bias are crucial considerations (Nebeker et al., 2019). Data transparency about how data is being used and by whom is essential. Privacy and Security concerns exist as Digital health technologies collect vast amounts of personal health data. Robust security measures are essential to prevent data breaches, and clear communication with patients about data collection and usage is crucial. This is a developing field, and new regulations and best practices are emerging. It is important to stay informed about the latest developments to ensure the ethical and responsible use of data in AI and digital health.

Hospital Patient Data and Stigmatization

To achieve how patients are made aware of and have control over how their medical data are used, the proposed policy changes include mandatory reporting where all hospitals (and potentially other medical institutions) are required to report standardized patient data to a government agency or a public authority created specifically for this purpose. These data should include diagnoses, procedures, prescribed medications, billing information, quality metrics, and other relevant details from patient records. Clinicians, such as doctors, should also be mandated to report data on drug prescriptions and dispensing. All entities involved (hospitals, clinics, and doctors) must submit the same standardized data to the government agency. This ensures consistency and allows easier analysis. The aggregated data should be made publicly available in a way that allows researchers, policymakers, and the public to analyze healthcare trends by factors such as hospital, doctor, diagnosis, treatment, and prescribed medications.

The author argues that public ownership of patient data offers several benefits such as privacy protection. Public oversight could potentially provide stronger safeguards for patient privacy than private ownership. Public access to aggregated data can be used for public health research and initiatives, leading to better overall health outcomes. While publicly owned, the data could still be used for private research and development, potentially leading to new treatments and innovations. This suggests that physicians should advocate for these policies, as they align with core medical values.

Health Insurance Claims

The argument against private ownership of patient data for privacy reasons highlights existing examples where public ownership allows for data sharing while protecting privacy. The argument is that private ownership does not necessarily provide better protection for patient privacy. As an example, it mentions the Health Insurance Portability and Accountability Act (HIPAA) amendment of 2003. While HIPAA protects some patient information, it also allows healthcare providers to share anonymized data with related businesses. Additionally, these raise concerns that for-profit companies can sell anonymized data with minimal oversight.

Public ownership does not guarantee complete privacy. However, there are examples in which public ownership allows data sharing for specific purposes while protecting individual identities. Existing laws require doctors to report certain illnesses and health data relevant to public health to government agencies. This finding demonstrates that public data collection can coexist with safeguards for sensitive information. Medicare, a public health insurance program, requires hospitals to report the cost data used to determine reimbursement. These data were then made public. This is an example of public ownership that facilitates transparency, without compromising individual privacy.

Several states in New England allow for the analysis of data from all payers (insurance companies). This suggests that public ownership can enable valuable research while protecting individuals' identities. California mandates hospitals to report patient discharge data, which are then made available for a fee. This exemplifies public ownership, allowing access to anonymized data for research or other purposes. These examples highlight that public ownership, with appropriate regulations, can balance data sharing for public health or research while protecting patient privacy. This approach contrasts with private ownership, which may not offer stronger privacy safeguards.[9]

Fragmented Data in Social Care

The rise of patient-generated health data, from wearables to online platforms, presents a challenge for healthcare data analysis. This data is often scattered and not unified, existing in separate databases ("silos") even within the same healthcare provider. This fragmentation makes it difficult to get a complete picture of a patient's health. The research community often forgets that social care data, which is relevant, is largely unintegrated with the health data in most countries. Out of a few, the UK is moving towards the integration of health and social care data. An integrated approach to health data, where information is unified across various sources, is crucial for understanding how factors like race, gender, and socioeconomic class (intersectionality) impact access to and utilization of healthcare resources. This is because fragmented data can obscure these underlying patterns. As healthcare services continue to evolve, it is vital to consider all types of relevant data while still prioritizing individual patient privacy and needs.

BEST PRACTICES IN HEALTH DATA REGULATION

Organizations should be transparent about the data they collect, how it will be used, and who will have access to it. Informed consent should be obtained from individuals before collecting their data. Only necessary data should be collected, and robust security measures should be implemented to protect data. Clear policies and procedures for data management should be established. Data should not be used in ways that perpetuate bias or discrimination. Organizations should be accountable for the outcomes of data analysis. AI algorithms should be fair, transparent, and unbiased. Data should be accurate, complete, and reliable before being used for analysis. Privacy should be protected when sharing data, especially sensitive and identifiable information. Data anonymization should be considered to remove personally identifiable information and protected health information. Clear agreements should be in place when sharing data with external parties.

Organizational Culture and Training

Organizations should foster a culture of ethical data practices and provide training to employees on data ethics principles. An ethics committee may be established to review and provide guidance on data-related decisions. Organizations should comply with the General Data Protection Regulation (GDPR) and other relevant data privacy laws. Industry-specific standards and guidelines for data ethics should be adhered to.

Data Governance Over Ownership for Equitable Health Data Use

It is essential to strike a balance between individual rights and societal benefits. A potential solution lies in a hybrid approach, granting individuals greater control over their data while ensuring responsible data governance. This could involve an opt-in mechanism where individuals could choose to share their data for specific research purposes or public health initiatives. Strong data governance frameworks like clear guidelines for data collection, storage, access, and usage are crucial. These frameworks should incorporate robust security measures and ensure data anonymization, when necessary. Transparency regarding data use and clear communication with patients regarding data collection practices are essential.

Limitations of Private and Public Ownership

The limitations of both models are in achieving effective, ethical, and equitable data use. There are ethical concerns as much as practical issues. Neither private nor public ownership perfectly addresses core ethical principles, such as justice and autonomy. Additionally, there are challenges in mitigating the risks associated with data use under both ownership models. Both models face technological, economic, and conceptual hurdles. For instance, private ownership might struggle with data fragmentation, whereas public ownership might incur high aggregation costs.

CONCLUSION

A shift in focus from ownership to data governance is elevated. This means prioritizing the institutional structures and procedures that manage data access and use, like establishing Data Access Committees. These committees can function as gatekeepers, balancing the need for data access for research with individual privacy and other ethical considerations. It can review requests, considering potential benefits and risks to patients, communities, researchers, and society overall. This nuanced approach to data governance is more effective than simply assigning ownership and can balance the strengths and weaknesses of private and public ownership frameworks, ultimately leading to the more equitable and ethical use of individual health data (Piasecki & Cheah, 2022)

A sound democracy is necessary for quality public health (Anand et al., 2004, p. 75). Only until the purpose and method of using health and social care data are transparent, can the data be considered for a public good. Ethical research using personal data can only be made possible with sufficient legal support and ethical guidelines from relevant regulatory agencies, as the field of health data ownership is evolving rapidly. Future research could explore this like the development of effective data governance models and examining how data governance frameworks can be tailored to specific contexts. The impact of data ownership models on research and innovation can also assess the potential benefits and drawbacks of different ownership approaches to healthcare advancement. Finally, the role of public education should investigate how to empower individuals to make informed decisions about their personal health data.

REFERENCES

Anand, S., Peter, F., & Sen, A. (Eds.). (2004). *Public health, ethics, and equity*. Oxford University Press. DOI: 10.1093/oso/9780199276363.001.0001

Andrews, L. B. (2002). Genes and patent policy: Rethinking intellectual property rights. *Nature Reviews. Genetics*, 3(10), 803–808. DOI: 10.1038/nrg909 PMID: 12360238

Ashcroft, R. E. (2006). Individual freedom versus collective responsibility: An ethicist's perspective. *Emerging Themes in Epidemiology*, 3(1), 11. DOI: 10.1186/1742-7622-3-11 PMID: 16999864

Ballantyne, A., & Schaefer, G. O. (2018). Consent and the ethical duty to participate in health data research. *Journal of Medical Ethics*, 44(6), 392–396. DOI: 10.1136/medethics-2017-104550 PMID: 29358219

Beauchamp, T. L., & Childress, J. F. (2001). *Principles of Biomedical Ethics*. Oxford University Press.

Bharadwaj, P., Pai, M. M., & Suziedelyte, A. (2017). Mental health stigma. *Economics Letters*, 159, 57–60. DOI: 10.1016/j.econlet.2017.06.028

Braun, M., & Hummel, P. (2022). Data justice and data solidarity. *Patterns*, 3(3), 100427. DOI: 10.1016/j.patter.2021.100427 PMID: 35510188

Claw, K. G., Anderson, M. Z., Begay, R. L., Tsosie, K. S., Fox, K., Garrison, N. A., Bader, A. C., Bardill, J., Bolnick, D. A., Brooks, J., Cordova, A., Malhi, R. S., Nakatsuka, N., Neller, A., Raff, J. A., Singson, J., TallBear, K., Vargas, T., & Yracheta, J. M. (2018). A framework for enhancing ethical genomic research with Indigenous communities. *Nature Communications*, 9(1), 2957. DOI: 10.1038/s41467-018-05188-3 PMID: 30054469

Commissioner, O. of the. (2024, May 29). *Food and Drug Administration Amendments Act (FDAAA) of 2007*. FDA; FDA. https://www.fda.gov/regulatory-information/selected-amendments-fdc-act/food-and-drug-administration-amendments-act-fdaaa-2007

Fishbein, E. A. (1991). Ownership of research data. *Academic Medicine*, 66(3), 129–133. DOI: 10.1097/00001888-199103000-00002 PMID: 1997022

Floridi, L., & Taddeo, M. (2016). What is data ethics? *Philosophical Transactions. Series A, Mathematical, Physical, and Engineering Sciences*, 374(2083), 20160360. DOI: 10.1098/rsta.2016.0360 PMID: 28336805

Grossman, R. L. (2023). Ten lessons for data sharing with a data commons. *Scientific Data*, 10(1), 1. Advance online publication. DOI: 10.1038/s41597-023-02029-x PMID: 36878917

Hafen, E., Kossmann, D., & Brand, A. (2014). Health data cooperatives—Citizen empowerment. *Methods of Information in Medicine*, 53(2), 82–86. DOI: 10.3414/ME13-02-0051 PMID: 24514946

Hudson, M., Garrison, N. A., Sterling, R., Caron, N. R., Fox, K., Yracheta, J., Anderson, J., Wilcox, P., Arbour, L., Brown, A., Taualii, M., Kukutai, T., Haring, R., Te Aika, B., Baynam, G. S., Dearden, P. K., Chagné, D., Malhi, R. S., Garba, I., & Carroll, S. R. (2020). Rights, interests and expectations: Indigenous perspectives on unrestricted access to genomic data. *Nature Reviews. Genetics*, 21(6), 377–384. DOI: 10.1038/s41576-020-0228-x PMID: 32251390

Humbly, C. (2006). *Data is the New Oil. ANA Senior Marketer's Summit*. Kellogg School.

Johnson, S. B., Slade, I., Giubilini, A., & Graham, M. (2020). Rethinking the ethical principles of genomic medicine services. *European Journal of Human Genetics*, 28(2), 147–154. DOI: 10.1038/s41431-019-0507-1 PMID: 31534213

Kukutai, T., & Taylor, J. (Eds.). (2016). *Indigenous Data Sovereignty: Toward an agenda* (Vol. 38). ANU Press. https://www.jstor.org/stable/j.ctt1q1crgf

Laurie, G., Dove, E., Ganguli-Mitra, A., McMillan, C., Postan, E., Sethi, N., & Sorbie, A. (Eds.). (2021). *The Cambridge Handbook of Health Research Regulation* (1st ed.). Cambridge University Press., DOI: 10.1017/9781108620024

Lee, N. (2014). The Lancet Technology: November, 2014. *Lancet*, 384(9958), 1917. DOI: 10.1016/S0140-6736(14)62267-4 PMID: 25478615

Mayer-Schönberger, V., & Cukier, K. (2013). *Big Data: A revolution that will transform how we live, work, and think*. Houghton Mifflin Harcourt.

Nebeker, C., Torous, J., & Bartlett Ellis, R. J. (2019). Building the case for actionable ethics in digital health research supported by artificial intelligence. *BMC Medicine*, 17(1), 137. DOI: 10.1186/s12916-019-1377-7 PMID: 31311535

Obermeyer, Z., Powers, B., Vogeli, C., & Mullainathan, S. (2019). Dissecting racial bias in an algorithm used to manage the health of populations. *Science*, 366(6464), 447–453. DOI: 10.1126/science.aax2342 PMID: 31649194

Open Data Institute. (2022, March 1). *Health data governance: A playbook for non-technical leaders*. https://open-data-institute.gitbook.io/data-governance-playbook/

Palmer, B. W., & Harmell, A. L. (2016). Assessment of Healthcare Decision-making Capacity. *Archives of Clinical Neuropsychology*, 31(6), 530–540. DOI: 10.1093/arclin/acw051 PMID: 27551024

Piasecki, J., & Cheah, P. Y. (2022). Ownership of individual-level health data, data sharing, and data governance. *BMC Medical Ethics*, 23(1), 104. DOI: 10.1186/s12910-022-00848-y PMID: 36309719

Prainsack, B., & Buyx, A. (2012). Solidarity in contemporary bioethics—Towards a new approach. *Bioethics*, 26(7), 343–350. DOI: 10.1111/j.1467-8519.2012.01987.x PMID: 22827315

Prainsack, B., El-Sayed, S., Forgó, N., Szoszkiewicz, Ł., & Baumer, P. (2022). Data solidarity: A blueprint for governing health futures. *The Lancet Digital Health*, 4(11), e773–e774. DOI: 10.1016/S2589-7500(22)00189-3 PMID: 36307191

Rodwin, M. A. (2009). The Case for Public Ownership of Patient Data. *Journal of the American Medical Association*, 302(1), 86. DOI: 10.1001/jama.2009.965 PMID: 19567445

Ross, M. W., Iguchi, M. Y., & Panicker, S. (2018). Ethical aspects of data sharing and research participant protections. *The American Psychologist*, 73(2), 138–145. DOI: 10.1037/amp0000240 PMID: 29481107

Stephenson, N., Smith, C., & Vajdic, C. M. (2022). Health and public sector data sharing requires social licence negotiations. *Australian and New Zealand Journal of Public Health*, 46(4), 426–428. DOI: 10.1111/1753-6405.13261 PMID: 35679055

Taylor, L. (2017). What is data justice? The case for connecting digital rights and freedoms globally. *Big Data & Society*, 4(2), 2053951717736335. DOI: 10.1177/2053951717736335

Wilkinson, M. D., Dumontier, M., Aalbersberg, I., Appleton, G., Axton, M., Baak, A., Blomberg, N., Boiten, J.-W., da Silva Santos, L. B., Bourne, P. E., Bouwman, J., Brookes, A. J., Clark, T., Crosas, M., Dillo, I., Dumon, O., Edmunds, S., Evelo, C. T., Finkers, R., & Mons, B. (2016). The FAIR Guiding Principles for scientific data management and stewardship. *Scientific Data*, 3(1), 160018. DOI: 10.1038/sdata.2016.18 PMID: 26978244

World Medical Association. (2013). World Medical Association Declaration of Helsinki: Ethical principles for medical research involving human subjects. *Journal of the American Medical Association*, 310(20), 2191–2194. DOI: 10.1001/jama.2013.281053 PMID: 24141714

ADDITIONAL READING

D'Ignazio, C., & Klein, L. (2020). *Data feminism*. MIT Press. DOI: 10.7551/mit-press/11805.001.0001

Laurie, G., Dove, E., Ganguli-Mitra, A., McMillan, C., Postan, E., Sethi, N., & Sorbie, A. (Eds.). (2021). *The Cambridge handbook of health research regulation* (1st ed.). Cambridge University Press. DOI: 10.1017/9781108620024

Sen, A. (2004). *Public health, ethics, and equity*. Oxford University Press.

Singer, P. A., & Viens, A. M. (Eds.). (2015). *The Cambridge textbook of bioethics*. Cambridge University Press.

Steinbock, B. (Ed.). (2009). *The Oxford handbook of bioethics*. Oxford University Press. DOI: 10.1093/oxfordhb/9780199562411.001.0001

World Health Organization. (2021). *WHO: Ethics and governance of artificial intelligence for health*. https://www.who.int/publications/i/item/9789240029200

KEY TERMS AND DEFINITIONS

Bioethics: Concept of socio-legal ethics in biomedicine or biomedical research.

Data Cooperatives: A data governance model which is socially oriented to avert ethical crisis in the management of data.

Data Ethics: Various noble principles that need to be applied in data science while handling data of human participants.

Data Feminism: Expressing feminist thoughts during different applications of data science.

Data Justice: Issues of Justice that evolve and revolve around the use of data, more often during research.

Data Science: An innovative way of scientific discovery, which uses data as the primary element, and analyses it to bring meaningful insights.

Ethics: Affirming cultural legitimacy and what is rightfully right, independent of its legality

Health Data: Data generated from a human (or in some context, animal) participant either, during delivery of healthcare or health research.

ABBREVIATIONS

AI -: Artificial Intelligence
DL –: Deep Learning
FDA: – Food and Drug Administration (USA)
HIPAA: - Health Insurance Portability and Accountability Act
ICH: - International Council for Harmonization of Technical Requirements for Pharmaceuticals for Human Use
ML –: Machine Learning
NHS –: National Health Service, United Kingdom
UK –: United Kingdom
WHO –: World Health Organization

ENDNOTES

1 Carnegie UK Trust. (2018). *Data for Public Benefit Report* (p. 6). https://understandingpatientdata.org.uk/sites/default/files/2018-04/Data%20for%20public%20good_0.pdf

2 NHS Transformation Directorate. *Data saves lives*. https://transform.england.nhs.uk/key tools-and-info/data-saves-lives/

3 US National Commission for the Protection of Human Subjects of Biomedical and Behavioral Research. (1978). The Belmont Report: Ethical principles and guidelines for the protection of human subjects of research (Vol. 1): Department of Health, Education, and Welfare, National Commission for the Protection of Human Subjects of Biomedical and Behavioral Research

4 US National Commission for the Protection of Human Subjects of Biomedical and Behavioral Research. (1978). The Belmont Report: Ethical principles and guidelines for the protection of human subjects of research (Vol. 1): Department of Health, Education, and Welfare, National Commission for the Protection of Human Subjects of Biomedical and Behavioral Research

5 ICH Official website. https://www.ich.org/page/ich-guidelines

6 Beauchamp, T. L., & Childress, J. F. (2001). *Principles of biomedical ethics*. Oxford University Press.

7 D'Ignazio, C. & Klein, L. F. (2020). *Data feminism*. MIT Press.

8 Health Insurance Portability and Accountability Act of 1996. (1996). Code of Federal Regulations (CFR) title 45, part 160.

9 Security and Privacy: Other requirements relating to uses and disclosures of protected health information. 45 CFR § 164.514 (2018). https://www.ecfr.gov/current/title-45/subtitle-A/subchapter-C/part-164

10 Powles, J. (2017, July 5). Why are we giving away our most sensitive health data to Google? *The Guardian*. https://www.theguardian.com/commentisfree/2017/jul/05/sensitive-health-information-deepmind-google

11 Vibrator maker ordered to pay out C$4m for tracking users' sexual activity. (2017). *The Guardian*. https://www.theguardian.com/technology/2017/mar/14/we-vibe-vibrator-tracking-users-sexual-habits

APPENDIX I

Case Studies

DeepMind Issue[10]

The phrase "DeepMind scandal" describes ethical issues or controversies involving DeepMind, a research lab for artificial intelligence (AI) that Google purchased in 2014. Concerns concerning the handling of patient data and data sharing for research purposes without appropriate consent were brought up in relation to DeepMind's partnership with the UK's National Health Service (NHS) in 2015. This was one noteworthy issue. Concerns regarding data security and privacy also surfaced when it was revealed that DeepMind, as part of its partnership with hospitals, had access to millions of patients' private medical records. These incidents have sparked discussions about the ethical implications of AI in healthcare and the appropriate use of personal data, despite DeepMind's stated commitment to privacy and transparency. It delves deeply into the morality of developing AI with patient data. In order to test an app, Google's DeepMind collaborated with a hospital to gain access to 1.6 million patient records. Important processes like patient consent, public debate, and ethical approval were circumvented in this collaboration. Critics voiced concerns about ownership of NHS (National Health Service, UK) data, patient expectations for data management, and Google's possible commercialization of this data.

Cambridge Analytica Issue

The now-defunct political consulting firm Cambridge Analytica came to light for its involvement in a number of political campaigns, such as the Brexit campaign in the UK and the 2016 US presidential election. It was alleged that Cambridge Analytica improperly collected information without authorization, from millions of Facebook users and used it to target political advertisements. Significant questions concerning data protection, privacy, the impact of technology on democratic processes, and data ownership and misuse have been brought up by this controversy.

We-Vibe Issue[11]

Canadian vibrator manufacturer We-Vibe surreptitiously gathered user information about vibration settings and usage durations and connected it to email addresses. Without the users' knowledge or consent, this data was used for product development. This begs the question of how businesses manage private customer information that is gathered from individual devices.

Samaritans Radar Issue

It examines the difficulties in applying algorithms to mental health assistance. A UK mental health charity called The Samaritans released an app(Lee, 2014) that looked for indications of distress on Twitter. However, user worries about privacy and feeling constantly watched led to the app's rapid removal. This incident emphasizes how important it is to carefully assess the ethical implications of using personal data, particularly when dealing with delicate subjects like mental health.

Organizations Fostering Data Ethics

With the growing importance of data ethics, several organizations in the United Kingdom are tackling these challenges:

The Open Data Institute helps in ethical data management of organizations using practical tools like the "Data Ethics Canvas".

The Alan Turing Institute focuses on real-world applications of data science and AI, developing data ethics frameworks around data-driven technologies. It is also the national institute of Data Science and Artificial Intelligence of the United Kingdom.

The Ada Lovelace Institute, an independent research body, investigates how data and AI can be used to maximize social well-being.

The Nuffield Council on Bioethics is a leading bioethics institution, with dedicated programs on biological/health data and AI in healthcare and research in the UK.

Compilation of References

Aboelenien, A., & Naaman, K. (2023). Ethical considerations in marketing: Academics, practitioners, and other marketplace actors. In *Research Handbook on Accounting and Ethics* (pp. 328–342). Edward Elgar Publishing. DOI: 10.4337/9781800881020.00033

Adams, D. P., & Miles, T. P. (2013). The application of Belmont Report principles to policy development. *Journal of Gerontological Nursing*, 39(12), 16–21. DOI: 10.3928/00989134-20131028-07 PMID: 24219075

Adashi, E. Y., Walters, L. B., & Menikoff, J. A. (2018). The Belmont Report at 40: Reckoning with time. *American Journal of Public Health*, 108(10), 1345–1348. DOI: 10.2105/AJPH.2018.304580 PMID: 30138058

Ademola, O. E. (2020). An ethical approach to understanding cyber security. *Cyber Security Practitioner's Guide*, 475-502.

Aderibigbe, N. A. (2021). Synopsis on cyberethics behaviour: A literature review. *Inkanyiso*, 13(2), 273–290. DOI: 10.4102/ink.v13i2.8

Ahmed, S. K. (2024). The pillars of trustworthiness in qualitative research. *Journal of Medicine, Surgery, and Public Health*, 2, 100051. Advance online publication. DOI: 10.1016/j.glmedi.2024.100051

Ainembabazi, P., Castelnuovo, B., Okoboi, S., Arinaitwe, W., Sebuyira, L., Parkes-Ratanshi, R., & Byakika-Kibwika, P. (2023). Building Capacity for Research Ethics Committees in Uganda. *Journal of Empirical Research on Human Research Ethics*, 18(3), 91–98. DOI: 10.1177/15562646231177410 PMID: 37272030

Ajwang, S. O., & Ikoha, A. P. (2024). Publish or perish in the era of artificial intelligence: Which way for the Kenyan research community? *Library Hi Tech News*. Advance online publication. DOI: 10.1108/LHTN-04-2024-0065

Alder, S. (2022). What is considered PHI under HIPAA? *HIPAA Journal*. https://www.hipaajournal.com/considered-phi-hipaa

Allahrakha, N. (2023). Balancing cyber-security and privacy: Legal and ethical considerations in the digital age. *Legal Issues in the Digital Age*, 4(2), 78–121. DOI: 10.17323/10.17323/2713-2749.2023.2.78.121

Amaral da Luz, K. C., & Musial, N. T. K. (2023). Can organizational culture encourage corporate social responsibility and improve environmental performance in [B] companies? *Journal of Management and Sustainability*, 13(1), 215. DOI: 10.5539/jms.v13n1p215

Anand, S., Peter, F., & Sen, A. (Eds.). (2004). *Public health, ethics, and equity*. Oxford University Press. DOI: 10.1093/oso/9780199276363.001.0001

Anderson, E. E., Newman, S. B., & Matthews, K. A. (2017). Improving informed consent: Stakeholder views. *AJOB Empirical Bioethics*, 8(3), 178–188. DOI: 10.1080/23294515.2017.1362488 PMID: 28949896

Anderson, L. C. (2007). Institutional and IACUC responsibilities for animal care and use education and training programs. *ILAR Journal*, 48(2), 90–95. DOI: 10.1093/ilar.48.2.90 PMID: 17420530

Anderson, R. E. (2018). ACM Code of Ethics and Professional Conduct. *Communications of the ACM*, 35(5), 94–99. DOI: 10.1145/129875.129885

Andrews, J., Zhao, D., Thong, W., Modas, A., Papakyriakopoulos, O., & Xiang, A. (2024). Ethical considerations for responsible data curation. *InProceedings of the 37th International Conference on Neural Information Processing Systems, 36*, 55320-55360.

Andrews, L. B. (2002). Genes and patent policy: Rethinking intellectual property rights. *Nature Reviews. Genetics*, 3(10), 803–808. DOI: 10.1038/nrg909 PMID: 12360238

Anom, B. Y. (2020). Ethics of Big Data and artificial intelligence in medicine. *Ethics, Medicine, and Public Health*, 15, 2352–5525. DOI: 10.1016/j.jemep.2020.100568

Aoki, J., Iwahashi, K., Ishigooka, J., Fukamauchi, F., Numajiri, M., Ohtani, N., & Ohta, M. (2012). Evaluation of cerebral activity in the prefrontal cortex in mood [affective] disorders during animal-assisted therapy (AAT) by near-infrared spectroscopy (NIRS): A pilot study. *International Journal of Psychiatry in Clinical Practice*, 16(3), 205–213. DOI: 10.3109/13651501.2011.644565 PMID: 22486555

Ara, A., & Mifa, A. F. (2024). Integrating artificial intelligence and big data in mobile health: A systematic review of innovations and challenges in healthcare systems. *Global Mainstream Journal of Business, Economics, Development & Project Management, 3*(01), 01-16.

Artz, J. M. (2008). Addressing the central problem in cyber ethics through stories. In *Information Security and Ethics: Concepts, Methodologies, Tools, and Applications* (pp. 3824–3828). IGI Global. DOI: 10.4018/978-1-59904-937-3.ch260

Ashcroft, R. E. (2006). Individual freedom versus collective responsibility: An ethicist's perspective. *Emerging Themes in Epidemiology*, 3(1), 11. DOI: 10.1186/1742-7622-3-11 PMID: 16999864

Atabek, O. (2020). Experienced educators' suggestions for solutions to the challenges to technology integration. *Education and Information Technologies*, 25(6), 1–17. DOI: 10.1007/s10639-020-10243-y

Atapour-Abarghouei, A., McGough, A. S., & Wall, D. S. (2020). Resolving the cybersecurity data sharing paradox to scale up cybersecurity via a co-production approach towards data sharing. In *2020 IEEE International Conference on Big Data (Big Data)*, 3867-3876. IEEE. DOI: 10.1109/BigData50022.2020.9378014

Atenas, J., Havemann, L., & Timmermann, C. (2023). Reframing data ethics in research methods education: A pathway to critical data literacy. *International Journal of Educational Technology in Higher Education*, 20(1), 1–27. DOI: 10.1186/s41239-023-00380-y PMID: 36846483

Audi, R. (2022). The marketing of human images is a challenge to ethical leadership. In *Leadership and business ethics* (pp. 297–311). Springer Netherlands. DOI: 10.1007/978-94-024-2111-8_18

AVMA. (2007). Wellness Guidelines for Animals in Animal-Assisted Activity, Animal-Assisted Therapy and Resident Animal Programs. Available online from www.avma.org/issues/policy/animal_assisted_activity.asp

Awad, A., Trenfield, S. J., Pollard, T. D., Ong, J. J., Elbadawi, M., McCoubrey, L. E., Goyanes, A., Gaisford, S., & Basit, A. W. (2021). Connected healthcare: Improving patient care using digital health technologies. *Advanced Drug Delivery Reviews*, 178, 113958. DOI: 10.1016/j.addr.2021.113958 PMID: 34478781

Bailey, M., Dittrich, D., Kenneally, E., & Maughan, D. (2012). The Menlo Report. *IEEE Security and Privacy*, 10(2), 71–75. DOI: 10.1109/MSP.2012.52

Bakare, S. S., Adeniyi, A. O., Akpuokwe, C. U., & Eneh, N. E.Seun Solomon BakareAdekunle Oyeyemi AdeniyiChidiogo Uzoamaka AkpuokweNkechi Emmanuella Eneh. (2024). Data privacy laws and compliance: A comparative review of the EU GDPR and USA regulations. *Computer Science & IT Research Journal*, 5(3), 528–543. DOI: 10.51594/csitrj.v5i3.859

Ballantyne, A., & Schaefer, G. O. (2018). Consent and the ethical duty to participate in health data research. *Journal of Medical Ethics*, 44(6), 392–396. DOI: 10.1136/medethics-2017-104550 PMID: 29358219

Bandarian, R. (2020). Explaining the evolution stages of strategic roles of research and technology management. *International Journal of Research, Innovation and Commercialisation*. .DOI: 10.1504/IJRIC.2020.10031470

Bankert, E. A. (2022). Assessing informed consent communication. In Bankert, E. A., Gordon, B. G., Hurley, E. A., & Shriver, S. P. (Eds.), *Institutional Review Board: Management and function* (3rd ed., pp. 317–322). Jones & Bartlett Learning.

Barnard, D., & Feige, M. M. (2022). Documentation, policies, and procedures. In Bankert, E. A., Gordon, B. G., Hurley, E. A., & Shriver, S. P. (Eds.), *Institutional review board: Management and function* (3rd ed., pp. 45–54). Jones & Bartlett Learning.

Barnes, M., Carrithers, J., & Sugarman, J. (2020). Ethical and practical concerns about IRB restrictions on the use of research data. *Ethics & Human Research*, 42(6), 29–34. DOI: 10.1002/eahr.500072 PMID: 33136330

Barrett, C. (2018). *Everyday ethics for practicing planners*. Routledge. DOI: 10.4324/9781351179461

Barrett, C. (2019). Are the EU GDPR and the California CCPA becoming the de facto global standards for data privacy and protection? *Scitech Lawyer*, 15(3), 24–29.

Barrow, J. M., Brannan, G. D., & Khandhar, P. B. (2017). *Research ethics*. Europe PMC.

Battle, L., & Scheidegger, C. (2020). A Structured Review of Data Management Technology for Interactive Visualization and Analysis. *IEEE Transactions on Visualization and Computer Graphics*, 27(2), 1128–1138. DOI: 10.1109/TVCG.2020.3028891 PMID: 33031039

Bauer, T. N., Truxillo, D. M., Jones, M. P., & Brady, G. (2020). Privacy and cybersecurity challenges, opportunities, and recommendations: Personnel selection in an era of online application systems and big data. *Big Data in Psychological Research*, 393–409. DOI: 10.1037/0000193-018

Bazhin, A., Taverna, M., Lee, S., Mayer, B., & Werner, J. (2021). Development and assessment of a decentralized management system for doctoral student administration. *International Journal of Educational Administration, Management, and Leadership*. .DOI: 10.51629/ijeamal.v2i2.44

Beauchamp, T. L., & Childress, J. F. (2001). *Principles of Biomedical Ethics*. Oxford University Press.

Ben-Ari, A., & Enosh, G. (2013). Power relations and reciprocity: Dialectics of knowledge construction. *Qualitative Health Research, 23*(3), 422–429. DOI: 10.1177/1049732312470030

Berndsen, C. (2021). Teaching data management and literacy to support course-embedded research projects. *The FASEB Journal*, 35(S1). Advance online publication. DOI: 10.1096/fasebj.2021.35.S1.03442

Bernstein, M. S., Levi, M., Magnus, D., Rajala, B., Satz, D., & Waeiss, C. (2021). ESR: Ethics and society review of artificial intelligence research. https://arxiv.org/pdf/2106.11521

Bernstein, M. S., Levi, M., Magnus, D., Rajala, B., Satz, D., & Waeiss, C. (2021). ESR: Ethics and society review of artificial intelligence research. Preprint. https://doi.org//arXiv.2106.11521DOI: 10.48550

Bezuidenhout, L., Quick, R., & Shanahan, H. (2020). "Ethics When You Least Expect It": A Modular Approach to Short Course Data Ethics Instruction. *Science and Engineering Ethics*, 26(4), 2189–2213. Advance online publication. DOI: 10.1007/s11948-020-00197-2 PMID: 32067185

Bhandari, R. S., & Bansal, A. (2018). A comparison of psychological and socio-economic factors affecting individual user's social media behaviour. *FIIB Business Review*, 7(2), 126–134. DOI: 10.1177/2319714518789759

Bharadwaj, P., Pai, M. M., & Suziedelyte, A. (2017). Mental health stigma. *Economics Letters*, 159, 57–60. DOI: 10.1016/j.econlet.2017.06.028

Bhima, B., Zahra, A., & Nurtino, T. (2023). *Enhancing Organizational Efficiency through the Integration of Artificial Intelligence in Management Information Systems. APTISI Transactions on Management*. ATM., DOI: 10.33050/atm.v7i3.2146

Bhupathi, P. A., & Ravi, G. R. (2017). Comprehensive format of informed consent in research and practice: A tool to uphold the ethical and moral standards. *International Journal of Clinical Pediatric Dentistry*, 10(1), 73–81. DOI: 10.5005/jp-journals-10005-1411 PMID: 28377660

Biable, S. E., Garcia, N. M., Midekso, D., & Pombo, N. (2022). Ethical issues in software requirements engineering. *Software*, 1(1), 31–52. DOI: 10.3390/software1010003

Birch, M., & Miller, T. (2002). Encouraging participation: ethics and responsibilities. In *Ethics in Qualitative Research* (pp. 92–106). SAGE. DOI: 10.4135/9781849209090.n5

Blackman, R. (2021). If your company uses AI, it needs an institutional review board. *Harvard Business Review*.https://hbr.org/2021/04/if-your-company-usesai-it-needs-an-institutional-review-board

Blackman, R. (2021). If your company uses AI, it needs an institutional review board. *Harvard Business Review*.https://hbr.org/2021/04/if-your-company-uses-ai-it-needs-an-institutional-review-board

Bloch, C. (2020). Heterogeneous impacts of research grant funding. *Research Evaluation*. Advance online publication. DOI: 10.1093/reseval/rvaa025

Blood, B. (2003). The Belmont Report: Ethical principles and guidelines for the protection of human subjects research. *Bracketing Interview, 58*, 115. *Foundations for Research*, 59, 421.

Bodnar, O., & Horishna, O. (2021). Mechanisms for overcoming resistance to changes in general secondary education institutions. *Educational Horizons*, (2), 180–189. DOI: 10.25128/2415-3605.20.2.24

Bombard, Y., Baker, G. R., Orlando, E., Fancott, C., Bhatia, P., Casalino, S., Onate, K., Denis, J. L., & Pomey, M. P. (2018). Engaging patients to improve quality of care: A systematic review. *Implementation Science : IS*, 13(1), 98. Advance online publication. DOI: 10.1186/s13012-018-0784-z PMID: 30045735

Bommu, R. (2022). Ethical considerations in the development and deployment of AI-powered medical device software: Balancing innovation with patient welfare. *Journal of Innovative Technologies*, 5(1), 1–7.

Borde, P., Arora, R., & Kakoty, S. (2022). Transformational health-care leaders in collaborative entrepreneurial model to achieve UNSDG: a qualitative study. *Leadership in health services*, ahead-of-print ahead-of-print. https://doi.org/.DOI: 10.1108/LHS-03-2022-0032

Bouderhem, R. (2024). Shaping the future of AI in healthcare through ethics and governance. *Humanities & Social Sciences Communications*, 11(1), 1–12. DOI: 10.1057/s41599-024-02894-w

Bourgeois, F. T., & Espinoza, J. C. (2023). Advancing equity in medical device development for children. *JAMA Pediatrics*, 177(6), 561–562. DOI: 10.1001/jamapediatrics.2023.0790 PMID: 37093608

Bracken-Roche, D., Bell, E., Macdonald, M. E., & Racine, E. (2017). The concept of "vulnerability" in research ethics: An in-depth analysis of policies and guidelines. *Health Research Policy and Systems*, 15(1), 8–8. DOI: 10.1186/s12961-016-0164-6 PMID: 28173859

Braun, M., & Hummel, P. (2022). Data justice and data solidarity. *Patterns*, 3(3), 100427. DOI: 10.1016/j.patter.2021.100427 PMID: 35510188

Brelsford, V. L., Dimolareva, M., Gee, N. R., & Meints, K. (2020). Best practice standards in animal-assisted interventions: How the LEAD risk assessment tool can help. *Animals (Basel)*, 10(6), 974. DOI: 10.3390/ani10060974 PMID: 32503309

Brennan, R. (2021). Business-to-business (industrial) marketing ethics. In *The Sage Handbook of marketing ethics*, 238-247. SAGE

Brighouse, H., Ladd, H. F., Loeb, S., & Swift, A. (2018). *Educational goods: Values, evidence, and decision-making*. University of Chicago Press. DOI: 10.7208/chicago/9780226514208.001.0001

Brønstad, A., Newcomer, C. E., Decelle, T., Everitt, J. I., Guillen, J., & Laber, K. (2016). Current concepts of harm–benefit analysis of animal experiments–report from the AALAS–FELASA working group on harm–benefit analysis–part 1. *Laboratory Animals*, 50(1, _suppl), 1–20. DOI: 10.1177/0023677216642398 PMID: 27188275

Brothers, K. B., Rivera, S. M., Cadigan, R. J., Sharp, R. R., & Goldenberg, A. J. (2019). A Belmont reboot: Building a normative foundation for human research in the 21st century. *The Journal of Law, Medicine & Ethics*, 47(1), 165–172. DOI: 10.1177/1073110519840497 PMID: 30994072

Brown, G., & Peterson, R. S. (2022). The Subordinated Board: Facebook. In *Disaster in the Boardroom* (pp. 89-101). Palgrave Macmillan, Cham.

Brown, N. (2023). Research ethics in a changing social sciences landscape. *Research Ethics*, 19(2), 157–165. DOI: 10.1177/17470161221141011

Bryant, A. (2019). Indigeneity. In *The varieties of grounded theory* (pp. 92–103). SAGE. DOI: 10.4135/9781529716542.n6

Buchanan, B. (2020). A national security research agenda for cybersecurity and artificial intelligence. *Center for Security and Emerging Technology Issue Brief, 7.*

Budda, M. L., & Pritt, S. L. (2020). Evaluating IACUCs: Previous research and future directions. *Journal of the American Association for Laboratory Animal Science; JAALAS*, 59(6), 656–664. DOI: 10.30802/AALAS-JAALAS-20-000077 PMID: 32928341

Burgett-Moreno, M., Cartmell-McGlyn, A., & Mongeon, B. (2024, May 21–23). HRPP/IRB staffing: Onboarding, education, and empowerment [Conference presentation]. AAHRPP Annual Conference, San Diego, CA, United States. Burr, J. S., Johnson, A. R., Vasenina, V., Bisping, S., Coleman, R. W., Botkin, J. R., & Dean, J.

Burles, M. C., & Bally, J. M. G. (2018). Ethical, practical, and methodological considerations for unobtrusive qualitative research About personal narratives shared on the internet. *International Journal of Qualitative Methods*, 17(1), 1. Advance online publication. DOI: 10.1177/1609406918788203

Burr, J. S., Johnson, A. R., Vasenina, V., Bisping, S., Coleman, R. W., Botkin, J. R., & Dean, J. M. (2019). Implementing a central IRB model in a multicenter research network. *Ethics & Human Research*, 41(3), 23–28.

Bynum, T. W. (2001). Computer ethics: Its birth and its future. *Ethics and Information Technology*, 3(2), 109–112. DOI: 10.1023/A:1011893925319

Capili, B., & Anastasi, J. K. (2024). Ethical research and the institutional review board: An introduction. *The American Journal of Nursing*, 124(3), 50–54. DOI: 10.1097/01.NAJ.0001008420.28033.e8 PMID: 38386835

Carline, J. D., O'Sullivan, P. S., Gruppen, L. D., & Richardson-Nassif, K. (2007). Crafting successful relationships with the IRB. *Academic Medicine*, 82(10, Suppl), S57–S60. DOI: 10.1097/ACM.0b013e31813e66d4 PMID: 17895692

Carlson, G., & Hecht, S. (2024). Advocating for learners in telecollaboration exchanges: The role of ethical training and IRB approval. Japan Association for Language Teaching, *Computer-Assisted Language Learning Conference*. DOI: 10.37546/JALTSIG.CALL.PCP2023-03

Carpenter, D. (2018). Ethics, reflexivity and virtue. In *The SAGE Handbook of Qualitative Research Ethics* (pp. 35–50). SAGE. DOI: 10.4135/9781526435446.n3

Chapman, C. R., Quinn, G. P., Natri, H. M., Berrios, C., Dwyer, P., Owens, K., Heraty, S., & Caplan, A. L. (2023). Consideration and disclosure of group risks in genomics and other data-centric research: Does the Common Rule need revision? *The American Journal of Bioethics*, 1–14. DOI: 10.1080/15265161.2023.2276161 PMID: 38010648

Chapman, J., Chinnaswamy, A., & Garcia-Perez, A. (2018, January). The severity of cyber attacks on education and research institutions: a function of their security posture. In *Proceedings of ICCWS 2018 13th International Conference on Cyber Warfare and Security. Academic Conferences and Publishing Limited* (pp. 111-9).

Chen, J., Henry, E., & Jiang, X. (2023). Is cybersecurity risk factor disclosure informative? Evidence from disclosures following a data breach. *Journal of Business Ethics*, 187(1), 199–224. DOI: 10.1007/s10551-022-05107-z

Chen, Y., & Esmaeilzadeh, P. (2024). Generative AI in medical practice: In-depth exploration of privacy and security challenges. *Journal of Medical Internet Research*, 26, e53008. DOI: 10.2196/53008 PMID: 38457208

Chimes, M., & Sankar, P. (2014). Confidential and proprietary information. *The IACUC Handbook*, 524-559. CRC Press.

Choi, Y. B. (2021). Organizational cyber data breach analysis of Facebook, Equifax, and Uber cases. [IJCRE]. *International Journal of Cyber Research and Education*, 3(1), 58–64. DOI: 10.4018/IJCRE.2021010106

Chong, W. K., & Patwa, N. (2023). The value of integrity: Empowering SMEs with ethical marketing communication. *Sustainability*. 15(15), 11673. DOI: 10.3390/su151511673

Christen, M., Gordijn, B., & Loi, M. (2020). *The ethics of cybersecurity*. Springer Nature., DOI: 10.1007/978-3-030-29053-5

Christl, W., Kopp, K., & Riechert, P. U. (2017). Corporate surveillance in everyday life. *Cracked Labs*, 6, 2017–10.

Ciampa, K., & Wolfe, Z. (2019). Preparing for dissertation writing: Doctoral education students' perceptions. *Studies in Graduate and Postdoctoral Education*, 10(2), 86-108.

Ciampa, K., & Wolfe, Z. (2019). Preparing for dissertation writing: Doctoral education students' perceptions. *Studies in Graduate and Postdoctoral Education, 10*(2), 86-108. DOI: 10.1108/SGPE-03-2019-0039

Clark, B., Lapid, M., & Wright, R. (2019). Institutional review boards: What clinician researchers need to know. *Mayo Clinic Proceedings Concise Review for Clinicians*, 515-525

Claw, K. G., Anderson, M. Z., Begay, R. L., Tsosie, K. S., Fox, K., Garrison, N. A., Bader, A. C., Bardill, J., Bolnick, D. A., Brooks, J., Cordova, A., Malhi, R. S., Nakatsuka, N., Neller, A., Raff, J. A., Singson, J., TallBear, K., Vargas, T., & Yracheta, J. M. (2018). A framework for enhancing ethical genomic research with Indigenous communities. *Nature Communications*, 9(1), 2957. DOI: 10.1038/s41467-018-05188-3 PMID: 30054469

Commissioner, O. of the. (2024, May 29). *Food and Drug Administration Amendments Act (FDAAA) of 2007*. FDA; FDA. https://www.fda.gov/regulatory-information/selected-amendments-fdc-act/food-and-drug-administration-amendments-act-fdaaa-2007

Costello, E., Brunton, J., Bolger, R., Soverino, T., & Juillerac, C. (2023). Massive omission of consent (MOOC): Ethical research in educational big data studies. *Online Learning: the Official Journal of the Online Learning Consortium*, 27(2). Advance online publication. DOI: 10.24059/olj.v27i2.3759

Council of American Survey Research Organizations (2001). CASRO code of standards and ethics for survey research.

Cox, D. J., Suarez, V. D., & Marya, V. (2023). Ethical principles and values guiding modern scientific research. In *Research ethics in behavior analysis* (pp. 35–61). Academic Press. DOI: 10.1016/B978-0-323-90969-3.00008-6

Crane, K., Blatch-Jones, A., & Fackrell, K. (2023). The post-award effort of managing and reporting on funded research: a scoping review. *F1000Research*, 12. https://doi.org/DOI: 10.12688/f1000research.133263.2

Creswell, J. W., & Poth, C. N. (2016). *Qualitative inquiry and research design: Choosing among five approaches*. Sage publications.

D'Agostino, S. (2023). *In class, some colleges overlook technology's dark side*. Inside Higher Ed. https://www.insidehighered.com/news/tech-innovation/teaching-learning/2023/11/16/some-colleges-overlook-technologys-dark-side

D'Antonoli, T. A. (2020). Ethical considerations for artificial intelligence: An overview of the current radiology landscape. *Diagnostic and Interventional Radiology (Ankara, Turkey)*, 26(5), 504–511. DOI: 10.5152/dir.2020.19279 PMID: 32755879

Dalal, R. S., Howard, D. J., Bennett, R. J., Posey, C., Zaccaro, S. J., & Brummel, B. J. (2022). Organizational science and cybersecurity: Abundant opportunities for research at the interface. *Journal of Business and Psychology*, 37(1), 1–29. DOI: 10.1007/s10869-021-09732-9 PMID: 33564206

Dalkiran, A (2024). Guidelines for scientific research and publication ethics. *International Journal of Aviation Science and Technology (IJAST)*, 1-14.

Dalton-Brown, S. (2020). The ethics of medical AI and the physician-patient relationship. *Cambridge Quarterly of Healthcare Ethics*, 29(1), 115–121. DOI: 10.1017/S0963180119000847 PMID: 31858938

Daswani, N., & Elbayadi, M. (2021). The Yahoo breaches of 2013 and 2014. In *Big Breaches* (pp. 155–169). Apress. DOI: 10.1007/978-1-4842-6655-7_7

Davison, R. M., Chughtai, H., Nielsen, P., Marabelli, M., Iannacci, F., van Offenbeek, M., & Panteli, N. (2024). *The ethics of using generative AI for qualitative data analysis*. Wiley., DOI: 10.1111/isj.12504

De Oliveira Neto, G. C., Pinto, L. F. R., Amorim, M. P. C., Giannetti, B. F., & de Almeida, C. M. V. B. (2018). A framework of actions for strong sustainability. *Journal of Cleaner Production*, 196, 1629–1643. DOI: 10.1016/j.jclepro.2018.06.067

De Roche, M., & Silver, R. C. (2024). Ethical issues in research with human subjects. *How Science Engages with Ethics and Why It Should. An Interdisciplinary Approach*, 61, 61–78. Advance online publication. DOI: 10.1515/9783111142463-005

Denison, T. L., Sorensen, K. U., Blanton, M. P., Johnson, L., Byrd, T., Pass, S. E., Philips, L., Miller, J., McMahon, L. R., & Cherry, B. (2024, October). (2024). The Quality Improvement Review Board: An innovative approach to oversight of projects that do not meet criteria of human subject research. *Quality Management in Health Care*, 33(4), 269–277. Advance online publication. DOI: 10.1097/QMH.0000000000000446 PMID: 38414295

Developer Staff, I. B. M. (2018, November 14). AI fairness 360. *IBM Developer*. https://www.ibm.com/opensource/projects/ai-fairness-360/

Dhablia, A. (2024). The role of cybersecurity policies in ensuring compliance with US federal and state network security laws. *Network Security*, 2024(7), 14–21.

Di Nucci, E. (2019). Should we be afraid of medical AI? *Journal of Medical Ethics*, 45(8), 556–558. DOI: 10.1136/medethics-2018-105281 PMID: 31227547

Doerr, M., & Meeder, S. (2022). Big health data research and group harm: The scope of IRB review. *Ethics & Human Research*, 44(4), 34–38. DOI: 10.1002/eahr.500130 PMID: 35802789

Doorn, N. (2022). Wastewater research and surveillance: An ethical exploration. *Environmental Science. Water Research & Technology*, 8(11), 2431–2438. DOI: 10.1039/D2EW00127F PMID: 36353217

Duan, W., & Zhang, J. (2021). The comparative Performance of online referral channels in e-commerce. *Journal of Management Information Systems*, 38(3), 828–854. DOI: 10.1080/07421222.2021.1962598

Duhigg, C. (2012, February 19). Psst, You in Aisle 5. *New York Times,* MM30.

Dumitrescu, C., Shaw Hughner, R., & Shultz, C. J.II. (2016). Policy and marketing changes to help curb childhood obesity: Government ban vs. industry self-regulation. *International Journal of Consumer Studies*, 40(5), 519–526. DOI: 10.1111/ijcs.12296

Dunn Cavelty, M. (2018). Cybersecurity research meets science and technology studies. *Politics and Governance*, 6(2), 22–30. DOI: 10.17645/pag.v6i2.1385

Dupuis, M., & Renaud, K. (2021). Scoping the ethical principles of cybersecurity fear appeals. *Ethics and Information Technology*, 23(3), 265–284. DOI: 10.1007/s10676-020-09560-0

Eagle, L., & Dahl, S. (2018). Product placement in old and new media: Examining the evidence for concern. *Journal of Business Ethics*, 147(3), 605–618. DOI: 10.1007/s10551-015-2955-z

Edquist, A., Grennan, L., Griffiths, S., & Rowshankish, K. (2022). Data ethics: What it means and what it takes. *McKinsey Report, 23.*

Eeckhout, D., Aelbrecht, K., & Ver Der Straeten, C. (2023). Informed consent: Research staff's perspectives and practical recommendations to improve research staff-participant communication. *Journal of Empirical Research on Human Research Ethics; JERHRE*, 18(1–2), 3–12. DOI: 10.1177/15562646221146043 PMID: 36562147

Elhajjar, S. (2023). The current and future state of the marketing management profession. *Journal of Marketing Theory and Practice*, 1–18.

Emanuel, E. J., Wendler, D., & Grady, C. (2008). An ethical framework for biomedical research. *The Oxford textbook of clinical research ethics*, 123-135. Oxford University Press.

Emanuel, E. J., Wendler, D., & Grady, C. (2000). What makes clinical research ethical? *Journal of the American Medical Association*, 283(20), 2701–2711. DOI: 10.1001/jama.283.20.2701 PMID: 10819955

Esmer, Y., & Arıbaş, A. N. (2022). Information ethics in the context of current developments. In *Handbook of Research on Digital Violence and Discrimination Studies* (pp. 685–707). IGI Global. DOI: 10.4018/978-1-7998-9187-1.ch030

Eto, T. (2022). Conducting an effective IRB review of artificial intelligence human subjects research (AI HSR). https://techinhsr.com/wp-content/uploads/2022/08/AI-HSR-WHITE-PAPER-TechInHSR-08.2022-1.pdf

Eto, T. (2022). *Conducting an effective IRB review of artificial intelligence human subjects research (AI HSR)*. Tech In HSR. https://techinhsr.com/wp-content/uploads/2022/08/AI-HSR-WHITE-PAPER-TechInHSR-08.2022-1.pdf

Eto, T. (2022). *Conducting an effective IRB review of artificial intelligence human subjects research (AI HSR)*. Tech InHSR. https://techinhsr.com/wp-content/ uploads/2022/08/AI-HSR-WHITE-PAPER-TechInHSR-08.2022-1.pdf

Eto, T., Lifson, M., & Vidal, D. (2024). A novel, streamlined approach to the IRB review of artificial intelligence human subjects research (AI HSR). Preprint. https://purl.stanford.edu/zj025zw1714

Evans, N., & Gray, C. (2012). The practice and ethics of animal-assisted therapy with children and young people: Is it enough that we don't eat our co-workers? *British Journal of Social Work*, 42(4), 600–617. DOI: 10.1093/bjsw/bcr091

Faden, R. R., & Beauchamp, T. L. (1986). *A history and theory of informed consent*. Oxford University Press.

Faden, R. R., Beauchamp, T. L., & Kass, N. E. (2014). Informed consent, comparative effectiveness, and learning health care. *The New England Journal of Medicine*, 370(8), 766–768. DOI: 10.1056/NEJMhle1313674 PMID: 24552325

Faden, R. R., & Kass, N. E. (2022). Learning health care and the obligation to participate in research. *The Hastings Center Report*, 52(3), 29–31. DOI: 10.1002/hast.1393 PMID: 35763203

Faden, R. R., Kass, N. E., Goodman, S. N., Pronovost, S. T., & Beauchamp, T. L. (2013). An ethics framework for a leaning health care system: A departure from traditional research ethics and clinical ethics. *The Hastings Center Report*, 43(s1), S16–S27. DOI: 10.1002/hast.134 PMID: 23315888

Faheem, H., & Dutta, S. (2023). Artificial intelligence failure at IBM 'Watson for Oncology'. *IUP Journal of Knowledge Management*, 21(3), 47–75.

Familoni, B. T.Babajide Tolulope Familoni. (2024). Cybersecurity challenges in the age of AI: Theoretical approaches and practical solutions. *Computer Science & IT Research Journal*, 5(3), 703–724. DOI: 10.51594/csitrj.v5i3.930

Feil, H. (2021). The cancer of corruption and World Bank project performance: Is there a connection? *Development Policy Review*, 39(3), 381–397. DOI: 10.1111/dpr.12503

Ferdowsian, H., Johnson, L. S. M., Johnson, J., Fenton, A., Shriver, A., & Gluck, J. (2020). A Belmont Report for animals? *Cambridge Quarterly of Healthcare Ethics*, 29(1), 19–37. DOI: 10.1017/S0963180119000732 PMID: 31581963

Ferretti, A., Ienca, M., Hurst, S., & Vayena, E. (2020). Big data, biomedical research, and ethics review: New challenges for IRBs. *Ethics & Human Research*, 42(5), 17–28. DOI: 10.1002/eahr.500065 PMID: 32937036

Ferretti, A., Ienca, M., Velarde, M. R., Hurst, S., & Vayena, E. (2022). The challenges of big data for research ethics committees: A qualitative Swiss study. *Journal of Empirical Research on Human Research Ethics; JERHRE*, 17(1-2), 129–143. DOI: 10.1177/15562646211053538 PMID: 34779661

Fiesler, C., Garrett, N., & Beard, N. (2020). What do we teach when we teach tech ethics?: A syllabi analysis. *Proceedings of the 51st ACM Technical Symposium on Computer Science Education*, 289–295. DOI: 10.1145/3328778.3366825

Fiesler, C., Zimmer, M., Proferes, N., Gilbert, S., & Jones, N. (2024). Remember the human: A systematic review of ethical considerations in Reddit research. *Proceedings of the ACM on Human-Computer Interaction, 8*(GROUP), 1-33. DOI: 10.1145/3633070

Fine, A. H., & Griffin, T. C. (2022, January). Protecting animal welfare in animal-assisted intervention: Our ethical obligation. Seminars in Speech and Language (Vol. 43, No. 01, pp. 008-023). Thieme Medical Publishers, Inc.

Finkelstein, J. A., Brickman, A. L., Capron, A., Ford, D., Gombosev, A., Greene, S. M., Iafrate, R. P., Kolaczkowski, L., Pallin, S. C., Pletcher, M. J., Staman, K. L., Vazquez, M., & Sugarman, J. (2015). Oversight on the borderline: Quality improvement and pragmatic research. *Clinical Trials*, 12(5), 457–466. DOI: 10.1177/1740774515597682 PMID: 26374685

Finn, M., & Shilton, K. (2023). Ethics governance development: The case of the Menlo Report. *Social Studies of Science*, 53(3), 315–340. DOI: 10.1177/03063127231151708 PMID: 36802925

Fiscella, K., Tobin, J. N., Carroll, J. K., He, H., & Ogedegbe, G. (2015). Ethical oversight in quality improvement and quality improvement research: New approaches to promote a learning health care system. *BMC Medical Ethics*, 16(1), 63. Advance online publication. DOI: 10.1186/s12910-015-0056-2 PMID: 26383770

Fischer, C., Xu, D., Rodriguez, F., Denaro, K., & Warschauer, M. (2020). Effects of course modality in summer session: Enrollment patterns and student performance in face-to-face and online classes. *The Internet and Higher Education*, 45, 100710. DOI: 10.1016/j.iheduc.2019.100710

Fishbein, E. A. (1991). Ownership of research data. *Academic Medicine*, 66(3), 129–133. DOI: 10.1097/00001888-199103000-00002 PMID: 1997022

Fiske, S. T., & Hauser, R. M. (2014). Protecting human research participants in the age of big data. *Proceedings of the National Academy of Sciences of the United States of America*, 111(38), 13675–13676. DOI: 10.1073/pnas.1414626111 PMID: 25157175

Floridi, L., & Taddeo, M. (2016). What is data ethics? *Philosophical Transactions. Series A, Mathematical, Physical, and Engineering Sciences*, 374(2083), 20160360. DOI: 10.1098/rsta.2016.0360 PMID: 28336805

Formosa, P., Wilson, M., & Richards, D. (2021). A principlist framework for cybersecurity ethics. *Computers & Security*, 109, 102382. DOI: 10.1016/j.cose.2021.102382

Fox, G., Lynn, T., & Rosati, P. (2022). Enhancing consumer perceptions of privacy and trust: A GDPR label perspective. *Information Technology & People, 35*(8), 181–204. DOI: 10.1108/ITP-09-2021-0706

Friesen, P., Douglas-Jones, R., Marks, M., Pierce, R., Fletcher, K., Mishra, A., Lorimer, J., Véliz, C., Hallowell, N., Graham, M., Chan, M. S., Davies, H., & Sallamuddin, T. (2021). Governing AI-driven health research: Are IRBs up to the task? *Ethics & Human Research*, 43(2), 35–42. DOI: 10.1002/eahr.500085 PMID: 33683015

Friesen, P., Gelinas, L., Kirby, A., Strauss, D. H., & Bierer, B. E. (2023). IRBs and the protection-inclusion dilemma: Finding a balance. *The American Journal of Bioethics*, 23(6), 75–88. DOI: 10.1080/15265161.2022.2063434 PMID: 35482887

Friesen, P., Kearns, L., Redman, B., & Caplan, A. L. (2017). Rethinking the Belmont Report? *The American Journal of Bioethics*, 17(7), 15–21. DOI: 10.1080/15265161.2017.1329482 PMID: 28661753

Frohmann, B. (2000). Cyber ethics: Bodies or bytes? *The International Information & Library Review*, 32(3-4), 423–435. DOI: 10.1080/10572317.2000.10762528

Fuller, D. A. (1999). *Sustainable marketing: Managerial-ecological issues*. Sage publications. DOI: 10.4135/9781452220611

Gangjee, D. S. (2024). Law (s) against unfair competition: The legitimacy of the UK approach. *Journal of Intellectual Property Law and Practice*, 19(2), 90–100. DOI: 10.1093/jiplp/jpad105

Gauthier, B., Goyette, J., & Kouamé, W. A. (2021). Why do firms pay bribes? Evidence on the demand and supply sides of corruption in developing countries. *Journal of Economic Behavior & Organization*, 190, 463–479. DOI: 10.1016/j.jebo.2021.06.041

Gehlhar, M. (2021). *The fashion designer survival guide: start and run your own fashion business*. Simon and Schuster.

Geist, T. S. (2011). Conceptual framework for animal assisted therapy. *Child & Adolescent Social Work Journal*, 28(3), 243–256. DOI: 10.1007/s10560-011-0231-3

Gerke, S., Minssen, T., & Cohen, G. (2020). Ethical and legal challenges of artificial intelligence-driven healthcare. In *Artificial intelligence in healthcare* (pp. 295–336). Academic Press. DOI: 10.1016/B978-0-12-818438-7.00012-5

Gibson, S., Hsu, M. K., & Zhou, X. (2022). Convenience stores in the digital age: A focus on the customer experience and revisit intentions. *Journal of Retailing and Consumer Services*, 68, 103014. DOI: 10.1016/j.jretconser.2022.103014

Giuffrida, I., & Hall, A. (2023). Technology integration in higher education and student privacy beyond learning environments - A comparison of the UK and US perspective. *British Journal of Educational Technology*, 54(6), 1587–1603. DOI: 10.1111/bjet.13375

Godfrey, P. C., & Jacobus, L. E. (2022). *Ethics and business: An integrated approach for business and personal success*. Wiley.

Godskesen, T., Bjork, J., & Juth, N. (2023). Challenges regarding informed consent in recruitment to clinical research: A qualitative study of clinical research nurses' experiences. *Trials*, 24(1), 801. Advance online publication. DOI: 10.1186/s13063-023-07844-6 PMID: 38082434

Goksel, N., & Bozkurt, A. (2019). Artificial intelligence in education: Current insights and future perspectives. In *Handbook of research on learning in the Age of Transhumanism* (pp. 224–236). IGI Global., DOI: 10.4018/978-1-5225-8431-5.ch014

Gordon, B. G. (2020). Vulnerability in research: Basic ethical concepts and general approach to review. *The Ochsner Journal*, 20(1), 34–38. DOI: 10.31486/toj.19.0079 PMID: 32284680

Gouthier, M. H., Nennstiel, C., Kern, N., & Wendel, L. (2022). The more the better? Data disclosure between the conflicting priorities of privacy concerns, information sensitivity and personalization in e-commerce. *Journal of Business Research*, 148, 174–189. DOI: 10.1016/j.jbusres.2022.04.034

Grady, C., & Wendler, D. (2013). Making the transition to a learning health care system. *The Hastings Center Report*, 43(1), S32–S33. DOI: 10.1002/hast.137 PMID: 23315892

Greene, T., Martens, D., & Shmueli, G. (2021). Barriers for academic data science research in the new realm of behavior modification by digital platforms. https://papers.ssrn.com/sol3/ papers.cfm?abstract_id=3946116

Gross, M. S., & Miller, R. C.Jr. (2019). Preserving privacy & promoting learning: Technology to empower Belmont 2.0. *SSRN*. https://papers.ssrn.com/sol3/papers.cfm?abstract_id=3390979

Grossman, R. L. (2023). Ten lessons for data sharing with a data commons. *Scientific Data*, 10(1), 1. Advance online publication. DOI: 10.1038/s41597-023-02029-x PMID: 36878917

Guillemin, M., & Gillam, L. (2004). Ethics, reflexivity, and "ethically important moments" in research. *Qualitative Inquiry*, 10(2), 261–280. DOI: 10.1177/1077800403262360

Gunathilake, S. (2024). *The role of ethics in shaping modern scientific research.* UVA Clinical Research.

Gurwitz, D. (2020). Repurposing current therapeutics for treating COVID-19: A vital role of prescription records data mining. *Drug Development Research*, 81(7), 777–781. DOI: 10.1002/ddr.21689 PMID: 32420637

Hafen, E., Kossmann, D., & Brand, A. (2014). Health data cooperatives—Citizen empowerment. *Methods of Information in Medicine*, 53(2), 82–86. DOI: 10.3414/ME13-02-0051 PMID: 24514946

Hagenbuch, D. J., & Mgrdichian, L. M. (2020). Mindful marketing: A strategy-based, branded Approach for encouraging ethical marketing. *Marketing Education Review*, 30(1), 15–28. DOI: 10.1080/10528008.2019.1686993

Hall, S., Lee, V., & Haase, K. (2020). Exploring the challenges of ethical conduct in quality improvement projects. *Canadian Oncology Nursing Journal*, 30(1), 64–68. PMID: 33118977

Hamby, A., Orazi, D., & Moreau, P. (2024). Whose story is this? Source reveal as a communication tactic to increase consumers' advocacy for social causes. *Journal of Business Research*, 170, 114309. DOI: 10.1016/j.jbusres.2023.114309

Hammersley, M., & Traianou, A. (2016). Ethics in qualitative research: Controversies and contexts. SAGE.

Hammond, W. (Ed.). (2016). *Informed consent: procedures, ethics and best practices*. Nova Publishers.

Hancock, D. R., Algozzine, B., & Lim, J. H. (2021). *Doing case study research: A practical guide for beginning researchers* (4th ed.). Teachers College Press.

Haneef, I., & Agrawal, M. (2024). Ethical issues in educational research. *Asian Research Journal of Arts & Social Sciences*, 22(5), 29–38. DOI: 10.9734/arjass/2024/v22i5535

Hansen, B. C., Gografe, S., Pritt, S., Jen, K. L. C., McWhirter, C. A., Barman, S. M., Comuzzie, A., Greene, M., McNulty, J. A., Michele, D. E., Moaddab, N., Nelson, R. J., Norris, K., Uray, K. D., Banks, R., Westlund, K. N., Yates, B. J., Silverman, J., Hansen, K. D., & Redman, B. (2017). Ensuring due process in the IACUC and animal welfare setting: Considerations in developing noncompliance policies and procedures for institutional animal care and use committees and institutional officials. *The FASEB Journal*, 31(10), 4216–4225. DOI: 10.1096/fj.201601250R PMID: 28821637

Hardy, J. (2018). Advertising regulation. In *The advertising handbook* (pp. 123–140). Routledge. DOI: 10.4324/9781315558646-9

Haubenhofer, D. K., & Kirchengast, S. (2007). 'Dog handlers' and dogs' emotional and cortisol secretion responses associated with animal-aassisted therapy sessions. *Society & Animals*, 15(2), 127–150. DOI: 10.1163/156853007X187090

Hawamleh, A. M. A., Alorfi, A. S. M., Al-Gasawneh, J. A., & Al-Rawashdeh, G. (2020). Cyber security and ethical hacking: The importance of protecting user data. *Solid State Technology*, 63(5), 7894–7899.

Head, G. (2020). Ethics in educational research: Review boards, ethical issues, and researcher development. *European Educational Research Journal, 19*(1), 72–83. DOI: 10.1177/1474904118796315

Hennessy, M., Dennehy, R., Doherty, J., & O'Donoghue, K. (2022). Outsourcing transcription: Extending ethical considerations in qualitative research. *Qualitative Health Research*, 32(7), 1197–1204. DOI: 10.1177/10497323221101709 PMID: 35599615

Herington, J., Connelly, K., & Illes, J. (2023). Ethical imperatives for working with diverse populations in digital research. *Journal of Medical Internet Research*, 25, e47884. DOI: 10.2196/47884 PMID: 37721792

Herrmann, D., & Pridöhl, H. (2020). Basic concepts and models of cybersecurity. *The Ethics of Cybersecurity*, 11-44. https://doi.org/DOI: 10.7196/SAJBL.2024.v17i2.1671

Hesse, A., Glenna, L., Hinrichs, C., Chiles, R., & Sachs, C. (2019). Qualitative research ethics in the Big Data Era. *The American Behavioral Scientist*, 63(5), 560–583. DOI: 10.1177/0002764218805806

Hicks, R. W., Hines, K., & Henson, B. (2021). Demystifying the Institutional Review Board. *AORN Journal*, 114(4), 309–318. DOI: 10.1002/aorn.13498 PMID: 34586652

Hilário, A. P., Augusto, F. R., Hilário, A. P., & Augusto, F. R. (2020). *Practical and ethical dilemmas in researching sensitive topics with populations considered vulnerable*. MDPI.

Hill, K. (2012, February 16). How Target figured out a teen girl was pregnant before her father did. *Forbes*.

Hine, C. (2021). Evaluating the prospects for university-based ethical governance in artificial intelligence and data-driven innovation. *Research Ethics*, 17(4), 464–479. DOI: 10.1177/17470161211022790

HIPAA Privacy Rule, 45 CFR §164.514(a) (2013).

Hirsch, D. D., Bartley, T., Chandrasekaran, A., Norris, D., Parthasarathy, S., & Turner, P. N. (2021). Business data ethics: Emerging trends in the governance of advanced analytics and AI. https://papers.ssrn.com/sol3/papers.cfm?abstract_id=3828239

Hite, R. L., Shin, S., & Lesley, M. (2022). Reflecting on responsible conduct of research: A self study of a research-oriented university community. *Journal of Academic Ethics*, 20(3), 399–419. DOI: 10.1007/s10805-021-09418-0 PMID: 34104126

Hopkins, C. D., Ferrell, O. C., Ferrell, L., & Hopkins, K. H. (2021). Changing perceptions of marketing ethics and social responsibility in principles of marketing. *Journal of Marketing Education*, 43(2), 244–259. DOI: 10.1177/0273475321995553

Hörnle, J., Schmidt-Kessen, M., Littler, A., & Padumadasa, E. (2019). Regulating online advertising for gambling–once the genie is out of the bottle…. *Information & Communications Technology Law*, 28(3), 311–334. DOI: 10.1080/13600834.2019.1664001

Hosseini, M., Wieczorek, M., & Gordijn, B. (2022). Ethical issues in social science research employing big data. *Science and Engineering Ethics*, 28(3), 1–21. DOI: 10.1007/s11948-022-00380-7 PMID: 35705883

Hsieh, H. F., & Shannon, S. E. (2005). Three approaches to qualitative content analysis. *Qualitative Health Research*, 15(9), 1277–1288. DOI: 10.1177/1049732305276687 PMID: 16204405

Hudson, M., Garrison, N. A., Sterling, R., Caron, N. R., Fox, K., Yracheta, J., Anderson, J., Wilcox, P., Arbour, L., Brown, A., Taualii, M., Kukutai, T., Haring, R., Te Aika, B., Baynam, G. S., Dearden, P. K., Chagné, D., Malhi, R. S., Garba, I., & Carroll, S. R. (2020). Rights, interests and expectations: Indigenous perspectives on unrestricted access to genomic data. *Nature Reviews. Genetics*, 21(6), 377–384. DOI: 10.1038/s41576-020-0228-x PMID: 32251390

Huh-Yoo, J., & Rader, E. (2020). It's the Wild, Wild West: Lessons learned from IRB members' risk perceptions toward digital research data. *Proceedings of the ACM on Human-Computer Interaction, 4*(CSCW1), (pp. 1-22). ACM. DOI: 10.1145/3392868

Humbly, C. (2006). *Data is the New Oil. ANA Senior Marketer's Summit.* Kellogg School.

Hunjan, U. G., & Reddy, K. J. (2023). Effect of Animal-Assisted Therapy On Depression, Memory, Attention, And Emotion Regulation. *Human Research in Rehabilitation*, 13(1), 120–129. DOI: 10.21554/hrr.042314

Hunt, D. F., Dunn, M., Harrison, G., & Bailey, J. (2021). Ethical considerations in quality improvement: Key questions and a practical guide. *BMJ Open Quality*, 10(3), e001497. Advance online publication. DOI: 10.1136/bmjoq-2021-001497 PMID: 34404685

Hurley, E., & Pierce, H. H. (2019, July 30). The Belmont Report at 40 – A moment of admiration. *Ampersand*. https://blog.primr.org/belmont-report-40/

Hussain, M. I., Figueiredo, M. C., Tran, B. D., Su, Z., Molldrem, S., Eikey, E. V., & Chen, Y. (2020). A scoping review of qualitative research in JAMIA: Past contributions and opportunities for future work. *Journal of the American Medical Informatics Association: JAMIA*, 28(2), 402–413. DOI: 10.1093/jamia/ocaa179 PMID: 33225361

IBM. (2023). 5725-W51 IBM Watson for oncology. https://www.ibm.com/docs/en/announcements/watson-oncology?region=US

IEEE. (2020). *IEEE Code of Ethics*. www.ieee.org

IEEE. (n.d.). *IEEE Code of Ethics*. Retrieved May 24, 2024, from https://www.ieee.org/about/corporate/governance/p7-8.html

Iglesias, O., Markovic, S., Bagherzadeh, M., & Singh, J. J. (2020). Co-creation: A key link between corporate social responsibility, customer trust, and customer loyalty. *Journal of Business Ethics*, 163(1), 151–166. DOI: 10.1007/s10551-018-4015-y

Information Commissioner's Office (ICO). (2023, May 19). *A guide to the data protection principles*. A Guide to the Data Protection Principles; ICO. https://ico.org.uk/for-organisations/uk-gdpr-guidanceand-resources/data-protection-principles/a-guide-to-the-data-protection-principles/

Institute of Medicine. (2013). *Best care at lower cost: The path to continuously learning health care in America*. The National Academies Press., DOI: 10.17226/13444

Institutional Animal Care and Use Committee Guidebook. (2002). https://grants.nih.gov/grants/olaw/guidebook.pdf

Jach, E., Gloeckner, G., & Kohashi, C. (2020). Social and behavioral research with undocumented immigrants: Navigating an IRB committee. *Hispanic Journal of Behavioral Sciences*, 42(1), 3–17. DOI: 10.1177/0739986319899979

Joanna, K. (2021). Cyber ethics: Historical outline and basic problems. *Редакційна колегія*, 21.

Johann, M. (2015). Marketing Communications in the Network Organizations. *Communication and Information Technology in Society: Volume 2 Information and Communication Technologies (ICT) in Management, 2*, 35.

Johnson, A. R., Pautler, M., Burr, J. S., Abdelsamad, N., VanBuren, J. M., Rigtrup, L. M., Dean, J. M., & Rothwell, E. (2022). Using single IRB consultations to meet the educational needs of investigative teams. *Contemporary Clinical Trials Communications*, 29, 100971. DOI: 10.1016/j.conctc.2022.100971 PMID: 36033361

Johnson, F. (2015). SC Johnson's CEO on doing the right thing, even when it hurts business. *Harvard Business Review*, •••, 33–36.

Johnson, J. L., Bottorff, J. L., Browne, A. J., Grewal, S., Hilton, B. A., & Clarke, H. (2004, April). Othering and being othered in the context of health care services. *Health Communication*, 16(2), 255–271. DOI: 10.1207/S15327027HC1602_7 PMID: 15090288

Johnson, N., Vermeulen, L., & Smith, K. M. (2006). A survey of academic medical centers to distinguish between quality improvement and research activities. *Quality Management in Health Care*, 15(4), 215–220. DOI: 10.1097/00019514-200610000-00003 PMID: 17047495

Johnson, P. D., & Besselsen, D. G. (2002). Practical aspects of experimental design in animal research. *ILAR Journal*, 43(4), 202–206. DOI: 10.1093/ilar.43.4.202 PMID: 12391395

Johnson, S. (2007). SC Johnson builds business at the base of the pyramid. *Global Business and Organizational Excellence*, 26(6), 6–17. DOI: 10.1002/joe.20170

Johnson, S. B., Slade, I., Giubilini, A., & Graham, M. (2020). Rethinking the ethical principles of genomic medicine services. *European Journal of Human Genetics*, 28(2), 147–154. DOI: 10.1038/s41431-019-0507-1 PMID: 31534213

Johnson, S. L. (2019). AI, machine learning, and ethics in health care. *Journal of Legal Medicine*, 39(4), 427–441. DOI: 10.1080/01947648.2019.1690604 PMID: 31940250

Jones, R. D., Krenz, C., Gornick, M., Griffith, K. A., Spence, R., Bradbury, A. R., De Vries, R., Hawley, S. T., Hayward, R. A., Zon, R., Bolte, S., Sadeghi, N., Schilsky, R. L., & Jagsi, R. (2020). Patient preferences regarding informed consent models for participation in a learning health care system for oncology. *JCO Oncology Practice*, 16(9), e977–e990. DOI: 10.1200/JOP.19.00300 PMID: 32352881

Kagendo-Kiarie, R. (2022). Ethical marketing: Concepts and reasons. In *Products for conscious consumers: Developing, marketing and selling ethical products* (pp. 15–33). Emerald Publishing Limited. DOI: 10.1108/978-1-80262-837-120221002

Kara, H. (2024). Ethics for independent researchers. *NCIS Guide for Independent Scholars*, 98.

Kass, N. E., Faden, R. R., Goodman, S. N., Pronovost, P., Tunis, S., & Beauchamp, T. L. (2013). The research-treatment distinction: A problematic approach for determining which activities should have ethical oversight. *The Hastings Center Report*, 43(1), S4–S15. DOI: 10.1002/hast.133 PMID: 23315895

Kavak, H., Padilla, J. J., Vernon-Bido, D., Diallo, S. Y., Gore, R., & Shetty, S. (2021). Simulation for cybersecurity: State of the art and future directions. *Journal of Cybersecurity*, 7(1), 1–13. DOI: 10.1093/cybsec/tyab005

Kaye, K. (2015, October 26). The $24 Billion Data Business That Telcos Don't Want to Talk About. *Advertising Age*, 12–14.

Kayla, C., Jennifer, P., Sandra, B., & Johnston, K. (2023). Building our research administrator workforce as our clinical and translational research programs become increasingly complex. *Frontiers in Pharmacology*, 14, 1295255. Advance online publication. DOI: 10.3389/fphar.2023.1295255 PMID: 37942484

Kennedy, A. M., Jones, K., & Williams, J. (2019). Children as vulnerable consumers in online environments. *The Journal of Consumer Affairs*, 53(4), 1478–1506. DOI: 10.1111/joca.12253

Kerulo, G., Kargas, N., Mills, D. S., Law, G., VanFleet, R., Faa-Thompson, T., & Winkle, M. Y. (2020). Animal-assisted interventions: Relationship between standards and qualifications. *People and Animals : The International Journal of Research and Practice*, 3(1), 4.

Khurana, D., Koli, A., Khatter, K., & Singh, S. (2023). Natural language processing: State of the art, current trends and challenges. *Multimedia Tools and Applications*, 82(3), 3713–3744. DOI: 10.1007/s11042-022-13428-4 PMID: 35855771

Kilkenny, C., Browne, W. J., Cuthill, I. C., Emerson, M., & Altman, D. G. (2014). Improving bioscience research reporting: The arrive guidelines for reporting animal research. *Animals (Basel)*, 4(1), 35–44. DOI: 10.3390/ani4010035

Kimmelman, J. (2020). What is human research for? reflections on the omission of scientific integrity from the Belmont report. *Perspectives in Biology and Medicine*, 63(2), 63. DOI: 10.1353/pbm.2020.0017 PMID: 33416651

Kimmons, R., & Veletsianos, G. (2018). Public internet data mining methods in instructional design, educational technology, and online learning research. *TechTrends*, 62(5), 492–500. DOI: 10.1007/s11528-018-0307-4

Klitzman, R., Pivovarova, E., & Lidz, C. W. (2017). Single IRBs in multisite trials: Questions posed by the new NIH policy. *Journal of the American Medical Association*, 317(20), 2061–2062. DOI: 10.1001/jama.2017.4624 PMID: 28445582

Knight, S., Shibani, A., & Vincent, N. (2024). Ethical AI governance: Mapping a research ecosystem. *AI and Ethics*, 1–22. DOI: 10.1007/s43681-023-00416-z

Koali, S., Khasoane, N., & Mongezi, M. (2024). Promoting research integrity through the lens of virtue ethics and deontological approach. *South African Journal of Bioethics and Law*, 17(2), 78–82.

Koester, J. V. (2003). Legal briefing: Advertising to children in the USA. *Young Consumers*, 4(1), 65–71. DOI: 10.1108/17473610310813735

Komić, D., Marušić, S. L., & Marušić, A. (2015). Research integrity and research ethics in professional codes of ethics: Survey of terminology used by professional organizations across research disciplines. *PLoS One*, 10(7), e0133662. Advance online publication. DOI: 10.1371/journal.pone.0133662 PMID: 26192805

Konda, S. R. (2022). Ethical considerations in the development and deployment of AI-driven software systems. *International Journal of Computer Science and Technology*, 6(3), 86–101.

Kraus, S., Breier, M., Lim, W. M., Dabić, M., Kumar, S., Kanbach, D., Mukherjee, D., Corvello, V., Piñeiro-Chousa, J., Liguori, E., Palacios-Marqués, D., Schiavone, F., Ferraris, A., Fernandes, C., & Ferreira, J. J. (2022). Literature reviews as independent studies: Guidelines for academic practice. *Review of Managerial Science*, 16(8), 2577–2595. DOI: 10.1007/s11846-022-00588-8

Kruger, K. A., & Serpell, J. A. (2010). Animal-assisted interventions in mental health: Definitions and theoretical foundations. In *Handbook on animal-assisted therapy* (pp. 33–48). Academic Press. DOI: 10.1016/B978-0-12-381453-1.10003-0

Kukutai, T., & Taylor, J. (Eds.). (2016). *Indigenous Data Sovereignty: Toward an agenda* (Vol. 38). ANU Press. https://www.jstor.org/stable/j.ctt1q1crgf

Kurfess, F., Vasilaky, K., Cheuk, T., Jenkins, R., Nolan, G., Hajrasouliha, A., & St John, E. (2022, August). Assessment of ethics and social justice aspects in data science and artificial intelligence. In *2022 ASEE Annual Conference & Exposition*. DOI: 10.18260/1-2--41550

Laber, K., & Smith, A. (2014). Submission and Maintenance of IACUC Protocols. *The IACUC Handbook*, 111.

Lachapelle, A. (2022). Data protection enforcement in the era of the Directive on Whistleblowers: Towards a collective approach? In *Research Handbook on EU Data Protection Law* (pp. 600–619). Edward Elgar Publishing. DOI: 10.4337/9781800371682.00032

Lahman, M. (2018). Becoming ethically responsive researchers: introduction. In *Ethics in Social Science Research: Becoming Culturally Responsive* (pp. 3-12). SAGE. DOI: 10.4135/9781071878750.n1

Langlais, P. J., & Bent, B. J. (2018). Effects of training and environment on graduate students' self-rated knowledge and judgments of responsible research behavior. *Ethics & Behavior*, 28(2), 133–153. DOI: 10.1080/10508422.2016.1260014

Lantos, J. D. (2020). The Belmont Report and innovative clinical research. *Perspectives in Biology and Medicine*, 63(2), 389–400. DOI: 10.1353/pbm.2020.0026 PMID: 33416660

Lapid, M. I., Ouellette, Y., Drake, M. T., & Clarke, B. L. (2023). Institutional Review Board (IRB): US perspectives. In *Handbook of Bioethical Decisions. Volume II: Scientific Integrity and Institutional Ethics*, 219-240. Springer International. DOI: 10.1007/978-3-031-29455-6_15

Laurie, G., Dove, E., Ganguli-Mitra, A., McMillan, C., Postan, E., Sethi, N., & Sorbie, A. (Eds.). (2021). *The Cambridge Handbook of Health Research Regulation* (1st ed.). Cambridge University Press., DOI: 10.1017/9781108620024

Lee, N. (2014). The Lancet Technology: November, 2014. *Lancet*, 384(9958), 1917. DOI: 10.1016/S0140-6736(14)62267-4 PMID: 25478615

Lee, W. W. (2020). Ethical computing for data protection. [IJT]. *International Journal of Technoethics*, 11(1), 43–58. DOI: 10.4018/IJT.2020010104

Lehtonen, T. (2021). Ethics of security: From personal safety to cyber security. In *Multidisciplinary Approaches to Ethics in the Digital Era* (pp. 44–59). IGI Global. DOI: 10.4018/978-1-7998-4117-3.ch004

Leonidou, L. C., Aykol, B., & Eteokleous, P. (2020). Internal and external drivers of an ethical international marketing strategy: Implications on reputational advantage and performance. *The SAGE Handbook of marketing ethics*, 20.

Levine, G. (2008). *Realism, ethics and secularism: essays on Victorian literature and science*. Cambridge University Press. DOI: 10.1017/CBO9780511484872

Levitt, H. M., Morrill, Z., Collins, K. M., & Rizo, J. L. (2021). The methodological integrity of critical qualitative research: Principles to support design and research review. *Journal of Counseling Psychology*, 68(3), 357–370. DOI: 10.1037/cou0000523 PMID: 34043379

Lewis, C., & Throne, R. (2021). Autoethnography and other self-inquiry methods for practice-based doctoral research. In *Practice-based and practice-led research for dissertation development* (pp. 87-107). IGI Global.

Liamputtong, P. (2007). Researching the vulnerable. *SAGE*. DOI: 10.4135/9781849209861

Libaque-Sáenz, C. F., Wong, S. F., Chang, Y., & Bravo, E. R. (2021). The effect of fair information practices and data collection methods on privacy-related behaviors: A study of mobile apps. *Information & Management*, 58(1), 103284. DOI: 10.1016/j.im.2020.103284

Lincoln, Y. S., & Guba, E. G. (1985). Naturalistic inquiry. *SAGE.*

Link, A., & Wagner, C. (2021). The publicness of publicly funded research. *Science & Public Policy*, 48(5), 757–762. Advance online publication. DOI: 10.1093/scipol/scab050

Lin, X. (2022, April). Big data analysis of personalized recommendation in e-commerce. In *2022 International Conference on Social Sciences and Humanities and Arts (SSHA 2022)* (pp. 768-771). Atlantis Press. DOI: 10.2991/assehr.k.220401.147

Loi, M., & Christen, M. (2020). Ethical frameworks for cybersecurity. *The Ethics of Cybersecurity*, 73-95.

Lonsdale, D. J. (2020). The ethics of cyber attack: Pursuing legitimate security and the common good in contemporary conflict scenarios. *Journal of Military Ethics*, 19(1), 20–39. DOI: 10.1080/15027570.2020.1764694

Löser, F. (2015). *Strategic information systems management for environmental sustainability: enhancing firm competitiveness with Green IS* (Vol. 6). Universitätsverlag der TU Berlin.

Lucas, G. R. (2017). *Ethics and cyber warfare: the quest for responsible security in the age of digital warfare*. Oxford University Press. DOI: 10.1093/acprof:oso/9780190276522.001.0001

Lu, J., Ren, L., Qiao, J., Yao, S., Strielkowski, W., & Streimikis, J. (2019). Corporate social responsibility and corruption: Implications for the sustainable energy sector. *Sustainability*, 11(15), 4128. DOI: 10.3390/su11154128

Lynch, S., & Kuntz, A. (2019). A critical autoethnography of a doctoral students' research journey: Learning to take risks in the academy. *Curriculum Studies in Health and Physical Education*, 10(2), 156–171. DOI: 10.1080/25742981.2019.1588762

Maalem Lahcen, R. A., Caulkins, B., Mohapatra, R., & Kumar, M. (2020). Review and insight on the behavioral aspects of cybersecurity. *Cybersecurity*, 3(1), 1–18. DOI: 10.1186/s42400-020-00050-w

Macnamara, J. (2024). Human research ethics review challenges in the social sciences: A case for review. *Journal of Academic Ethics*, 1-17. https://doi/org/DOI: 10.1007/s10805-024-09532-9

Macnish, K., & Van der Ham, J. (2020). Ethics in cybersecurity research and practice. *Technology in Society*, 63, 101382. DOI: 10.1016/j.techsoc.2020.101382

Madan, A., Rosca, L. D., Dumitru, I., & Canda, A. (2022). The value of art in persuasive marketing communication and its sustainable effect on the country of origin. *Sustainability (Basel)*, 14(3), 1228. DOI: 10.3390/su14031228

Mahfood, S., Astuto, A., Olliges, R., & Suits, B. (2005). Cyberethics: Social ethics teaching in educational technology programs. *Communication Research Trends*, 24(4), 3–23.

Makridis, C. A., Boese, A., Fricks, R., Workman, D., Klote, M., Mueller, J., Hildebrandt, I. J., Kim, M., & Alterovitz, G. (2023). Informing the ethical review of human subjects research utilizing artificial intelligence. *Frontiers of Computer Science*, 5, 1235226. DOI: 10.3389/fcomp.2023.1235226

Mamaysky, I. (2021). Understanding ethics and compliance: A practitioner's guide to effective corporate compliance programs. *Loy. U. Chi. J. Reg. Compl.*, 6, 58.

Mandal, P. C. (2019). Ethical issues and concerns in collection of marketing information and marketing intelligence: Ethical issue in collection of information. [IJBIR]. *International Journal of Business Intelligence Research*, 10(2), 16–28. DOI: 10.4018/IJBIR.2019070102

Mandal, P. C. (2021). Socially-responsible marketing communications: Concerns, strategies, and initiatives. [JMME]. *Journal of Media Management and Entrepreneurship*, 3(1), 1–17. DOI: 10.4018/JMME.290301

Manjikian, M. (2017). *Cybersecurity ethics: an introduction*. Routledge. DOI: 10.4324/9781315196275

Manton, A., Wolf, L. A., Baker, K. M., Carman, M. J., Clark, P. R., Henderson, D., & Zavotsky, K. E. (2014). Ethical considerations in human subjects' research. *Journal of Emergency Nursing: JEN*, 40(1), 92–94. DOI: 10.1016/j.jen.2013.11.001 PMID: 24384483

Marinelli, L., Normando, S., Siliprandi, C., Salvadoretti, M., & Mongillo, P. (2009). Dog assisted interventions in a specialized centre and potential concerns for animal welfare. *Veterinary Research Communications*, 33(S1), 93–95. DOI: 10.1007/s11259-009-9256-x PMID: 19578960

Markman, K. M., Weicker, N. P., Klein, A. K., & Sege, R. (2023). Community-engaged training in informed consent. *Journal of Clinical and Translational Science*, 7(1), e108. Advance online publication. DOI: 10.1017/cts.2023.534 PMID: 37313379

Marmo, R. (2021). Social media mining. In *Encyclopedia of organizational knowledge, administration, and technology* (pp. 2153-2165). DOI: 10.4018/978-1-7998-3473-1.ch149

Martinez-Martin, N., Insel, T. R., Dagum, P., Greely, H. T., & Cho, M. K. (2018). Data mining for health: Staking out the ethical territory of digital phenotyping. *NPJ Digital Medicine*, 1(1), 1–5. DOI: 10.1038/s41746-018-0075-8 PMID: 31211249

Mathis, D. (2024). Preparation of an ethics application. *Arthroskopie*, 1-4(4), 254–257. Advance online publication. DOI: 10.1007/s00142-024-00684-9

Mayer-Schönberger, V., & Cukier, K. (2013). *Big Data: A revolution that will transform how we live, work, and think.* Houghton Mifflin Harcourt.

Mbinjama-Gamatham, A., & Olivier, B. (2020). 'Dark technology', aggressiveness and the question of cyber-ethics. *Acta Academica*, 52(1), 99–120. DOI: 10.18820/24150479/aa52i1/1

McCallister, E., Grance, T., & Scarfone, K. (2010). Guide to protecting the confidentiality of personally identifiable information (PII). Special Publication (NIST SP), National Institute of Standards and Technology. https://tsapps.nist.gov/publication/get_pdf.cfm?pub_id=904990

McCarry, M. (2012). Who benefits? A critical reflection of children and young people's participation in sensitive research. *International Journal of Social Research Methodology*, 15(1), 55–68. DOI: 10.1080/13645579.2011.568196

McCutchan, S. (2022). *Research data governance for global health.* Duke Global Health Institute.

McGraw, D., & Mandl, K. D. (2021). Privacy protections to encourage use of health-relevant digital data in a learning health system. *NPJ Digital Medicine*, 4(1), 2. DOI: 10.1038/s41746-020-00362-8 PMID: 33398052

McKee, K. R. (2024). Human participants in AI research: Ethics and transparency in practice. *IEEE Transactions on Technology and Society*, 5(3), 279–288. DOI: 10.1109/TTS.2024.3446183

Menih, H. (2013). Applying ethical principles in researching a vulnerable population: Homeless women in Brisbane. *Current Issues in Criminal Justice*, 25(1), 527–539. DOI: 10.1080/10345329.2013.12035979

Miller, R. B. (2003). How the Belmont report fails. *Essays in Philosophy*, 4(2), 119–134. DOI: 10.5840/eip2003423

Miller, S., & Bossomaier, T. (2024). *Cybersecurity, ethics, and collective responsibility*. Oxford University Press. DOI: 10.1093/oso/9780190058135.001.0001

Minssen, T., Gerke, S., Aboy, M., Price, N., & Cohen, G. (2020). Regulatory responses to medical machine learning. *Journal of Law and the Biosciences*, 7(1), 1–18. DOI: 10.1093/jlb/lsaa002 PMID: 34221415

Miteu, G. D. (2024). Ethics in scientific research: A lens into its importance, history, and future. *Annals of Medicine and Surgery (London)*, 86(5), 2395–2398. DOI: 10.1097/MS9.0000000000001959 PMID: 38694399

Mogaji, E., & Mogaji, E. (2021). Brand Extension. *Brand Management: An Introduction through Storytelling*, 181-205.

Mohamed, K. (2013). Evaluation of societal marketing (sustainable marketing) in maintaining ethics in marketing goods or services to the customers. *International Journal of Management Research and Reviews*, 3(11), 3703.

Mohan, S., & Huneke, R. (2019). The role of IACUCs in responsible animal research. *ILAR Journal*, 60(1), 43–49. DOI: 10.1093/ilar/ilz016 PMID: 31711237

Mokha, A. K. (2018). Social responsibility and ethics in marketing. *IITM Journal of Management and IT*, 9(2), 50–56.

Morain, S. R., Kraft, S. A., Wilfond, B. S., McGuire, A., Dickert, N. W., Garland, A., & Sugarman, J. (2022). Toward meeting the obligation of *respect for persons* in pragmatic clinical trials. *The Hastings Center Report*, 52(3), 9–17. DOI: 10.1002/hast.1391 PMID: 35763201

Morgan, G., & Gordijn, B. (2020). A care-based stakeholder approach to ethics of cybersecurity in business. *The Ethics of Cybersecurity*, 119-138.

Morley, J., Machado, C. C., Burr, C., Cowls, J., Joshi, I., Taddeo, M., & Floridi, L. (2020). The ethics of AI in health care: A mapping review. *Social Science & Medicine*, 260, 113172. DOI: 10.1016/j.socscimed.2020.113172 PMID: 32702587

Morton, F., & Treviño, T. (2021). Targeting kids in the digital age: The ethics of online marketing towards children. In *Humanistic management in Latin America* (pp. 50–71). Routledge. DOI: 10.4324/9780429351174-4

Morton, L. G.II, & Cunningham-Williams, R. M. (2009). The capacity to give informed consent in a homeless population with developmental disabilities. *Community Mental Health Journal*, 45(5), 341–348. DOI: 10.1007/s10597-009-9184-9 PMID: 19263221

Mozersky, J., Walsh, H., Parsons, M., McIntosh, T., Baldwin, K., & DuBois, J. M. (2020). Are we ready to share qualitative research data? Knowledge and preparedness among qualitative researchers, IRB members, and data repository curators. *IASSIST Quarterly*, 43(4), 1–23. Advance online publication. DOI: 10.29173/iq952 PMID: 32205903

Muckle, J., & Lasikiewicz, N. (2017). An exploration of the benefits of animal-assisted activities in undergraduate students in Singapore. *Asian Journal of Social Psychology*, 20(2), 75–84. DOI: 10.1111/ajsp.12166

Nadeem, W., Tan, T. M., Tajvidi, M., & Hajli, N. (2021). How do experiences enhance brand relationship performance and value co-creation in social commerce? The role of consumer engagement and self brand-connection. *Technological Forecasting and Social Change*, 171, 120952. DOI: 10.1016/j.techfore.2021.120952

Nagai, H., Nakazawa, E., & Akabayashi, A. (2022). The creation of the Belmont Report and its effect on ethical principles: A historical study. *Monash Bioethics Review*, 40(2), 157–170. DOI: 10.1007/s40592-022-00165-5 PMID: 36357708

Najafi, K., Heidarzadeh Hanzaee, K., & Khounsiavash, M. (2021). Exploration of the role of visual merchandising (mannequin) on women's shopping behavior through the moderating role of type of store and customers' intention: An experimental research approach. *Journal of Business and Management*, 13(1), 247–272.

Narayanan, A., & Felten, E. W. (2014). No silver bullet: De-identification still doesn't work. *White Paper, 8*. https://www.cs.princeton.edu/~arvindn/publications/no-silver-bullet-de-identification.pdf

Narvekar, H. N., & Ghani, S. (2022). Animal-Assisted Therapy for the Promotion of Social Competence: a Conceptual Framework. *Human Arenas*, 1-10.

National Commission for the Protection of Human Subjects in Biomedical and Behavioral Research. (1979). *The Belmont report: Ethical principles and guidelines for the protection of human subjects in biomedical and behavioral research.* U.S. Department of Health and Human Services. https://www.hhs.gov/ohrp/regulations-and-policy/belmont-report

National Commission for the Protection of Human Subjects of Biomedical and Behavioral Research. (1978). *The Belmont Report: Ethical principles and guidelines for the protection of human subjects of research.* https://www.hhs.gov/ohrp/sites/default/files/ the-belmont-report-508c_FINAL.pdf

National Commission for the Protection of Human Subjects of Biomedical and Behavioral Research. (1979). *The Belmont Report: Ethical principles and guidelines for the protection of human subjects of research.* U.S. Department of Health and Human Services. https://www.hhs.gov/ohrp/regulations-andpolicy/belmont-report/read-the-belmont-report/index.html

National Institute of Standards and Technology. (n.d.). Information security. https://csrc.nist.gov/glossary/term/information_security

National Institutes of Health (US). Office for Protection from Research Risks. (1986). Public Health Service policy on humane care and use of laboratory animals. Office for Protection from Research Risks (OPRR), National Institutes of Health.

Nazarenko, T. (2023). Strategic resources for Digital transformation of Economic activity and Public administration in the field of Landscaping. *Artificial societies.* .DOI: 10.18254/S207751800025960-5

Nebeker, C., Torous, J., & Bartlett Ellis, R. J. (2019). Building the case for actionable ethics in digital health research supported by artificial intelligence. *BMC Medicine*, 17(1), 1–7. DOI: 10.1186/s12916-019-1377-7 PMID: 31311535

Negri, F., & Ieva, M. (2022). The interplay between privacy failure, recovery and crisis communication management: An integrative review and research agenda. *Sinergie Italian Journal of Management*, 40(1), 77–101. DOI: 10.7433/s117.2022.05

Nerenz, D. R., Stoltz, P. K., & Jordan, J. (2003). Quality improvement and the need for IRB review. *Quality Management in Health Care*, 12(3), 159–170. DOI: 10.1097/00019514-200307000-00006 PMID: 12891960

Newcomer, C. E., & Greer, W. (2000). General composition of the IACUC and specific roles of the IACUC members. *The IACUC handbook.*

Ngiam, K. Y., & Khor, W. (2019). Big data and machine learning algorithms for health-care delivery. *The Lancet Oncology*, 20(5), e262–e273. DOI: 10.1016/S1470-2045(19)30149-4 PMID: 31044724

Ng, Z. (2021). Strategies to assessing and enhancing animal welfare in animal-assisted interventions. In *The welfare of animals in animal-assisted interventions: Foundations and best practice methods* (pp. 123–154). Springer International Publishing. DOI: 10.1007/978-3-030-69587-3_6

Ng, Z., Morse, L., Albright, J., Viera, A., & Souza, M. (2019). Describing the use of animals in animal-assisted intervention research. *Journal of Applied Animal Welfare Science*, 22(4), 364–376. DOI: 10.1080/10888705.2018.1524765 PMID: 30295540

Nicolaides, A. (2016). Bioethical considerations, the common good approach and some shortfalls of the Belmont Report. *Medical Technology SA*, 30(1), 15–24.

Nii Laryeafio, M., & Ogbewe, O. C. (2023). Ethical consideration dilemma: Systematic review of ethics in qualitative data collection through interviews. *Journal of Ethics in Entrepreneurship and Technology*, 3(2), 94–110. DOI: 10.1108/JEET-09-2022-0014

Nill, A., & Laczniak, G. (2022). Direct-to-consumer genetic testing and its marketing: Emergent ethical and public policy implications. *Journal of Business Ethics*, 175(4), 669–688. DOI: 10.1007/s10551-020-04632-z

Nucera, S., Tartarisco, G., Epasto, A., Smeriglio, D., Mazzeo, A., Pioggia, G., & Anastasi, A. (2018). Ubiquitous, wearable, mobile: Paradigm shifts in e-learning and diffusion of knowledge. In *Handbook of research on mobile devices and smart gadgets in K-12 education* (pp. 286–307). IGI Global. DOI: 10.4018/978-1-5225-2706-0.ch018

Nuremburg Code. (1949). *Trials of war criminals before the Nuremburg Military Tribunals under Control Council Law No. 10.* (Vol. 2); 181-182. U.S. Government Printing Office. https://history.nih.gov/display/history/Nuremburg+Code Office of Civil Rights. (1996). Health Insurance Portability and Accountability Act. https://www.hhs.gov/hipaa/for-professionals/privacy/laws-regulations/index.html

Nwokeji, J., & Iwanenko, W. (2021). Requirements analysis and design of IRB system. *American Journal of Operations Management and Information Systems*, 6(4), 75–85. DOI: 10.11648/j.ajomis.20210604.12

Obermeyer, Z., Powers, B., Vogeli, C., & Mullainathan, S. (2019). Dissecting racial bias in an algorithm used to manage the health of populations. *Science*, 366(6464), 447–453. DOI: 10.1126/science.aax2342 PMID: 31649194

Office for Human Research (OHRP). (2018, January 15). *Read the Belmont Report [Text].* OHRP. https://www.hhs.gov/ohrp/regulations-and-policy/belmont-report/read-the-belmont-report/index.html

Office for Human Research Protections (OHRP). (2019, September 19). Privacy and health research in a data-driven world [Workshop]. https://www.hhs.gov/ohrp/sites/default/files/report-privacy-and-health-10-31-19.pdf

Office for Human Research Protections. (2022, October 17). *The Belmont Report: Ethical principles and guidelines for the protection of human subjects of research.* U.S. Department of Health & Human Services. https://www.hhs.gov/ohrp/regulations-and-policy/belmont-report/index.html

Office of Human Research Protection. (2024). *Participant-centered informed consent training video* [Video]. U.S. Department of Health and Human Services. https://www.hhs.gov/ohrp/education-and-outreach/human-research-protection- training/participant-centered-informed-consent-training/index.html

Office of Human Subjects Research Protections. (n.d.). *Understanding and submitting for an exempt determination*. National Institutes of Health.

Onyancha, O. B. (2015). An informetrics view of the relationship between internet ethics, computer ethics and cyberethics. *Library Hi Tech*, 33(3), 387–408. DOI: 10.1108/LHT-04-2015-0033

Open Data Institute. (2022, March 1). *Health data governance: A playbook for non-technical leaders*. https://open-data-institute.gitbook.io/data-governance-playbook/

Opsal, T., Wolgemuth, J., Cross, J., Kaanta, T., Dickmann, E., Colomer, S., & Erdil-Moody, Z. (2016). (2016). "There are no known benefits. . ." Considering the risk/benefit ratio of qualitative research. *Qualitative Health Research*, 26(8), 1137–1150. DOI: 10.1177/1049732315580109 PMID: 25857654

Oravec, J. A. (2018). "Don't be evil" and beyond for high tech organizations: Ethical statements and mottos (and responsibility). In *Handbook of research on civic engagement and social change in contemporary society* (pp. 220–237). IGI Global. DOI: 10.4018/978-1-5225-4197-4.ch013

Palad, J. (2023). Strategies for Improving Organizational Efficiency, Productivity, and Performance through Technology Adoption. *Journal of Management and Administration Provision*. .DOI: 10.55885/jmap.v2i3.230

Palley, L. S., O'Rourke, P. P., & Niemi, S. M. (2010). Mainstreaming animal-assisted therapy. *ILAR Journal*, 51(3), 199–207. DOI: 10.1093/ilar.51.3.199 PMID: 21131720

Palmer, B. W., & Harmell, A. L. (2016). Assessment of Healthcare Decision-making Capacity. *Archives of Clinical Neuropsychology*, 31(6), 530–540. DOI: 10.1093/arclin/acw051 PMID: 27551024

Pandya, M., & Desai, C. (2013). Compensation in clinical research: The debate continues. *Perspectives in Clinical Research*, 4(1), 70–74. DOI: 10.4103/2229-3485.106394 PMID: 23533986

Panicker, S., & Huh-Yoo, J. (2022, October 18). Research ethics and oversight in the digital age. PRIM&R Webinar. https://www.pathlms.com/primr/courses/47358

Paraskevas, A. (2020). Human trafficking and modern slavery in Europe's hotels. In *Slavery and liberation in hotels, restaurants and bars* (pp. 70–86). Routledge. DOI: 10.4324/9780367855383-6

Parthasarathy, S., Panigrahi, P. K., & Subramanian, G. H. (2024). A framework for managing ethics in data science projects. *Engineering Reports*, 6(3), e12722. DOI: 10.1002/eng2.12722

Pater, J., Fiesler, C., & Zimmer, M. (2022). No humans here: Ethical speculation on public data, unintended consequences, and the limits of institutional review. *Proceedings of the ACM on Human-Computer Interaction, 6*(GROUP), 1-13. DOI: 10.1145/3492857

Pattison, J. (2020). From defence to offence: The ethics of private cybersecurity. *European Journal of International Security*, 5(2), 233–254. DOI: 10.1017/eis.2020.6

Pecora, N. (2022). Media policies for children: Issues and histories in the US. In *The Routledge International Handbook of children, adolescents, and media* (pp. 425–431). Routledge. DOI: 10.4324/9781003118824-54

Peled-Raz, M., Tzafrir, S. S., Enosh, G., Efron, Y., & Doron, I. (2021). Ethics review boards for research with human participants: Past, present, and future. *Qualitative Health Research*, 31(3), 590–599. DOI: 10.1177/1049732320972333 PMID: 33213280

Pendergraft, G. (2019). Controversy over gender differences and free speech at Google. *Sage (Atlanta, Ga.)*. Advance online publication. DOI: 10.4135/9781526488664

Perlman, D. (2006). Putting the" ethics" back into research ethics: A process for ethical reflection for human research protection. *The Journal of Research Administration*, 37(1/2), 13.

Perneger, T. V. (2004). Why we need ethical oversight of quality improvement projects. *International Journal for Quality in Health Care : Journal of the International Society for Quality in Health Care*, 16(5), 343–344. DOI: 10.1093/intqhc/mzh075 PMID: 15375092

Petermann, M., Tempini, N., Kherroubi Garcia, I., Whitaker, K., & Strait, A. (2022). *Looking before we leap: Expanding ethical review processes for AI and data science research.* Ada Lovelace Institute.

Peterson, M., Minton, E. A., Liu, R. L., & Bartholomew, D. E. (2021). Sustainable marketing and consumer support for sustainable businesses. *Sustainable Production and Consumption*, 27, 157–168. DOI: 10.1016/j.spc.2020.10.018

Piasecki, J., & Cheah, P. Y. (2022). Ownership of individual-level health data, data sharing, and data governance. *BMC Medical Ethics*, 23(1), 104. DOI: 10.1186/s12910-022-00848-y PMID: 36309719

Pietrzykowski, T., & Smilowska, K. (2021). The reality of informed consent: Empirical studies on patient comprehension – Systematic review. *Trials*, 22(1), 57. Advance online publication. DOI: 10.1186/s13063-020-04969-w PMID: 33446265

Pike, E. R. (2019). Defending data: Toward ethical protections and comprehensive data governance. *Emory Law Journal*, 69, 687.

Pinillos, R. G., Appleby, M. C., Manteca, X., Scott-Park, F., Smith, C., & Velarde, A. (2016). One Welfare-a platform for improving human and animal welfare. *The Veterinary Record*, 179(16), 412–413. DOI: 10.1136/vr.i5470 PMID: 27770094

Pirani, S. (2024). Navigating research ethics: Strategies for preventing and addressing research misconduct. *International Journal of Multidisciplinary Research & Reviews*, 3(02), 96–104. DOI: 10.56815/IJMRR.V3I2.2024/96-104

Plattner, H., & Leukert, B. (2015). Quality time at work. In *The In-memory revolution* (pp. 222–242). Springer. DOI: 10.1007/978-3-319-16673-5_10

Pousti, H., Urquhart, C., & Linger, H. (2021). Researching the virtual: A framework for reflexivity in qualitative social media research. *Information Systems Journal*, 31(3), 356–383. DOI: 10.1111/isj.12314

Pradeep, R., & Deora, K. [REMOVED HYPERLINK FIELD]Danca, E., Young, T., Mott, K., Chezik, A., Rocca, N., & Kudelka, A. P. (August 8-10, 2024). *Unveiling the impact: Exploring malnutrition rates in patients with cancer and its association with anticancer treatments* [poster]. ASCO Breakthrough Meeting, Yokohama, Japan. https://conferences.asco.org/breakthrough/welcome

Prainsack, B., & Buyx, A. (2012). Solidarity in contemporary bioethics—Towards a new approach. *Bioethics*, 26(7), 343–350. DOI: 10.1111/j.1467-8519.2012.01987.x PMID: 22827315

Prainsack, B., El-Sayed, S., Forgó, N., Szoszkiewicz, Ł., & Baumer, P. (2022). Data solidarity: A blueprint for governing health futures. *The Lancet Digital Health*, 4(11), e773–e774. DOI: 10.1016/S2589-7500(22)00189-3 PMID: 36307191

Prentice, E. D., Oki, G. S., & Mann, M. D. (2014). General concepts of protocol review. *The IACUC Handbook*, 107-147. CRC Press.

Preston, C. (2016). Pre-school children and marketing communications. *International Journal of Consumer Studies*, 40(5), 618–623. DOI: 10.1111/ijcs.12307

Pritt, S. L., & Hammer, R. E. (2017). The interplay of ethics, animal welfare, and IACUC oversight on the reproducibility of animal studies. *Comparative Medicine*, 67(2), 101–105. PMID: 28381309

Protection of Human Subjects, 45 CFR §46.102(e)(5) (2018).

Protection of Human Subjects. 45 C.F.R. § 46 (2009). http://www.hhs.gov/ohrp/regulations-and-policy/regulations/regulatory-text/index.html

Protection of Human Subjects. 45 C.F.R. § 46 (2018). http://www.hhs.gov/ohrp/regulations-and-policy/regulations/45-cfr-46/revised-common-rule-text/index.html

Quach, S., Thaichon, P., Martin, K. D., Weaven, S., & Palmatier, R. W. (2022). Digital technologies: Tensions in privacy and data. *Journal of the Academy of Marketing Science*, 50(6), 1–25. DOI: 10.1007/s11747-022-00845-y PMID: 35281634

Rafi, N., & Snyder, B. (2015). *Ethics in research: How to collect data ethically. Doing Research in Emergency and Acute Care: Making Order Out of Chaos*. Wiley.

Råheim, M., Magnussen, L. H., Sekse, R. J. T., Lunde, Å., Jacobsen, T., & Blystad, A. (2016). Researcher-researched relationship in qualitative research: Shifts in positions and researcher vulnerability. *International Journal of Qualitative Studies on Health and Well-being*, 11(1), 30996–12. DOI: 10.3402/qhw.v11.30996 PMID: 27307132

Rajasekharaiah, K. M., Dule, C. S., & Sudarshan, E. (2020). Cyber security challenges and its emerging trends on latest technologies. [). IOP Publishing.]. *IOP Conference Series. Materials Science and Engineering*, 981(2), 022062. DOI: 10.1088/1757-899X/981/2/022062

Raman, R., Utts, J., Cohen, A. I., & Hayat, M. J. (2023). Integrating ethics into the guidelines for assessment and instruction in statistics education (GAISE). *The American Statistician*, 77(3), 1–8. DOI: 10.1080/00031305.2022.2156612

Ramírez, K. (2024). Ethical principles in dental research of behavioral and social factors. *Odovtos-International Journal of Dental Sciences*, 20-25. DOI: 10.15517/ijds.2024.59408

Raul, A. C. (Ed.). (2021). *The privacy, data protection and cybersecurity law review*. Law Business Research Limited.

Rawat, S. R., Bhatia, K., Hegde, M., Bhat, N., & Tewari, S. (2015). The importance of ethical marketing practices. *Journal of Business Management and Economics*, 3(2), 23–30.

Raymond, N. (2019). Reboot ethical review for the age of big data. *Nature*, 568(7752), 277. DOI: 10.1038/d41586-019-01164-z PMID: 30988466

Reddy, P., Chandu, V., Ryali, N., Gangula, A., Mishra, D., & Mishra, S. (2023). Ethical Business Marketing: The Practices to Compete Successfully in Business. In *Data-Driven Approaches for Effective Managerial Decision Making* (pp. 259-276). IGI Global.

Reddy, S., Allan, S., Coghlan, S., & Cooper, P. (2020). A governance model for the application of AI in health care. *Journal of the American Medical Informatics Association : JAMIA*, 27(3), 491–497. DOI: 10.1093/jamia/ocz192 PMID: 31682262

Redmiles, E. M. (2022). The need for respectful technologies: Going beyond privacy. In Werthner, H., Prem, E., Lee, E. A., & Ghezzi, C. (Eds.), *Perspectives on Digital Humanism* (pp. 309–313). Springer. DOI: 10.1007/978-3-030-86144-5_42

Reinhardt, A., & Ray, L. (2003). Differentiating quality improvement from research. *Applied Nursing Research*, 16(1), 2–8. DOI: 10.1053/apnr.2003.59000 PMID: 12624857

Reisig, M. D., Flippin, M., & Holtfreter, K. (2022). Toward the development of a perceived IRB violation scale. *Accountability in Research*, 29(5), 309–323. DOI: 10.1080/08989621.2021.1920408 PMID: 33877941

Resnik, D. B. (2012). *Environmental health ethics*. Cambridge University Press. DOI: 10.1017/CBO9781139161848

Resnik, D. B., Antes, A., & Mozersky, J. (2024). Should researchers destroy audio or video recordings? *Ethics & Human Research*, 46(2), 30–35. DOI: 10.1002/eahr.500205 PMID: 38446098

Rezníčková, M., & Zaušková, A. (2020, July). Use of Social Media in the Presentation of Eco-Innovations of Slovak Businesses. In *7th European Conference on Social Media ECSM 2020* (p. 232).

Richards, D., Formosa, P., Ryan, M., Hitchens, M., & McEwan, M. (2020). A proposed AI-enhanced serious game for cybersecurity ethics training. In *Conference of the Australasian Institute of Computer Ethics (9th: 2020)* (pp. 1-9). Australasian Institute of Computer Ethics (AiCE).

Ritchie, K. L. (2021). Using IRB protocols to teach ethical principles for research and everyday life: A high-impact practice. *The Journal of Scholarship of Teaching and Learning*, 21(1), 120–130. DOI: 10.14434/josotl.v21i1.30554

Rochel, J. (2023). Learning from the ethics of AI - A research proposal on soft law and ethics of AI. *Tilburg Law Review*, 27(1). Advance online publication. DOI: 10.5334/tilr.297

Rodwin, M. A. (2009). The Case for Public Ownership of Patient Data. *Journal of the American Medical Association*, 302(1), 86. DOI: 10.1001/jama.2009.965 PMID: 19567445

Rosenfeld, S. J., Shaler, G., & Hickey, R. (2023). The Harraseeket Conference–Revisiting systems for ethics oversight of research with human participants. *Research Ethics*, 19(3), 231–249. DOI: 10.1177/17470161231157053

Rossi, A., Arenas, M. P., Kocyigit, E., & Hani, M. (2022, June). Challenges of protecting confidentiality in social media data and their ethical import. In *2022 IEEE European Symposium on Security and Privacy Workshops (EuroS&PW)* (pp. 554-561). IEEE. DOI: 10.1109/EuroSPW55150.2022.00066

Ross, M. W., Iguchi, M. Y., & Panicker, S. (2018). Ethical aspects of data sharing and research participant protections. *The American Psychologist*, 73(2), 138–145. DOI: 10.1037/amp0000240 PMID: 29481107

Sakka, G., & Spyrou, I. (2015). CyberEthics Case Study. In *Human Rights and Ethics: Concepts, Methodologies, Tools, and Applications* (pp. 646-658). IGI Global. DOI: 10.4018/978-1-4666-6433-3.ch036

Salazar, T. (2023). A new frontier: Human subject research ethics in an artificial intelligence world. *The Health Lawyer*, 35(4), 23–52.

Samuel, G., Chubb, J., & Derrick, G. (2021). Boundaries between research ethics and ethical research use in artificial intelligence health research. *Journal of Empirical Research on Human Research Ethics; JERHRE*, 16(3), 325–337. DOI: 10.1177/15562646211002744 PMID: 33733915

Samuel, G., Derrick, G. E., & Van Leeuwen, T. (2019). The ethics ecosystem: Personal ethics, network governance and regulating actors governing the use of social media research data. *Minerva*, 57(3), 317–343. DOI: 10.1007/s11024-019-09368-3 PMID: 31501635

Sánchez Abril, P., Oliva Blázquez, F., & Martínez Evora, J. (2018). The Right of Withdrawal in Consumer Contracts: A comparative analysis of American and European law. *InDret*, 3, 18–13.

Santaniello, A., Dicé, F., Claudia Carratú, R., Amato, A., Fioretti, A., & Menna, L. F. (2020). Methodological and terminological issues in animal-assisted interventions: An umbrella review of systematic reviews. *Animals*, 10(5), 759. DOI: 10.3390/ani10050759 PMID: 32349351

Saritas, O., Bakhtin, P., Kuzminov, I., & Khabirova, E. (2021). Big data augmentated business trend identification: The case of mobile commerce. *Scientometrics*, 126(2), 1553–1579. DOI: 10.1007/s11192-020-03807-9 PMID: 33424052

Sattar, M. M., Barr, J. C., & Sfodera, F. (2023). Cause-Related Marketing. In *Cultural Marketing and Metaverse for Consumer Engagement* (pp. 71–95). IGI Global. DOI: 10.4018/978-1-6684-8312-1.ch006

Saura, J. R., Palacios-Marqués, D., & Ribeiro-Soriano, D. (2021). Using data mining techniques to explore security issues in smart living environments in Twitter. *Computer Communications*, 179, 285–295. DOI: 10.1016/j.comcom.2021.08.021

Saura, J. R., Ribeiro-Soriano, D., & Palacios-Marqués, D. (2022). Assessing behavioral data science privacy issues in government artificial intelligence deployment. *Government Information Quarterly*, 39(4), 101679. DOI: 10.1016/j.giq.2022.101679

Sawka, K. (2020). The use and misuse of statistics. In Gruber, C. W. (Ed.), *The Theory of Statistics in Psychology* (pp. 95–110). Springer., DOI: 10.1007/978-3-030-48043-1_7

Saygılı, E., Ozturkoglu, Y., & Ozturkoglu, O. (2022). Corporate responsibilities and values in codes of ethics: An exploratory study of the Global Fortune companies. *European Journal of International Management*, 18(1), 76–101. DOI: 10.1504/EJIM.2022.123752

Schlehahn, E. (2020). Cybersecurity and the State. *The Ethics of Cybersecurity*, 205-225. Springer Open.

Schneble Butz, C. O. (2024). *Ethical and legal perspectives on big data in research* (Doctoral dissertation, University of Basel).

Schultz, R. A. (Ed.). (2005). *Contemporary issues in ethics and information technology*. IGI Global.

Schupmann, W., & Moreno, J. D. (2020). Belmont in context. *Perspectives in Biology and Medicine*, 63(2), 220–239. DOI: 10.1353/pbm.2020.0028 PMID: 33416649

Serpico, K. (2021). Making metrics meaningful: How human research protection programs can efficiently and effectively use their data. *Ethics & Human Research*, 43(5), 26–35. DOI: 10.1002/eahr.500102 PMID: 34496157

Serpico, K. (2024). Defining institutional review board application quality: Critical research gaps and future opportunities. *Research Ethics*, 20(1), 19–35. DOI: 10.1177/17470161231187099

Shahbazi Takabi, A., Amiri, H., & Karami, R. (2024). The Effect of Ethical Marketing in Auditing on Auditors' Job Performance. *International Journal of Finance & Managerial Accounting*, 9(35), 43–64.

Sharan, R., & Boruah, B. (2016). Ethical Concerns of Human-Being, Cyber-Being and Cybertariat: An Educational Perspective. *International Journal of Information Ethics*, 25, 25. DOI: 10.29173/irie107

Shaw, I. (2008). Ethics and the practice of qualitative research. *Qualitative Social Work: Research and Practice*, 7(4), 400–414. DOI: 10.1177/1473325008097137

Shaw, I. (2023). *Research and social work in time and place: Crossing boundaries* (1st ed.). Routledge., DOI: 10.4324/9781003306740

Shearer, E., Cho, M., & Magnus, D. (2021). Regulatory, social, ethical, and legal issues of artificial intelligence in medicine. In *Artificial intelligence in medicine* (pp. 457–477). Academic Press. DOI: 10.1016/B978-0-12-821259-2.00023-5

Shin, S. Y. (2021). Privacy protection and data utilization. *Healthcare Informatics Research*, 27(1), 1–2. DOI: 10.4258/hir.2021.27.1.1 PMID: 33611870

Shore, N. (2006). Re-conceptualizing the Belmont Report: A community-based participatory research perspective. *Journal of Community Practice*, 14(4), 5–26. DOI: 10.1300/J125v14n04_02

Shou, D. (2012). Ethical considerations of sharing data for cybersecurity research. In *Financial Cryptography and Data Security: FC 2011 Workshops* (pp. 169–177). Springer Berlin Heidelberg. DOI: 10.1007/978-3-642-29889-9_15

Shukla, S., George, J. P., Tiwari, K., & Kureethara, J. V. (2022). Data security. In *Data Ethics and Challenges* (pp. 41–59). Springer Singapore. DOI: 10.1007/978-981-19-0752-4_3

Shulman, Y., Kitkowska, A., & Meyer, J. (2022, July 5). *Informing users: Effects of notification properties and user characteristics on sharing attitudes*. ArXiv.Org. DOI: 10.1080/10447318.2022.2086592

Shykhnenko, K., & Nozhovnik, O. (2023). Innovations in professional training of research administrators in the USA and Ukraine: A technological approach. *Education. Innovation.Practice*, 11(9), 71–78. Advance online publication. DOI: 10.31110/2616-650X-vol11i9-010

Siddiqui, W., & Sharp, R. R. (2021). Beyond the Belmont Report. *The American Journal of Bioethics*, 21(10), 1–4. DOI: 10.1080/15265161.2021.1972649 PMID: 34554060

Silverman, J. (2015). The institutional animal care and use committee. In *Research Regulatory Compliance* (pp. 41–78). Academic Press. DOI: 10.1016/B978-0-12-420058-6.00003-4

Silverman, J., Suckow, M. A., & Murthy, S. (2014). *The IACUC handbook*. CRC Press. DOI: 10.1201/b17109

Singh, G., & Mishra, A. (2019). Ethics and Celebrity Advertising: Cases in the Indian Advertising Industry. In *Ethical Branding and Marketing* (pp. 173-184). Routledge.

Skorburg, J., & Friesen, P. (2020). Ethical issues in text mining for mental health. In *The atlas of language analysis in psychology* (in press). Guilford.

Slepchuk, A. N., Milne, G. R., & Swani, K. (2022). Overcoming privacy concerns in consumers' use of health information technologies: A justice framework. *Journal of Business Research*, 141, 782–793. DOI: 10.1016/j.jbusres.2021.11.073

Slovin, L. J., & Semenec, P. (2019). Thinking/writing within and outside the IRB box: Ethical disruptions of data in qualitative research. *Reconceptualizing Educational Research Methodology*, 10(1), 14–27. DOI: 10.7577/rerm.3241

Smale, M. A. (2010). Demystifying the IRB: Human subjects research in academic libraries. *portal. Portal (Baltimore, Md.)*, 10(3), 309–321. DOI: 10.1353/pla.0.0114

Smith, A. J., Clutton, R. E., Lilley, E., Hansen, K. E. A., & Brattelid, T. (2018). PREPARE: Guidelines for planning animal research and testing. *Laboratory Animals*, 52(2), 135–141. DOI: 10.1177/0023677217724823 PMID: 28771074

Soini, S. (2020). The GDPR, secondary research purposes and global data sharing—One-wheel too many. *European Journal of Human Genetics, 28*(6), 694. doi:DOI: 10.103841431-020-0608-x

Sontan, A. D., & Samuel, S. V.Adewale Daniel SontanSegun Victor Samuel. (2024). The Intersection of Artificial Intelligence and Cybersecurity: Challenges and Opportunities. *World Journal of Advanced Research and Reviews*, 21(2), 1720–1736. DOI: 10.30574/wjarr.2024.21.2.0607

Spinello, R. (2010). *Cyberethics: Morality and law in cyberspace*. Jones & Bartlett Publishers.

Stegenga, S. M., Steltenpohl, C. N., Lustick, H., Meyer, M. S., Renbarger, R., Standiford Reyes, L., & Lee, L. E. (2024). Qualitative research at the crossroads of open science and big data: Ethical considerations. *Social and Personality Psychology Compass*, 18(1), e12912. DOI: 10.1111/spc3.12912

Stephenson, N., Smith, C., & Vajdic, C. M. (2022). Health and public sector data sharing requires social licence negotiations. *Australian and New Zealand Journal of Public Health*, 46(4), 426–428. DOI: 10.1111/1753-6405.13261 PMID: 35679055

Stern, C., & Chur-Hansen, A. (2013). Methodological considerations in designing and evaluating animal-assisted interventions. *Animals (Basel)*, 3(1), 127–141. DOI: 10.3390/ani3010127 PMID: 26487313

Stevens, D. D., & Caskey, M. M. (2022). Building a foundation for a successful doctoral student journey: A scholarship of teaching and learning investigation. Innovative Higher Education, 1–23. PMID:36467630 14 Graduate Student Investigator.

Stewart, R. (2021). Big data and Belmont: On the ethics and research implications of consumer-based datasets. *Big Data & Society*, 8(2). Advance online publication. DOI: 10.1177/20539517211048183

Stiegler, M. P., & Tung, A. (2017). Is it quality improvement or is it research? Ethical and regulatory considerations. *Anesthesia and Analgesia*, 125(1), 342–344. DOI: 10.1213/ANE.0000000000001815 PMID: 28207587

Stoudt, S., Jernite, Y., Marshall, B., Marwick, B., Sharan, M., Whitaker, K., & Danchev, V. (2024). Ten simple rules for building and maintaining a responsible data science workflow. *PLoS Computational Biology*, 20(7), 1–19. DOI: 10.1371/journal.pcbi.1012232 PMID: 39024267

Surmiak, A., & Męcfal, S. (2024). Ethical challenges in contemporary social research [editorial]. *Diametros: a Journal of Philosophy*, 1–6. DOI: 10.33392/diam.1979

Tangudu, A., Jain, S., & Aggarwal, A. (2024). Best Practices for Ensuring Salesforce Application Security and Compliance. *Journal of Quantum Science and Technology*, 1(2), 88–101. DOI: 10.36676/jqst.v1.i2.18

Taquette, S. R., & Borges da Matta Souza, L. M. (2022). Ethical dilemmas in qualitative research: A critical literature review. *International Journal of Qualitative Methods*, •••, 21.

Taylor, C. R., Cho, Y. N., Anthony, C. M., & Smith, D. B. (2018). Photoshopping of models in advertising: A review of the literature and future research agenda. *Journal of Global Fashion Marketing*, 9(4), 379–398. DOI: 10.1080/20932685.2018.1511380

Taylor, L. (2017). What is data justice? The case for connecting digital rights and freedoms globally. *Big Data & Society*, 4(2), 2053951717736335. DOI: 10.1177/2053951717736335

The Secretary's Advisory Committee on Human Research Protections (SACHRP). (2022, July 21). Considerations for IRB review of research involving artificial intelligence. https://www.hhs.gov/ohrp/sachrp-committee/recommendations/attachment-e-july-25-2022-letter/index.html

The United States Federal Trade Commission. (2023, May 31). *FTC Says Ring employees illegally surveilled customers, failed to stop hackers from taking control of users' cameras*. Federal Trade Commission. https://www.ftc.gov/news-events/news/press-releases/2023/05/ftc-says-ring-employees-illegally-surveilled-customers-failed-stop-hackers-taking-control-users

Throne, R. (2023, August 28). Belmont Justice and fair, equitable, and inclusive research recruitment. Ampersand. https://blog.primr.org/belmont-justice-and-fairequitable-and-inclusive-research-recruitment/

Throne, R., & Hendon, M. (2024). Belmont 2.0: Ethical research expectations for PII in AI, ML, and data mining/scraping. In *Methodologies and ethics for social sciences research* (pp. 1-19). IGI Global.

Throne, R. (2022). Adverse trends in data ethics: The AI Bill of Rights and human subjects protections. Information Policy & Ethics eJournal. doi:DOI: 10.2139/ssrn.4279922

Throne, R., & Hendon, M. (2024). Belmont 2.0: Ethical research expectations for PII in AI, ML, and data mining/scraping. In Demircioğlu, A. (Ed.), (pp. 1–19). Advances in information quality and management. IGI Global., DOI: 10.4018/979-8-3693-1726-6.ch001

Throne, R., Hendon, M., & Kozinski, J. (2023). Graduate student investigator: Best practices for human research protections within online graduate research. In Walters, K. (Ed.), *Dynamic curriculum development and design strategies for effective online learning in higher education*. IGI Global. DOI: 10.4018/978-1-6684-8646-7.ch001

Tiwari, A. (2023). Regulatory consideration in AI-driven healthcare. *FOCUS WTO*, 25(4), 23–31.

Traianou, A., & Hammersley, M. (2024). Interrogating the concept of vulnerability in social research ethics. *Diametros : a Journal of Philosophy*, 1–16. Advance online publication. DOI: 10.33392/diam.1891

Trajkovski, G. (2023). *A more beautiful prompt: The art of human-centered prompt engineering* (1st ed., Kindle ed.). The Art of Living Large. ASIN: B0CJQZ68TM. https://beautifulprompt.expert

Trajkovski, G. (2024). Bridging the public administration-AI divide: A skills perspective. *Public Administration and Development*, 1–15. DOI: 10.1002/pad.2061

Tripathi, S., & Sharma, D. (2023). Green Marketing: Sustainable Strategies for Success. In *Climate Change Management and Social Innovations for Sustainable Global Organization* (pp. 76-92). IGI Global.

Tuhiwai-Smith, L. (2021). *Decolonizing methodologies: Research and indigenous peoples* (3rd ed.). Bloomsbury., DOI: 10.5040/9781350225282

U.S. Department of Health and Human Services [HHS]. (2022, October 25). Guidance regarding methods for de-identification of protected health information in accordance with the Health Insurance Portability and Accountability Act (HIPAA) Privacy Rule. Health Information Privacy. https://www.hhs.gov/hipaa/for-professionals/privacy/special-topics/de-identification/index.html

U.S. Department of Health and Human Services. (2017). Research. https://www.hhs.gov/hipaa/for-professionals/special-topics/research/index.html

U.S. Department of Health and Human Services. (2021). Exemptions (2018 requirements). Office for Human Research Protections. https://www.hhs.gov/ohrp/regulations-and-policy/regulations/45-cfr-46/common-rule-subpart-a-46104/index.html

U.S. Department of Health and Human Services. (n.d.). Quality improvement activities FAQs. https://www.hhs.gov/ohrp/regulations-and-policy/guidance/faq/quality-improvement-activities/index.html

U.S. Department of Homeland Security (DHS), Science and Technology, Cyber Security Division (CSD). (2012). *The Menlo Report: Ethical principles guiding information and communication technology research (No. 249;* p. 17). DHS. https://www.dhs.gov/sites/.../CSD-MenloPrinciplesCORE-20120803_1.pdf

U.S. Department of Homeland Security (DHS), Science and Technology, Cyber Security Division (CSD). (2012). The Menlo Report: Ethical principles guiding information and communication technology research (No. 249; p. 17). DHS. https://www.dhs.gov/sites/.../CSD-MenloPrinciplesCORE-20120803_1.pdf

U.S. Food and Drug Administration. (1998, January). Recruiting study subjects. https://www.fda.gov/regulatory-information/search-fda-guidance-documents/recruiting-study-subjects

U.S. Food and Drug Administration. (2021, January 29). Institutional Review Boards frequently asked questions. U.S. Department of Health and Human Services.

U.S. Food and Drug Administration. [FDA]. (2018, December 4). Software as a medical device (SaMD). https://www.fda.gov/medical-devices/digital-health-center-excellence/software-medical-device-samd

U.S. Food and Drug Administration. [FDA]. (2021, October 27). Good machine learning practice for medical device development: Guiding principles https://www.fda.gov/medical-devices/software-medical-device-samd/good-machine-learning-practice-medical-device-development-guiding-principles

U.S. Food and Drug Administration. [FDA]. (2024, March 15). Artificial intelligence and medical products: How CBER, CDER, CDRH, and OCP are working together. https://www.fda.gov/media/177030/download?attachment

U.S. Food and Drug Administration. [FDA]. (2024, May 13). Artificial intelligence and machine learning (AI/ML)-enabled medical devices. https://www.fda.gov/medical-devices/software-medical-device-samd/artificial-intelligence-and-machine-learning-aiml-enabled-medical-devices

Udegbe, F. C., Ebulue, O. R., Ebulue, C. C., & Ekesiobi, C. S.Francisca Chibugo UdegbeOgochukwu Roseline EbulueCharles Chukwudalu EbulueChukwunonso Sylvester Ekesiobi. (2024). The role of artificial intelligence in healthcare: A systematic review of applications and challenges. *International Medical Science Research Journal*, 4(4), 500–508. DOI: 10.51594/imsrj.v4i4.1052

Uğurlu, K. (2023). Personal Selling in Accommodation Businesses. In *International Handbook of Skill, Education, Learning, and Research Development in Tourism and Hospitality* (pp. 1-18). Springer Nature Singapore.

United States Office of Management and Budget (OMB). (2007). *OMB circular M-07-1616 compliance supplement* [Washington, D.C.]. Executive Office of the President, Office of Management and Budget.

United States Office of Management and Budget (OMB). (2019). *OMB circular M-10-23 compliance supplement* [Washington, D.C.]. Executive Office of the President, Office of Management and Budget.

Valado, T., & Amster, R. (2012). *Professional lives, personal struggles: ethics and advocacy in research on homelessness* (1st ed.). Rowman & Littlefield.

Valiyamattam, G., Yamamoto, M., Fanucchi, L., & Wang, F. (2018). Multicultural considerations in animal-assisted intervention. *Human Animal Interactions Bulletin*, hai.2018.0019. DOI: 10.1079/hai.2018.0019

Vallejo, M., Ramirez, M., Gurumendi, I., & Perdomo, T. (2020). Importance of research and drafting in the scientific production of the health area. *International Journal of Research*, 8(1), 83–89. DOI: 10.29121/granthaalayah.v8.i1.2020.251

Valls Martínez, M. D. C., Martín-Cervantes, P. A., & Peña Rodríguez, S. (2021). Ethical banking and poverty alleviation banking: The two sides of the same solidary coin. *Sustainability*, 13(21), 11977. DOI: 10.3390/su132111977

Van de Poel, I. (2020). Core values and value conflicts in cybersecurity: beyond privacy versus security. *The Ethics of Cybersecurity*, 45-71. Springer Open

Van den Hoven, J. (2010). The use of normative theories in computer ethics. *The Cambridge handbook of information and computer ethics*, 59-76. Cambridge University Press.

Varlamov, O. O., Chuvikov, D. A., Adamova, L. E., Petrov, M. A., Zabolotskaya, I. K., & Zhilina, T. N. (2019). Logical, philosophical and ethical aspects of AI in medicine. *International Journal of Machine Learning and Computing*, 9(6), 868–873. DOI: 10.18178/ijmlc.2019.9.6.885

Vearrier, L., & Henderson, C. M. (2021). Utilitarian Principlism as a framework for crisis healthcare ethics. *HEC Forum*, 33(1–2), 45–60. DOI: 10.1007/s10730-020-09431-7 PMID: 33449232

Vitak, J., Shilton, K., & Ashktorab, Z. (2016, February). Beyond the Belmont Principles: Ethical challenges, practices, and beliefs in the online data research community. In *Proceedings of the 19th ACM Conference on Computer-supported Cooperative Work & Social Computing* (pp. 941-953). DOI: 10.1145/2818048.2820078

Vollmer, S. H., & Howard, G. (2010). Statistical power, the Belmont Report, and the ethics of clinical trials. *Science and Engineering Ethics*, 16(4), 675–691. DOI: 10.1007/s11948-010-9244-0 PMID: 21063801

Wachter, S., & Mittelstadt, B. (2019). A right to reasonable inferences: Re-thinking data protection law in the age of big data and AI. *Columbia Business Law Review*, 2019(2), 494–620.

Warsaw. (2020). Personality Assessment For Dogs In AAI (PADA).

Weatherly, N. L. (2021). The ethics of organizational behavior management. *Journal of Organizational Behavior Management*, 41(3), 197–214. DOI: 10.1080/01608061.2021.1890664

White House Office of Science and Technology Policy (OSTP). (2022). A blueprint for an AI bill of rights. https://www.whitehouse.gov/ostp/ai-bill-of-rights/

White, M. G. (2020). Why human subjects research protection is important. *The Ochsner Journal*, 20(1), 16–33. DOI: 10.31486/toj.20.5012 PMID: 32284679

Wiafe, I., Yaokumah, W., & Kissi, F. A. (2020). Students' intentions on cyber ethics issues. In *Modern Theories and Practices for Cyber Ethics and Security Compliance* (pp. 105-121). IGI Global.

Wilkinson, M. D., Dumontier, M., Aalbersberg, I., Appleton, G., Axton, M., Baak, A., Blomberg, N., Boiten, J.-W., da Silva Santos, L. B., Bourne, P. E., Bouwman, J., Brookes, A. J., Clark, T., Crosas, M., Dillo, I., Dumon, O., Edmunds, S., Evelo, C. T., Finkers, R., & Mons, B. (2016). The FAIR Guiding Principles for scientific data management and stewardship. *Scientific Data*, 3(1), 160018. DOI: 10.1038/sdata.2016.18 PMID: 26978244

Williamson, S. M., & Prybutok, V. (2024). Balancing privacy and progress: A review of privacy challenges, systemic oversight, and patient perceptions in AI-driven healthcare. *Applied Sciences (Basel, Switzerland)*, 14(2), 675. DOI: 10.3390/app14020675

Williams, R. (2023). Think piece: Ethics for the virtual researcher. *Practice*, 5(1), 41–47. DOI: 10.1080 /25783858.2023.2179893

Winter, J. S., & Davidson, E. (2022). Harmonizing regulatory regimes for the governance of patient-generated health data. *Telecommunications Policy*, 46(5), 102285. DOI: 10.1016/j.telpol.2021.102285

Wolf, L. E., Ram, N., Contreras, J., & Beskow, L. M. (2024). Certificates of confidentiality: Privileging research data. *Journal of Law and the Biosciences*, 11(1), lsae003. Advance online publication. DOI: 10.1093/jlb/lsae003 PMID: 38405099

Wong, WC, P., Yu WM, R., Chau KY, K., Lee CY, J., & NGAI Tsz-kin, J. (2021). A responding to commentary: "Recognizing the role of animal-assisted therapies in addressing mental health needs during the COVID-19 pandemic". *Asian Journal of Psychiatry*. PMID: 33989958

World Medical Association. (2013). World Medical Association Declaration of Helsinki: Ethical principles for medical research involving human subjects. *Journal of the American Medical Association*, 310(20), 2191–2194. DOI: 10.1001/jama.2013.281053 PMID: 24141714

World Medical Association. (2018). *Declaration of Helsinki—Ethical principles for medical research involving human subjects*. www.wma.net/policies-post/wma-declaration-of-helsinki-ethical-principles-for-medical-research-involving-human-subjects/

Wylde, V., Rawindaran, N., Lawrence, J., Balasubramanian, R., Prakash, E., Jayal, A., & Platts, J. (2022). Cybersecurity, data privacy and blockchain: A review. *SN Computer Science*, 3(2), 127. DOI: 10.1007/s42979-022-01020-4 PMID: 35036930

Yang, A., Lee, D., Hwang, J., & Shin, J. (2013). The influence of regulations on the efficiency of telecommunications operators: A meta-frontier analysis. *Telecommunications Policy*, 37(11), 1071–1082. DOI: 10.1016/j.telpol.2013.02.004

Yaokumah, W., Rajarajan, M., Abdulai, J. D., Wiafe, I., & Katsriku, F. A. (Eds.). (2020). *Modern Theories and Practices for Cyber Ethics and Security Compliance*. IGI Global. DOI: 10.4018/978-1-7998-3149-5

Yonezawa, K., Gómez, M. I., & Richards, T. J. (2020). The Robinson–Patman act and vertical relationships. *American Journal of Agricultural Economics*, 102(1), 329–352. DOI: 10.1093/ajae/aaz049

Younis, H. A., Eisa, T. A. E., Nasser, M., Sahib, T. M., Noor, A. A., Alyasiri, O. M., Salisu, S., Hayder, I. M., & Younis, H. A. (2024). A systematic review and meta-analysis of artificial intelligence tools in medicine and healthcare: Applications, considerations, limitations, motivation and challenges. *Diagnostics*, 14(1), 109. DOI: 10.3390/diagnostics14010109 PMID: 38201418

Yu, K. H., Healey, E., Leong, T. Y., Kohane, I. S., & Manrai, A. K. (2024). Medical artificial intelligence and human values. *The New England Journal of Medicine*, 390(20), 1895–1904. DOI: 10.1056/NEJMra2214183 PMID: 38810186

Zhang, X., & Ghorbani, A. A. (2021). Human factors in cybersecurity: Issues and challenges in big data. *Research Anthology on Privatizing and Securing Data*, 1695-1725. IGI Global.

Zhao, H., & Yang, G. (2022). Information security and legal ethics of artificial intelligence medical devices. *Forest Chemicals Review*, 236-245.

Zhong, S., Zhao, X., & Song, J. (2023). MNEs' Ambidexterity Strategies and Moral Conflicts: The Case of Google in China. *Journal of Business Ethics*, 186(4), 781–796. DOI: 10.1007/s10551-023-05437-6

Znamirowski, B. (2021). *Getting a Grip on Research Data Management. Bulletin - Association of Canadian Map Libraries and Archives*. ACMLA., DOI: 10.15353/acmla.n169.4739

Zondag, A. G. M., Groenhof, K. J., van der Graf, R., van Solinge, W. W., Bots, M. L., & Haitjema, S.UCC-CVRM study group. (2023). Asking informed consent may lead to significant participation bias and suboptimal cardiovascular risk management in learning healthcare systems. *BMC Medical Research Methodology*, 23(1), 98. Advance online publication. DOI: 10.1186/s12874-023-01924-6 PMID: 37087415

Zywicki, T. J. (2007). Institutional Review Boards as academic bureaucracies: An economic and experiential analysis. *Northwestern University Law Review*, 101(2), 861–895.

About the Contributors

Robin Throne, PhD, is a human research protections professional and research methodologist. Her research agenda continues to consider doctoral researcher positionality and agency, and voice and land dispossession from various social justice research approaches.

***

Chad Becker holds the position of Sr. IRB Education Analyst in the HRPP and Sr. Core Part-Time Professor of Education at National University. He has served as a professor and dissertation chair at Concordia University – Portland, Northcentral University, and Indiana State University. Chad earned advanced degrees from Purdue University and Indiana University, Bloomington in Curriculum and Instruction with emphases on Curriculum Studies. His research interests include education law, equity in education, and ethical research.

Paula Garcia McAllister has been a research support professional for 20 years. She has a Ph.D. in Applied Linguistics, and a Certificate in Hospital Research Compliance. She has managed Institutional Review Boards in academic and healthcare organizations. Currently, she is Administrator of Research and Sponsored Programs at a community health system in Northern Arizona.

Michalina Hendon is a passionate computer enthusiast with a keen interest in the interaction between users and technology. As a determined problem solver, she constantly seeks out innovative approaches to harness technology for seamless information dissemination and lifestyle enhancement. With over a decade of experience in Higher Education, Dr. Hendon has dedicated her expertise to teaching Information Technology courses while, and for the last few years, providing ethical oversight in research endeavors. Outside of her professional endeavors, Dr. Hendon cherishes her time with loved ones, including her family, beloved English bulldog, and friends. She also devotes time to engaging in research activities, research ethics, and staying abreast of the latest technological advancements. Through her multifaceted pursuits, Dr. Hendon remains committed to harnessing the power of technology ethically to enhance the lives of individuals and society as a whole.

Pratap Chandra Mandal is an Assistant Professor (Marketing) at Indian Institute of Management, Shillong, India. He has completed graduate degree from the reputed Indian Institute of Technology, Kharagpur (IIT Kharagpur), India (Bachelor of Technology in Mechanical Engineering), post-graduate degree from Vinod Gupta School of Management, IIT Kharagpur (Masters in Business Adminstration), PhD (Marketing) from Vinod Gupta School of Management, IIT Kharagpur. His research concerns customer relationship management, customer satisfaction, services marketing, marketing intelligence, and qualitative methods in management. He is the editor-in-chief of two international journals and is on the editorial board of journals like Journal of Global Marketing. Pratap has won several prestigious scholarships and awards throughout his academic career.

Brianne Mongeon holds an Ed.D. in Organizational Change and Leadership from the University of Southern California. With a bachelor's and master's degree in Sociology, as well as a master's in Leadership, she brings a rich interdisciplinary perspective to her work. Over the past 15 years, Brianne has dedicated her career to working and consulting within the Institutional Review Board (IRB) space for various universities. Her research focuses on the study and improvement of IRBs, emphasizing the development of ethical standards and practices in academic research. A committed member of PRIM&R (Public Responsibility in Medicine and Research) since 2010, Brianne has consistently contributed to advancing ethical research oversight throughout her career.

Hemangi Narayan Narvekar is a licensed Clinical Psychologist currently working as an Assistant Professor at CHRIST University, Bangalore. She has worked in different clinical setups treating mental health issues. She has pursued her PhD, which was aimed at understanding the effectiveness of Animal-Assisted Therapy in adults with ASD. She has authored seven research publications and contributed three chapters in three book publications. She has presented six papers at three International and three national-level conferences. She is the recipient of two Best Paper Awards at National Conferences. She was also a participant in the Emerging Psychologist's Programme (EPP) 2024, associated with the Internal Congress of Psychology, Prague.

R.J. Reding-Waldrep earned a PhD in Information Technology from the University of the Cumberlands. She has worked in the information technology industry for more than three decades. Her primary domains of experience include software development, system integration, project management, and cybersecurity. She obtained a Security + certification. She is an ISSC(2) Certified Information System Security Professional (CISSP).

Muhammad Usman Tariq has more than 16+ year's experience in industry and academia. He has authored more than 200+ research articles, 100+ case studies, 60+ book chapters and several books other than 4 patents. He is founder and CEO of The Case HQ, a unique repository for courses, narrative and video case studies. He has been working as a consultant and trainer for industries representing six sigma, quality, health and safety, environmental systems, project management, and information security standards. His work has encompassed sectors in aviation, manufacturing, food, hospitality, education, finance, research, software and transportation. He has diverse and significant experience working with accreditation agencies of ABET, ACBSP, AACSB, WASC, CAA, EFQM and NCEAC. Additionally, Dr. Tariq has operational experience in incubators, research labs, government research projects, private sector startups, program creation and management at various industrial and academic levels. He is Certified Higher Education Teacher from Harvard University, USA, Certified Online Educator from HMBSU, Certified Six Sigma Master Black Belt, Lead Auditor ISO 9001 Certified, ISO 14001, IOSH MS, OSHA 30,

Sheetal Temara is an Information Security professional with an extensive background in Penetration Testing and Vulnerability Assessments spanning various applications and domains. She is currently pursuing a doctoral degree in Information Technology at University of the Cumberlands. Her academic journey also includes a Master of Science degree in Computer Engineering from Oklahoma City University and a Bachelor of Technology in Information Technology from India. In her current role, as a Cyber Security Research Scientist, Sheetal demonstrates her expertise in penetration testing and identifies security vulnerabilities for the organizational networks, application systems, hardware infrastructure and emerging technologies to improve the enterprise information security posture. She excels in conducting security risk assessments ensuring compliance with corporate information security policies and communicating risks effectively to stakeholders. Prior to her current position, Sheetal served as a Security Engineer where she conducted dynamic application security testing and provided expert guidance on security vulnerabilities and risks to development teams. She has extensive experience in utilizing various penetration testing tools and possesses advanced skills in programming languages. With a passion for cybersecurity and a dedication to advancing the field, Sheetal's research interests incline toward the development of innovative solutions to address emerging cyber threats. She actively contributes to the academic community through her research publications and publishes her academic work on platforms such as ResearchGate. Outside of her professional endeavors, Sheetal is committed to sharing her knowledge and expertise with others. As an author, she aims to empower readers with practical insights and strategies to enhance their understanding of cybersecurity and strengthen their defenses against evolving threats.

Goran Trajkovski, Ph.D., has 30 years of experience in educational design, focusing on learner-centered approaches. His expertise covers data science, artificial intelligence, machine learning, and learning analytics. He has developed programs and courses in these areas for various educational institutions. His research interests are in cognitive and developmental robotics, leading to the establishment of an NSF-funded research lab. Dr. Trajkovski has a significant publication record, including more than 20 books and over 300 articles, with his involvement in AI research starting in 1995. He prioritizes ongoing learning and skill development, consistently seeking to broaden his knowledge.

Index

A

B

C

D

L

O

P

Q

R

S

T

U

W

www.ingramcontent.com/pod-product-compliance
Lightning Source LLC
LaVergne TN
LVHW080609300926
849162LV00005B/1096